## Student Solutions Manual and Study Guide



# Chemistry for Engineering Stude

**THIRD EDITION**

**Lawrence S. Brown**
Texas A&M University

**Thomas A. Holme**
Iowa State University

Prepared by

**Steve Rathbone**
Blinn College, Bryan Campus

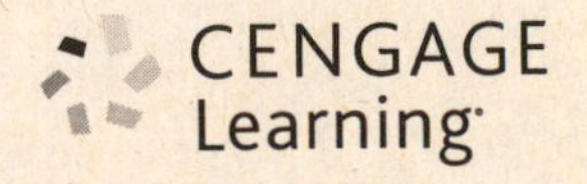


Australia • Brazil • Mexico • Singapore • United Kingdom • United States

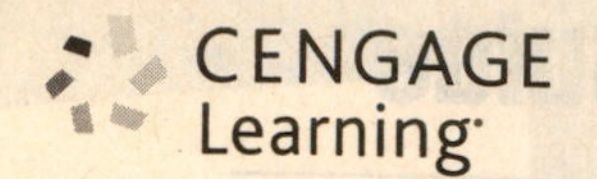



ISBN-13: 978-1-285-84524-1
ISBN-10: 1-285-84524-2

**Cengage Learning**
200 First Stamford Place, 4th Floor
Stamford, CT 06902
USA

Cengage Learning is a leading provider of customized learning solutions with office locations around the globe, including Singapore, the United Kingdom, Australia, Mexico, Brazil, and Japan. Locate your local office at: **www.cengage.com/global**.

Cengage Learning products are represented in Canada by Nelson Education, Ltd.

To learn more about Cengage Learning Solutions, visit **www.cengage.com**.

Purchase any of our products at your local college store or at our preferred online store **www.cengagebrain.com**.

Printed in the United States of America
1 2 3 4 5 6 7 17 16 15 14 13

# CONTENTS

# PREFACE

Learning chemistry can be an intimidating task. There is a tremendous quantity of information, sometimes very abstract, and much of it might seem unrelated. Most freshman chemistry courses follow a certain progression, and Chemistry for Engineering Students, 3rd Ed. is no exception. Try to visualize the "big picture" of chemistry describing different forms of matter, expressing chemical reactions and utilizing chemical equations, understanding how and why reactions occur (bonding), and studying characteristics of chemical reactions (energy changes, rate, and equilibrium). This clearly is a gross simplification of the course, but it helps sometimes to "see where you are going" in a subject such as chemistry.

Chemistry, like a sport or a musical instrument, requires practice. Just sitting in class taking notes is usually not sufficient to master this subject. Students should take an active role in their learning process. Chapters should be read prior to coming to class and read again after covering the material in class. Notes should be re-read and possibly re-written. Example problems in class should be scrutinized, and similar problems in the text found and worked. There is no doubt that all this chemistry practice (studying) requires a lot of time. If you are seeking the greatest success in your chemistry course you must be willing to commit a significant amount of time.

Use of this student solutions manual/ study guide can help make effective use of your time. Each chapter has four separate parts:

1. Each chapter starts with a list of study goals. These are the major concepts to be mastered in each section of the text chapters. End-of-chapter problem references are given to direct the student to specific problems relevant to each section.

2. Detailed solutions are then given for all odd-numbered problems in the text chapters. Emphasis is placed on solving problems in a logical, systematic fashion, an important skill for engineering students to acquire. The questions are included as well as the solutions to avoid having to look in both the text and the study guide while studying.

3. A chapter objective quiz follows. These are additional questions designed specifically to target the chapter objectives outlined in the text.

4. Finally, solutions to the chapter objective quiz are given.

After reading the text chapter and your class notes thoroughly, begin working the end-of-chapter problems. If some problems are particularly confusing, carefully study the solutions in this guide. Look for problems in your book similar to ones emphasized by your professor in class. If you still don't understand a solution to a problem, don't be shy about asking for help from your professor, a tutor, or a classmate. When you are preparing for an exam, you can use the chapter objective quiz to check your knowledge of the material. Alternatively, you may want to work the quiz problems just as extra practice.

If this sounds like a lot of work, you are right. But with hard work comes success: a good understanding of chemistry and solid study habits which will help you during your entire college career. Good luck!

I am indebted to all who have helped make this Student Solutions Manual and Study Guide possible. The third edition contains many reworked and new problems and some additional practice test questions. My thanks go to Brendan Killion, Assistant Editor of Cengage Learning, for his assistance in revising the third edition of the manual. I am also indebted to my colleagues at Blinn College for their insight, suggestions, and encouragement. I thank my students for their questions, curiosity, and enthusiasm, and for always motivating me to be a better teacher. Last, and certainly not least, thanks to my family for their love and support in my endeavors.

Steve Rathbone

# CHAPTER 1

## *Study Goals*

The study goals outline specific concepts to be mastered in this section of the text chapter. Related problems at the end of the text chapter will also be noted. Working the questions noted should aid in mastery of each study goal and also highlight any areas that you may need additional help in.

**Section 1.1 *INSIGHT INTO* Aluminum**

1. Recognize the significant properties of aluminum and how matter must be physically and chemically altered to meet the needs of society. *Work Problems 1–6.*

**Section 1.2 The Study of Chemistry**

2. Describe the difference between the three perspectives of chemistry: macroscopic, microscopic, and symbolic. *Work Problems 7–9, 11, 17.*
3. Distinguish between chemical and physical properties and between chemical and physical changes. *Work Problems 10, 13–16.*
4. Define several forms of matter: atoms, elements, molecules, and compounds. *Work Problems 12, 18.*

**Section 1.3 The Science of Chemistry: Observations and Models**

5. Describe the scientific method as it relates to observations, models and theories. *Work Problems 19–28.*

**Section 1.4 Numbers and Measurements in Chemistry**

6. Be familiar with common units of measurement in the SI system (International System of Units) and the metric system of prefixes. *Work Problems 29–39, 56–58.*
7. Express numbers using scientific notation with the correct number of significant figures. *Work Problems 43–52.*

**Section 1.5 Problem Solving in Chemistry and Engineering**

8. Use the factor-label method (Dimensional Analysis) technique of problem solving to perform unit conversions, density calculations, and simple computations. *Work Problems 40–42, 5 –55, 59–70, 73–76, 84–90.*
9. Recognize the importance of visualization in chemistry and how matter can be represented conceptually. *Work Problems 71, 72.*

Section 1.6 ***INSIGHT INTO* Material Selection and Bicycle Frames**

10. Using the example of a bicycle frame selection, discuss the importance of the properties of matter in design considerations. *Work Problems 77–83.*

## *Solutions to the Odd-Numbered Problems*

Problem 1.1 Use the web to determine the mass of a steel beverage can from the 1970s and find the mass of a modern aluminum can. How much more would a 12-pack of soda weigh in steel cans?

**Answers will vary according to websites found. One website stated the mass of an average steel can was 2.74 ounces. The mass of an aluminum soda can is approximately 15.0 grams. We can convert mass in ounces to grams and then calculate the difference in mass of a "12-pack".**

$$2.74 \text{ oz} \times \frac{28.35 \text{ g}}{1 \text{ oz}} = 77.7 \text{ g/steel can}$$

**The difference in mass is 77.7 g/steel can – 15.0 g/aluminum can = 62.7 g**

**A 12-pack of steel cans would have 12 × 62.7 g = 752 g more mass.**
**In pounds,**

$$752 \text{ g} \times \frac{1 \text{ lb}}{454 \text{ g}} = 1.66 \text{ lbs of extra weight}$$

Problem 1.3 Where does the scientific method "start?" What is the first step?

**Observation is the first step in the scientific method. The process of understanding natural phenomena must start by first making some observations.**

Problem 1.5 Use the web to find current prices offered for aluminum for recycling. Is there a variation for the price based on where in the United States the aluminum is returned?

**Answers will vary according to the websites found. Much of the variation in price is due to differences in demand for aluminum at various locations and transportation costs.**

Problem 1.7 When making observations in the laboratory, which perspective of chemistry are we normally using?

**We make observations in the laboratory using the macroscopic perspective of chemistry, unless very sophisticated instruments are used.**

Problem 1.9 Which macroscopic characteristics differentiate solids, liquids, and gases? (List as many as possible.)

**Solids maintain a definite shape; liquids and gases do not. Gases expand to completely occupy their container; liquids assume the shape of the container but do not fully occupy it. Solids tend to have high densities, liquids slightly lower, and gases typically have very low densities, comparatively.**

Problem 1.11 Some farmers use ammonia, $NH_3$, as a fertilizer. This ammonia is stored in liquid form. Use the particulate perspective to show the transition from liquid ammonia to gaseous ammonia.

**Ammonia, $NH_3$, in the liquid form will have the molecules very close together, in random orientation. As a gas, the molecules will be far apart.**

**Liquid:** **Gas:**

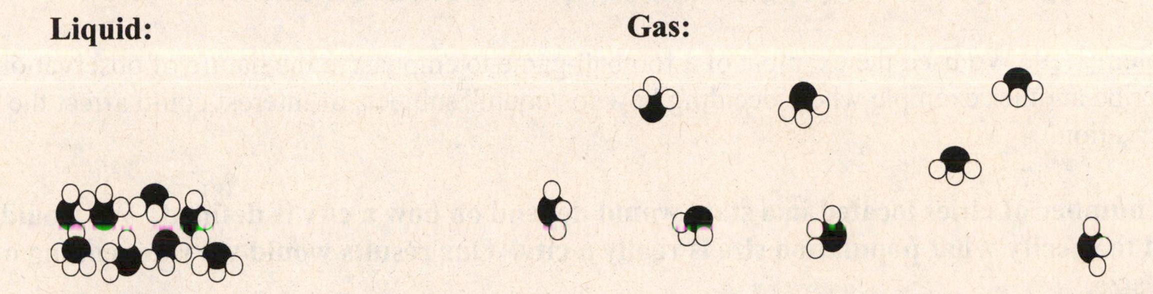

Problem 1.13 ■ Label each of the following as either a physical process or a chemical process: **(a)** rusting of an iron bridge, **(b)** melting of ice, **(c)** burning of a wooden stick, **(d)** digestion of a baked potato, **(e)** dissolving of sugar in water.

**A chemical change involves a change in the *composition* of matter; that is, some new compounds are formed. A physical change only involves a change in the *physical state* of matter; no new compounds are formed.**

(a) rusting of an iron bridge **Chemical, rust forms when iron and oxygen react chemically.**

(b) melting of ice **Physical, change from the solid to the liquid state.**

(c) burning of a wooden stick **Chemical, the molecules in the wood are changed into carbon dioxide and water during combustion.**

(d) digestion of a baked potato **Chemical, larger food molecules are changed into smaller ones and eventually oxidized ("burned") to produce energy.**

(e) dissolving of sugar in water **Physical, the sugar molecules are not changed they just become surrounded by water molecules in solution.**

Problem 1.15 Physical properties may change because of a chemical change. For example, the color of an egg "white" changes from clear to white because of a chemical change when it is cooked. Think of another common situation when a chemical change also leads to a physical change.

**The "rusting" of iron is a chemical change. A chemical reaction changes the metal to a new compound. The rust formed is brittle, orange-red compound with very different properties than iron.**

Problem 1.17 Use a molecular level description to explain why gases are less dense than liquids or solids.

**See Problem 1.11. There is a large amount of empty space between the molecules when a substance is in the gaseous state. Fewer molecules in a given volume means less mass per volume; meaning the density decreases. Solids and liquids have little space between the molecules, giving a relatively high mass per volume; much larger density.**

Problem 1.19 We used the example of a football game to emphasize the nature of observations. Describe another example where deciding how to "count" subjects of interest could affect the observation.

**The number of cities located in a state would depend on how a city is defined. We would need to specify what population size is really a city. Our results would vary depending on that size.**

Problem 1.21 Complete the following statement: Data that have a large systematic error are (a) accurate, (b) precise, (c) neither.

**(b) Precise. Data that have large systematic error may still be precise. Accuracy is how close our data is to the actual value. Large systematic error means the data will not be close to the actual value but they will be off by about the same amount. Precision is the repeatability of your data. The data will agree closely but will not be close the true value.**

Problem 1.23 Use your own words to explain the difference between deductive and inductive reasoning.

**In deductive reasoning, facts are considered and conclusions are drawn from this information. In inductive reasoning, you first infer what seems to be accurate or true, and then find ways to determine if later observations fit the inferred conclusion.**

Problem 1.25 When a scientist looks at an experiment and then predicts the results of other related experiments, which type of reasoning is she using? Explain your answer.

**This is inductive reasoning. Inductive reasoning is the reverse of deductive. A scientist makes predictions and then tries to prove the prediction by later observations. Deductive reasoning involves starting with facts and then drawing conclusions from those.**

Problem 1.27 Should the word *theory* and *model* be used interchangeably in the context of science? Defend your answer using web information.

**The word theory implies something more advanced and supported than a model. When a model makes predictions and all observations are consistent with these observations, then the model may be described as a theory.**

Problem 1.29 Describe a miscommunication that can arise because units are not included as part of the information.

**Discussing the time of day and omitting PM or AM. If you are supposed to meet someone at 9:00 but they didn't say in the evening or in the morning, the meeting might not occur. Another example might be negotiating a price while trying to buy something.**

Problem 1.31 Identify which of the following units are base units in the SI system: grams, meters, joules, liters, and amperes.

**Meter and ampere are base units. The other base units are: kilogram, second, kelvins, mole, and candela. Therefore grams are not a base unit. Joules and liters are both derived units.**

Problem 1.33 Rank the following prefixes in order of increasing size of the number they represent: centi–, giga–, nano–, and kilo–.

**In order of smallest to largest: nano– ($10^{-9}$), centi– ($10^{-2}$), kilo– ($10^{3}$), giga– ($10^{9}$).**

Problem 1.35 Historically, some unit differences reflected the belief that the quantity measured was different when it was later revealed to be a single entity. Use the web to look up the origins of the energy units *erg* and *calorie*, and describe how they represent an example of this type of historical development.

**The erg was an energy unit associated with work done, while the calorie was an energy unit associated with heat. At the time scientists thought heat and work were different things – instead of forms of energy used in different ways.**

Problem 1.37 How many $\mu$g are equal to one gram?

**The prefix $\mu$ stands for micro–, or $10^{-6}$. One $\mu$g is therefore equal to $\frac{1}{1{,}000{,}000}$ gram.**

**There are one million $\mu$g in one gram.**

Problem 1.39 How was the Fahrenheit temperature scale calibrated? Describe how this calibration process reflects measurement errors that were evident when the temperature scale was devised.

**The upper temperature, 100 degrees, was set by body temperature. The lower temperature was determined by the coldest temperature that could be achieved by adding salt to water and set at 0 degrees. This shows limitations of the measurement in that the accepted temperature of a human body is 98.6 $^{o}$F (although this can vary from individual).**

Problem 1.41 Express each of the following temperatures in kelvins:

**To convert from the Celsius scale to kelvins, we add 273. The unit size is the same, making this conversion easier. Note: we do not write the degree symbol with kelvins.**

(a) −10. °C **kelvins = −10. °C + 273 = 260 K**

(b) 0.00 °C **kelvins = 0 °C + 273 = 273 K**

(c) 280. °C **kelvins = 280 °C + 273 = 553 K**

(d) $1.4 \times 10^3$ °C **kelvins = 1400 °C + 273 = 1700 K**

Problem 1.43 Express each of the following numbers in scientific notation:

**Scientific notation expresses a number factoring out all powers of ten and writing a number between 1 and 10 multiplied by some power of 10. It makes writing very small or very large numbers much easier.**

(a) 62.13 **$6.213 \times 10^1$**

(b) 0.000414 **$4.14 \times 10^{-4}$**

(c) 0.0000051 **$5.1 \times 10^{-6}$**

(d) 871,000,000 **$8.71 \times 10^8$**

(e) 9100 **$9.1 \times 10^3$**

Problem 1.45 ■ How many significant figures are present in these measured quantities?

(a) 1374 kg **4 significant figures**
(b) 0.00348 s **3 significant figures (leading zeros only locate decimal point)**
(c) 5.619 mm **4 significant figures**
(d) $2.475 \times 10^{-3}$ cm **4 significant figures**
(e) 33.1 mL **3 significant figures**

Problem 1.47 ■ Calculate the following to the correct number of significant figures. Assume that all these numbers are measurements.

**In addition and subtraction, the number that has a doubtful digit in the largest decimal place determines the number of significant digits retained.**
**When multiplying and dividing, the number of significant digits retained is determined by whichever factor has the fewest significant digits.**

(a) $x = 17.2 + 65.18 - 2.4$ **= 80.0 (round to tenths place)**

(b) $x = 13.0217/17.10$ **= 0.7615 (round to four significant digits)**

(c) $x = (0.0061020)(2.0092)(1200.0)$ **= 14.712 (round to five significant digits)**

(d) $x = 0.0034 + \sqrt{(0.0034)^2 + 4(1.000)(6.3\times10^{-4})} \div (2)(1.000)$
**= 0.03 (round to one significant digit)**

<u>Problem 1.49</u> A student finds that the mass of an object is 4.131 g and its volume is 7.1 mL. What density should be reported in g/mL?

**Density is the ratio of mass per unit volume.**

$$\textbf{density} = \frac{\textbf{mass}}{\textbf{volume}} = \frac{\textbf{4.131 g}}{\textbf{7.1 mL}} = \textbf{0.58 g/mL (rounded to two significant digits)}$$

<u>Problem 1.51</u> A student weighs 10 quarters and finds their total mass is 56.63 grams. What should she report as the average mass of a quarter based on her data?

**The average mass of a quarter would be found by dividing the total mass by ten quarters.**

$$\textbf{Average mass of a quarter} = \frac{\textbf{56.63 g}}{\textbf{10 quarters}} = \textbf{5.663 g}$$

**The answer should retain all four significant digits because the ten quarters is an *exact* number.**

<u>Problem 1.53</u> A package of eight apples has a mass of 1.00 kg. What is the average mass of one apple in grams?

**1.00 kg = 1.00 × 10$^3$ g**

**The average mass of an apple would be found by dividing the total mass by eight apples.**

$$\textbf{Average mass of an apple} = \frac{\textbf{1.00}\times\textbf{10}^3\textbf{ g}}{\textbf{8 apples}} = \textbf{1.25}\times\textbf{10}^2\textbf{ g per apple}$$

**The answer should retain three significant digits because the eight apples is an *exact* number.**

Problem 1.55 A person measures 173 cm in height. What is this height in meters? feet and inches?

$$173\ \text{cm} \times \frac{1\ \text{m}}{100\ \text{cm}} = 1.73\ \text{m}$$

$$173\ \text{cm} \times \frac{1\ \text{in}}{2.54\ \text{cm}} = 68.1\ \text{inches} \times \frac{1\ \text{ft}}{12\ \text{in}} = 5.68\ \text{feet.} \qquad 0.68\ \text{ft} \approx \frac{8}{12}\ \text{inches,}$$

**Therefore, the height is 5 ft 8 in.**

Problem 1.57 Carry out the following unit conversions:

(a) $3.47 \times 10^{-6}$ g to $\mu$g

$$3.47 \times 10^{-6}\ \text{g} \times \frac{1\times10^{6}\ \mu\text{g}}{1\ \text{g}} = 3.47\ \mu\text{g}$$

(b) $2.73 \times 10^{-4}$ L to mL

$$2.73 \times 10^{-4}\ \text{L} \times \frac{1000\ \text{mL}}{1\ \text{L}} = 2.73 \times 10^{-1}\ \text{mL}$$

(c) 725 ns to s

$$725\ \text{ns} \times \frac{10^{-9}\ \text{s}}{1\ \text{ns}} = 7.25 \times 10^{-7}\ \text{s}$$

(d) 1.3 m to km

$$1.3\ \text{m} \times \frac{1\ \text{km}}{1000\ \text{m}} = 1.3 \times 10^{-3}\ \text{km}$$

Problem 1.59 ■ Convert 22.3 mL to

(a) liters

$$22.3\ \text{mL} \times \frac{1\ \text{L}}{1000\ \text{mL}} = 0.0223\ \text{L}$$

(b) $\text{in}^3$

$$22.3\ \text{mL} \times \frac{1\ \text{cm}^3}{1\ \text{mL}} \times \frac{(1\ \text{in})^3}{(2.54\ \text{cm})^3} = 1.36\ \text{in}^3$$

(c) quarts

$$22.3\ \text{mL} \times \frac{1\ \text{L}}{1000\ \text{mL}} \times \frac{1.057\ \text{qt}}{1\ \text{L}} = 0.0236\ \text{qt}$$

Problem 1.61 ■ A load of asphalt weighs 254 lbs and occupies a volume of 220.0 L. What is the density of this asphalt in g/L?

**Density is the ratio of mass per unit volume. The mass = 254 lbs ×** $\frac{454\ \text{g}}{1\ \text{lb}} = 1.15 \times 10^{5}\ \text{g}$

$$\text{density} = \frac{\text{mass}}{\text{volume}} = \frac{1.15\times10^{5}\ \text{g}}{220.0\ \text{L}} = 524\ \text{g/L}$$

**(The answer is rounded to 3 significant figures)**

Problem 1.63 ■ A sample of crude oil has a density of 0.87 g/mL. What volume in liters does a 3.6-kg sample of this oil occupy?

**Mass = 3.6 kg = 3600 g**

**Density is the ratio of mass per unit volume, density =** $\frac{\text{mass}}{\text{volume}}$**.**

**Rearranging,** $\text{volume} = \frac{\text{mass}}{\text{density}} = \frac{3600\text{ g}}{\frac{0.87\text{ g}}{\text{mL}}} = 4100\text{ mL or }4.1\text{ L}$

Problem 1.65 ■ The area of the contiguous states is $3.02 \times 10^6$ mi$^2$. Assume that these states are completely flat (no mountains and no valleys). What volume of water, in liters, would cover these states with a rainfall of two inches?

**The volume of water would be the area of the contiguous states multiplied by the depth of water.**

**Area:**

$$3.02 \times 10^6 \text{ mi}^2 \times \frac{(1.609\text{ km})^2}{(1\text{ mi})^2} \times \frac{(1000\text{ m})^2}{(1\text{ km})^2} \times \frac{(100\text{ cm})^2}{(1\text{ m})^2} = 7.82 \times 10^{16}\text{ cm}^2$$

**Depth:**

$$2\text{ in} \times \frac{2.54\text{ cm}}{1\text{ in}} = 5.08\text{ cm}$$

**Volume:**

$$7.82 \times 10^{16}\text{ cm}^2 \times 5.08\text{ cm} = 3.97 \times 10^{17}\text{ cm}^3 \times \frac{1\text{ L}}{1000\text{ cm}^3} = 3.97 \times 10^{14}\text{ L}$$

Problem 1.67 ■ Titanium is used in airplane bodies because it is strong and light. It has a density of 4.55 g/cm$^3$. If a cylinder of titanium is 7.75 cm long and has a mass of 153.2 g, calculate the diameter of the cylinder. ($V = \pi r^2 h$, where $V$ is the volume of the cylinder, $r$ is its radius, and $h$ is the height.

**Using the density, the volume can be calculated: volume =** $\frac{\text{mass}}{\text{density}} = \frac{153.2\text{ g}}{\frac{4.55\text{ g}}{\text{cm}^3}} = 33.7\text{ cm}^3$

**The radius may now be calculated by using the height of the cylinder and the formula for its volume:** $r = \sqrt{V/\pi h} = \sqrt{(33.7\text{ cm}^3)/(3.1416)(7.75\text{ cm})} = 1.18\text{ cm}$

**Finally, the diameter is two × $r$ = 2 × 1.18 cm = 2.36 cm**

Problem 1.69 ■ An industrial engineer is designing a process to manufacture bullets. The mass of each bullet must be within 0.25 % of 150 grains. What range of bullet masses, in mg, will meet this tolerance? 1 gr = 64.74891 mg

**First, calculate 0.25% of 150 grains. (Recall percentage is parts per 100 parts.)**

$$\frac{0.25}{100} \times 150 \text{ gr} = 0.375 \text{ gr}$$

**Next, calculate this mass in milligrams:** $0.375 \text{ gr} \times \frac{64.74891 \text{ mg}}{1 \text{gr}} = 24.3 \text{ mg}$

**The bullets must manufactured with a tolerance of ± 24.3 mg**

**Note: It is extremely important that bullets be manufactured with tight tolerances in mass and shape to produce precise shooting results.**

Problem 1.71 Draw a molecular scale picture to show how a crystal is different from a liquid.

**A crystalline solid has a regular repeating arrangement of the particles that make up the solid. A liquid has particles randomly arranged.**

**Alumina ($Al_2O_3$, crystal)**

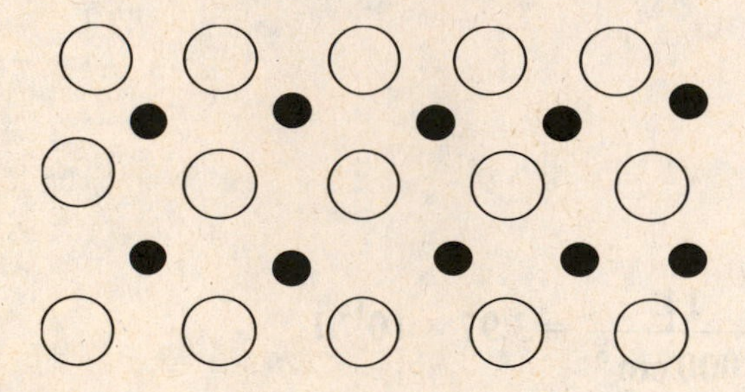

**Water ($H_2O$, liquid)**

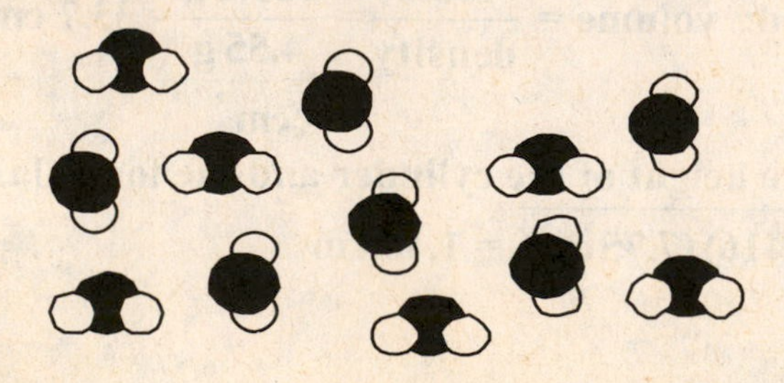

Problem 1.73 ■ On average, Earth's crust contains about 8.1% aluminum by mass. If a standard 12-ounce soft drink can contains approximately 15.0 g of aluminum, how many cans could be made from one ton of the Earth's crust?

**There are 2000 pounds in one ton.**
**First, calculate the mass of aluminum in one ton of the Earth's crust. (Recall percentage is parts per 100 parts.)**

$$2000 \text{ lbs Earth} \times \frac{8.1 \text{ lbs Al}}{100 \text{ lbs Earth}} \times \frac{454 \text{ g}}{1 \text{ lb}} = 73548 \text{ g of aluminum.}$$

**Next, divide by the mass of one can to find the number of cans that could be made.**

**73548 g Al ÷ 15.0 g Al per can = $4.90 \times 10^3$ cans (rounded to three significant figures)**

Problem 1.75 The "Western Stone" in Jerusalem is one of the largest stone building blocks ever to have been used. It has a mass of 517 metric tons, and measures 13.6 m long, 3.00 m high and 3.30 m wide. What is the density of this rock in $g/cm^3$? (1 metric ton = 1000 kg)

**Assuming the "Western Stone" is rectangular, the volume is calculated by multiplying length × height × width: 13.6 m × 3.00 m × 3.30 m = 134.64 $m^3$.**

**Converting to centimeters:** $134.64 \text{ m}^3 \times \frac{(100 \text{ cm})^3}{1 \text{ m}^3} = 1.3464 \times 10^8 \text{ cm}^3$

**The mass is** $517 \text{ metric tons} \times \frac{1000 \text{ kg}}{1 \text{ metric ton}} \times \frac{1000 \text{ g}}{1 \text{ kg}} = 5.17 \times 10^8 \text{ g}$

$$\text{Density} = \frac{\text{mass}}{\text{volume}} = \frac{5.17 \times 10^8 \text{ g}}{1.3464 \times 10^8 \text{ cm}^3} = 3.84 \text{ g/cm}^3$$

Problem 1.77 Suppose that a new material has been devised for a bicycle frame with an elastic modulus of $22.0 \times 10^6$ psi. Is this bike frame likely to be more or less stiff than an aluminum frame?

**The elastic modulus of aluminum is $10.0 \times 10^6$ psi. This is smaller than that of the new material, so aluminum is less stiff than the new material. The higher the value of elastic modulus, the less flexible that material is.**

Problem 1.79 Compare the strengths of aluminum, steel, and titanium. If high strength were needed for a particular design, would aluminum be a good choice?

**Yield strength is the property used to compare strength of metals. Yield strength is the amount of force required to cause failure of a standard-size amount of material. The higher the yield strength, the stronger the material is. Aluminum's yield strength (5–60 ×**

**$10^4$ psi) is much lower than that of either titanium (40–120 × $10^4$ psi) or iron (45–160 × $10^4$ psi). If strength were the most important feature of a bicycle's design, Al would not be a good choice.**

Problem 1.81 Use the web to research the differences in the design of steel-framed bicycles versus aluminum-framed bicycles. Write a report that details the similarities and differences you discover.

**Answers will vary according to the websites found. One of the primary differences is the diameter of the tubes in the frame. Aluminum bikes use larger diameter tubing than steel-frame bikes. Larger diameter tubes are stronger. Lightweight but weaker materials like aluminum require larger diameter tubing than steel-frame bikes.**

Problem 1.83 Use the web to research the relative cost of aluminum, steel, and titanium frames for bicycles. Speculate about how much of the relative cost is due to the costs of the materials themselves.

**Answers will vary according to the websites found. Titanium is the most expensive material and titanium-frame bikes are the most expensive. One could then conclude that the material of construction is closely related to the cost of the bike.**

Problem 1.85 Battery acid has a density of 1.285 g/mL and contains 38.0% sulfuric acid by mass. Describe how you would determine the mass of pure sulfuric acid in a car battery, noting which items(s) you would have to measure or look up.

**First you would measure the volume of battery acid and then using the density provided, calculate the mass of the acid solution:** $\text{V (mL)} \times \dfrac{1.285\text{ g}}{\text{mL}} = \text{mass (g)}$

**Next, using the percent sulfuric acid would allow the mass of pure sulfuric acid to be determined:** $\text{mass battery acid (g)} \times \dfrac{38.0\text{ g sulfuric acid}}{100\text{ g battery acid}} = \text{mass pure sulfuric acid (g)}$

Problem 1.87 A solution of ethanol in water has a volume of 54.2 mL and a mass of 49.6 g What information would you need to look up and how would you determine the percentage of ethanol?

**With the mass and volume given, the density of the solution can be determined. This density is the weighted average of the densities of pure ethanol and water, according to their percentages. Let $X$ = the fraction (percent/100) of ethanol and $(1 - X)$ = the fraction of water. Look up the density of pure ethanol and water. Multiply these fractions by their respective densities and these will sum to the density of the solution. Algebraically solve for $X$, the fraction of ethanol in the solution.**

Problem 1.89 Imagine you place a cork measuring 1.30 cm × 4.50 cm × 3.00 cm in a pan of water. On top of this cork you place a small cube of lead measuring 1.15 cm on a side. Describe how you would determine if the combination of the cork and the lead cube would still float in the water. Note any information you would need to look up to answer the question.

**The average density of the cork and the lead cube must be less than 1.0 g/mL for the combination to still float. The mass of each can be measured and the volume of each can be calculated. Dividing the total mass by the total volume will give the average density.**

## CHAPTER OBJECTIVE QUIZ

This quiz will test your understanding of the basic text chapter objectives and give you additional practice problems. You should work this quiz after completing the end-of-chapter questions. The solutions to these questions are found at the end of this chapter.

1. Which of the following statements about aluminum is not true?
   a. Aluminum does not occur naturally as a free metal.
   b. Aluminum is found in nature combined with other elements.
   c. Bauxite is the most common ore from which aluminum is produced.
   d. Only physical changes are necessary to extract the aluminum from the ore.
   e. In the smelting of aluminum, the metal is separated from other atoms by a chemical reaction.

2. A new form of matter is produced in the laboratory and its melting point, boiling point, and density are measured. Which perspective of chemistry are we using?

3. Why is the symbolic perspective of chemistry important in describing matter?

4. Which of the following are chemical changes?

   I. A nail rusts — II. A chocolate candy melts
   III. The dog's water evaporates — IV. Some milk goes sour

   a. I, II
   b. II, III
   c. I, IV
   d. I, II, IV
   e. All of them

5. Which of the following are physical properties of water?

   I. Density = 1.00 g/mL  II. Vapor pressure @ 25°C = 24 torr
   III. Decomposes into hydrogen and oxygen during electrolysis
   IV. Boiling point = 100 °C
   V. Forms a solution of calcium hydroxide when combined with calcium metal

   a. All of them
   b. I, IV
   c. I, II, IV
   d. III, V
   e. I, II

6. TRUE or FALSE: Chemical properties always involve a change in the composition of matter.

7. Use molecular pictures to describe the changes that occur as iron goes from a solid to a liquid to a gas.

8. TRUE or FALSE: Inductive reasoning begins with scientific observations and attempts to draw a universal conclusion from them.

9. Perform the following operations and express the answer in the proper number of significant figures.

   a. $0.01101 \times (5.08 \times 10^2) =$
   b. $17.44 - 11.7 =$

10. Express 38 miles per hour in kilometers per hour.

11. Express the density of titanium, 4.5 g/cm$^3$, in pounds per cubic feet.

12. Calculate the density of a copper cylinder that has a diameter of 1.25 cm, a length of 3.11 cm, and a mass of 34.0845 grams.

13. A quarter has a mass of 5.6579 g and a density of 7.89 g/cm$^3$. What is the volume of the quarter?

14. How would you go about determining the density of an unknown solid that is irregularly shaped? Of an unknown liquid?

15. A brass nut is placed in a graduated cylinder containing 17.45 mL of water. The volume reading of the cylinder rises to 19.89 mL. The mass of the nut is 17.0073 g. What is the density of the brass nut in g/cm$^3$?

# ANSWERS TO THE CHAPTER OBJECTIVE QUIZ

1. d

2. The macroscopic perspective. Melting point, boiling point, and density are all properties observed in large, tangible amounts of matter.

3. Matter is made of large numbers of extremely small particles (atoms, molecules) which cannot be seen and whose shapes can only be approximated. Symbols are very convenient for representing these particles and allow us to consider changes in matter taking place on either a microscopic or macroscopic scale.

4. c The formation of rust and the production of substances in the milk that cause a sour taste represent changes in the composition of matter. Chocolate melting and water becoming gas are just physical changes.

5. c No change in the composition of water occurs when we observe the density, boiling point, or the vapor pressure of water.

6. TRUE

7. The particles in solid iron are very close and orderly. In the liquid state, they are only slightly farther apart but random. In the gaseous state, the particles are very far apart and completely independent.

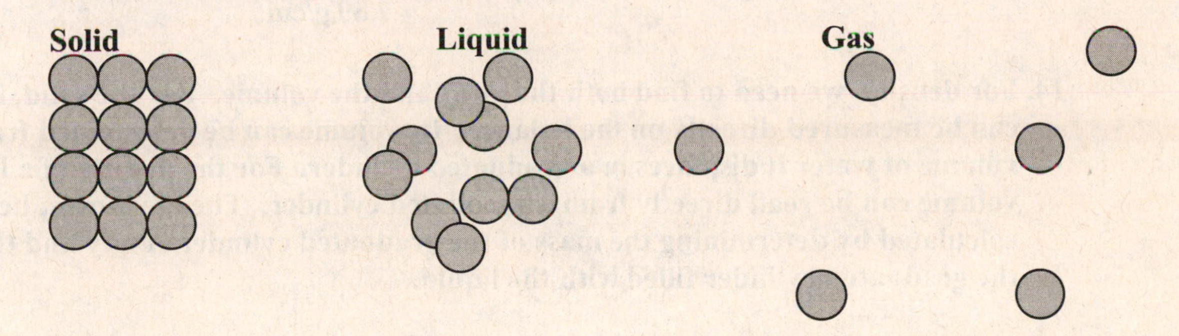


8. TRUE

9. When multiplying and dividing, the number of significant figures in the answer must be the same as the number of significant figures in the factor with the fewest significant figures. In subtracting and dividing, the number of significant figures depends on the position of the first doubtful digit. The answer should be rounded so that the last digit retained is the first uncertain digit.

   a. $0.01101 \times (5.08 \times 10^2) = ? = 5.59$
   b. $17.44 - 11.7 = ? = 5.7$

10. We need to convert miles to kilometers.

$$\frac{38 \text{ miles}}{\text{hour}} \times \frac{5280 \text{ feet}}{\text{miles}} \times \frac{12 \text{ inches}}{1 \text{ foot}} \times \frac{2.54 \text{ cm}}{1 \text{ inch}} \times \frac{1 \text{ meter}}{100 \text{ cm}} \times \frac{1 \text{ km}}{1000 \text{ m}} = 61 \text{ km/hr}$$

11. We need to convert from grams to pounds and from cubic centimeters to cubic feet.

$$\frac{4.5 \text{ g titanium}}{\text{cm}^3} \times \frac{1 \text{ pound}}{454 \text{ g}} \times \frac{1 \text{ cm}^3}{1 \text{ mL}} \times \frac{1000 \text{ mL}}{1 \text{ L}} \times \frac{28.316 \text{ L}}{1 \text{ ft}^3} = 280 \text{ lbs/ft}^3$$

12. Density is the ratio of mass per unit volume, density = $\frac{\text{mass}}{\text{volume}}$.

The volume of a cylinder is $V = \pi \times r^2 \times l$, $r = \frac{1.25 \text{ cm}}{2} = 0.625 \text{ cm}$, $l = 3.11 \text{ cm}$

$V = \pi \times (0.625 \text{ cm})^2 \times 3.11 \text{ cm} = 3.82 \text{ cm}^3$

$$\text{Density} = \frac{34.0845 \text{ g Cu}}{3.82 \text{ cm}^3} = 8.92 \text{ g Cu/cm}^3$$

13. Density = $\frac{\text{mass}}{\text{volume}}$, so, volume = $\frac{\text{mass}}{\text{density}}$

The volume of the quarter is: volume = $\frac{5.6579 \text{ g}}{7.89 \text{ g/cm}^3} = 0.717 \text{ cm}^3$

14. For density, we need to find both the mass and the volume. For the solid, its mass can be measured directly on the balance. Its volume can be determined from the volume of water it displaces in a graduated cylinder. For the density of a liquid, its volume can be read directly from a graduated cylinder. The mass must be calculated by determining the mass of the graduated cylinder empty and the mass of the graduated cylinder filled with the liquid.

15. The volume of the nut is the increase in water volume after it is added to the cylinder. The density is the mass divided by the volume. 1 cm$^3$ = 1 mL.

$V = 19.89 - 17.45 = 2.44 \text{ mL}$

$$\text{Density} = \frac{17.0073 \text{ g}}{2.44 \text{ mL}} \times \frac{1 \text{ mL}}{1 \text{ cm}^3} = 6.97 \text{ g/cm}^3$$

# CHAPTER 2

## *Study Goals*

The study goals outline specific concepts to be mastered in this section of the text chapter. Related problems at the end of the text chapter will also be noted. Working the questions noted should aid in mastery of each study goal and also highlight any areas that you may need additional help in.

**Section 2.1 *INSIGHT INTO* Polymers**

1. Understand how polymer molecules are formed and name at least three common polymers and give examples of their uses. *Work Problems 1–7.*

**Section 2.2 Atomic Structure and Mass**

2. Depict the general structure of an atom in terms of protons, neutrons, and electrons. *Work Problem 8.*
3. Describe isotopes in terms of atomic number and mass number, using atomic symbols. Understand how a mass spectrometer gives information about an atom's isotopes. *Work Problems 9–16.*
4. Calculate the atomic mass of an element using isotopic abundance data. *Work Problems 17–23, 84, 86–88.*

**Section 2.3 Ions**

5. Define monoatomic and polyatomic ions, cations and anions. Discuss the type of attractive forces involved between positive and negative ions. *Work Problems 24–34.*

**Section 2.4 Compounds and Chemical Bonds**

6. Define chemical and empirical formulas and be familiar with the common rules for writing chemical formulas. *Work Problems 35–36, 39, 40, 44–47, 85, 89*
7. Distinguish between ionic, metallic, and covalent bonds and be able to identify which chemical bonds are present in various forms of matter. *Work Problems 37, 38, 41–43.*

**Section 2.5 The Periodic Table**

8. List the basic features of the periodic table and discuss how its form is related to the periodic law. *Work Problems 47–51, 91, 92.*
9. Be able to identify special areas on the periodic table including the alkali metals, alkaline earth metals, noble gases, halogens, main group elements, transition metals, lanthanides, and actinides. *Work Problems 52, 53.*
10. Distinguish among metals, nonmetals, and metalloids. Be able to locate each element type on the periodic table. *Work Problems 54–60.*

**Section 2.6 Inorganic and Organic Chemistry**

11. Distinguish among organic and inorganic chemistry and which elements are involved in those areas. *Work Problems 61–64.*

12. Use line structures to represent larger organic molecules. *Work Problems 66–70.*
13. Recognize several common functional groups found in organic molecules. *Work Problem 65.*

**Section 2.7 Chemical Nomenclature**

14. Use the IUPAC rules of nomenclature to write correct names of simple molecular and ionic inorganic compounds. *Work Problems 71–79.*

**Section 2.8 *INSIGHT INTO* Polyethylene**

15. Recognize the importance of polyethylene in consumer products and how the properties of polyethylene depend largely on the physical properties of the individual polymer molecules. *Work Problems 80–83, 90, 93.*

## *Solutions to the Odd-Numbered Problems*

Problem 2.1 Define the terms *polymer* and *monomer* in your own words.

**A polymer is a large molecule made up of many small repeating units of atoms, called monomers, bonded together. A monomer is a small, unique group of atoms which can be bonded together many times to produce this large molecule (polymer) made of one repeating unit of atoms.**

Problem 2.3 Look around you and identify several objects that you think are probably made of polymers.

**Answers will vary. Some possible responses include clothes, carpet fibers, a calculator case, soda bottles, etc.**

Problem 2.5 The fact that a polymer's physical properties depend on its atomic composition is very important in making these materials so useful. Why do you think this would be so?

**Different monomers have different physical properties, because they are made with different combinations of atoms. By making polymers with different monomers, a scientist can design a polymer for a specific purpose.**

Problem 2.7 Use the web to research the amount of polyethylene produced annually in the United States. What are the three most common uses of this polymer?

**Answers will vary. According to the American Chemistry Council, Production for Polyethylene in 2007 was: LDPE (low-density polyethylene) 7.9 billion pounds, LLDPE (linear low-density polyethylene) 13.6 billion pounds, HDPE (high-density polyethylene) 18.2 billion pounds. Depending on how the usage is grouped, the top three uses are most likely (1) product containers like shampoo and motor oil bottles, (2) thin polymer films (for construction, greenhouses, etc.), and (3) grocery and garbage bags.**

Problem 2.9 Why is the number of protons called the atomic number?

**Each different element has a unique number of protons in its nucleus that distinguishes it from other elements. Therefore, that "atomic" number can identify an atom. The number of neutrons in the nucleus is not unique nor is the number of electrons, if ions are included.**

Problem 2.11 Define the term isotope.

**Isotopes are different atomic forms of an element that have different numbers of neutrons in the nucleus.**

Problem 2.13 ■ How many protons, neutrons, and electrons are there in each of the following atoms?

**The atomic number of the element gives the number of protons. This will equal the number of electrons in a neutral atom. The mass number is the sum of protons and neutrons; so subtracting the atomic number from the mass number gives the number of neutrons.**

a. magnesium−24, $^{24}Mg$ **12 protons, 12 electrons, 24 – 12 = 12 neutrons**

b. tin−119, $^{119}Sn$ **50 protons, 50 electrons, 119 – 50 = 69 neutrons**

c. thorium−232, $^{232}Th$ **90 protons, 90 electrons, 232 – 90 = 142 neutrons**

d. carbon−13, $^{13}C$ **6 protons, 6 electrons, 13 – 6 = 7 neutrons**

e. copper−63, $^{63}Cu$ **29 protons, 29 electrons, 63 – 29 = 34 neutrons**

f. bismuth−205, $^{205}Bi$ **83 protons, 83 electrons, 205 – 83 = 122 neutrons**

Problem 2.15 Mercury is 16.716 times more massive than carbon−12. What is the atomic mass of mercury? Remember to express your answer with the correct number of significant figures.

**The mass of a carbon−12 atom is *exactly* 12 atomic mass units (amu). The mass of a mercury atom is**

**$16.716 \times 12$ amu = 200.59 amu (rounded to 5 significant digits)**

Problem 2.17 Explain the concept of a "weighted" average in your own words.

**A weighted average is the average of several samples that also accounts for the abundance of individual samples.**

Problem 2.19 ■ The atomic mass of copper is 63.55 amu. There are only two isotopes of copper, $^{63}Cu$ with a mass of 62.94 amu and $^{65}Cu$ with a mass of 64.93 amu. What is the percentage abundance for each of these two isotopes?

**Since there are only two isotopes, the percent abundance must sum to 100, and their fractions must sum to one. We can let $X$ = the fraction of $^{63}Cu$ and $(1 - X)$ = the fraction of $^{65}Cu$. The given atomic mass of copper is the weighted average of the two isotopes. To calculate the "weighted" average mass of copper, we convert each percentage into a fraction, multiply each fraction by the mass of that isotope, and add them together.**

**$X$ (mass of $^{63}Cu$) + $(1 - X)$ (mass of $^{65}Cu$) = 63.55 amu (weighted average mass of copper)**
**$X$ (62.94 amu) + $(1 - X)$ (64.93 amu) = 63.55 amu**

**62.94 $X$ amu + 64.93 amu – 64.93 $X$ amu = 63.55 amu**

$$1.38 \text{ amu} = 1.99\, X \text{ amu} \quad \rightarrow \quad X = \frac{1.38 \text{ amu}}{1.99 \text{ amu}} = 0.693$$

**Therefore the percentage of $^{63}Cu$ is 69.3%.**

**The percentage of $^{65}Cu$ is 100% – 69.3% = 30.7%**

Problem 2.21 ■ Naturally occurring uranium consists of two isotopes, whose masses and abundances are shown below.

| Isotope | Abundance | Mass |
|---|---|---|
| $^{235}U$ | 0.720% | 235.044 amu |
| $^{238}U$ | 99.275% | 238.051 amu |

Only $^{235}U$ can be used as fuel in a nuclear reactor, so uranium for use in the nuclear industry must be enriched in this isotope. If a sample of enriched uranium has an average atomic mass of 235.684 amu, what percentage of $^{235}U$ is present?

**The percent abundances in the enriched sample are different from above, but they must still sum to 100% and the isotopic masses are still the same. We can let $X$ = the fraction of $^{235}U$ and $(1 - X)$ = the fraction of $^{238}U$. The given average atomic mass of uranium is the weighted average of the two isotopes. To calculate a "weighted" average mass of uranium, we convert each percentage into a fraction, multiply each fraction by the mass of that isotope, and add them together.**

**$X$ (mass of $^{235}U$) + $(1 - X)$ (mass of $^{238}U$) = 235.684 amu (weighted average mass of uranium)**
**$X$ (235.044 amu) + $(1 - X)$ (238.051 amu) = 235.684 amu**

**235.044$X$ amu + 238.051 amu – 238.051$X$ amu = 235.684 amu**

**2.367 amu = 3.007$X$ amu** $\rightarrow$ $X = \dfrac{2.367\text{ amu}}{3.007\text{ amu}} = 0.7872$

**Therefore the percentage of $^{235}U$ is 78.72%.**

Problem 2.23 Chlorine has only two isotopes, one with mass 35 and the other with mass 37. One is present at roughly 75% abundance and the atomic mass of chlorine on a periodic table is 35.45. Which must be the correct mass spectrum for chlorine?

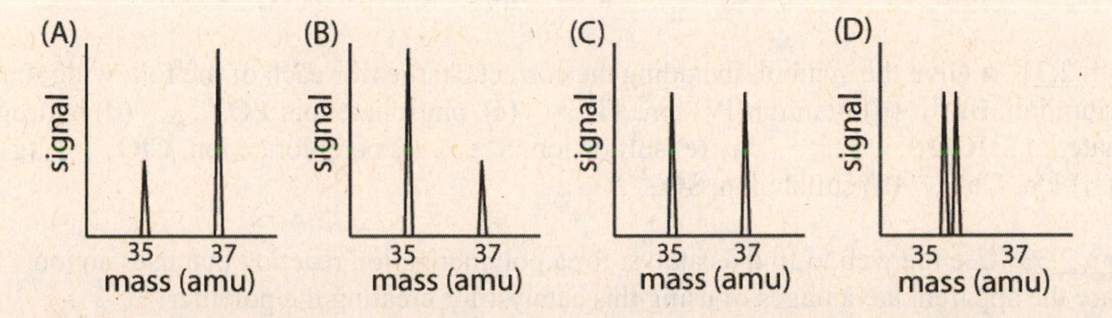


**The correct spectrum is (B), which shows the 35-amu isotope with approximately three times greater abundance (signal strength) than the 37-amu isotope. This corresponds to the 75% abundance of one isotope, and it has to be the one at 35 amu to yield an atomic mass of 35.45 amu:**

**0.75 (35 amu) + 0.25 (37 amu) = 35.5 amu**

Problem 2.25 What is the difference between cations and anions?

**A cation is an atom or group of atoms that have *lost* one or more electrons, making them positively charged. Anions are atoms or groups of atoms that have *gained* one or more electrons, making them negatively charged.**

Problem 2.27 ■ How many protons and electrons are in each of the following ions?

**The atomic number will give the number of protons for the ion. The charge of the ion indicates how many electrons were gained or lost. A positive charge means electrons (negative) were lost, a negative charge means electrons were gained. Subtract the charge number from the atomic number to get the number of electrons in a cation, or add the charge number to the atomic number to get the electrons in an anion.**

| | | | |
|---|---|---|---|
| (a) | $Na^+$ | **Atomic number = 11,** | **11 protons; 11 – 1 = 10 electrons** |
| (b) | $Al^{3+}$ | **Atomic number = 13,** | **13 protons; 13 – 3 = 10 electrons** |
| (c) | $S^{2-}$ | **Atomic number = 16,** | **16 protons; 16 + 2 = 18 electrons** |
| (d) | $Br^-$ | **Atomic number = 35,** | **35 protons; 35 + 1 = 36 electrons** |

Problem 2.29 ■ Write the atomic symbol for the element whose ion has a 2− charge, has 20 more neutrons than electrons, and has a mass number of 126.

**We'll use the symbols p, n, and e for the number of protons, neutrons and electrons. The mass number, 126 = p + n. The neutrons outnumber the electrons by 20, n = e + 20. And because of the 2− charge, there are two more electrons than protons, e = p + 2.**
**Substituting, $126 = (e - 2) + (e + 20)$, $126 = 2e + 18$, $e = 54$**
**Thus, $p = 54 - 2 = 52$**
**Therefore the atomic number is 52, making this the telluride ion, $^{126}_{52}Te^{2-}$.**

Problem 2.31 ■ Give the symbol, including the correct charge, for each of the following ions. **(a)** barium ion, **$Ba^{2+}$**, **(b)** titanium(IV) ion, **$Ti^{4+}$**, **(c)** phosphate ion, **$PO_4^{3-}$**, **(d)** hydrogen carbonate ion, **$HCO_3^-$**, **(e)** sulfide ion, **$S^{2-}$**, **(f)** perchlorate ion, **$ClO_4^-$**, **(g)** cobalt(II) ion, **$Co^{2+}$**, **(h)** sulfate ion, **$SO_4^{2-}$**.

Problem 2.33 Use the web to find a catalyst for a polymerization reaction that uses an ion. What are the apparent advantages of using this catalyst for creating the polymer?

**Answers will vary. One such catalyst used in ionic polymerization is butyl-lithium. This substance decomposes into a lithium ion (1+ charge) and a butyl ion (1− charge). One advantage of ionic polymerization is that the polymer formed is easier to process further to produce desirable properties.**

Problem 2.35 How many atoms of each element are represented in the formula, $Ba(OH)_2$?

**Subscripts give the numbers of atoms present in the formula. Subscripts outside parenthesis mean all atoms inside are multiplied by that number.**
**In one formula unit of $Ba(OH)_2$, there are one Ba atom, two O atoms, and two H atoms.**

Problem 2.37 In general, how are electrons involved in chemical bonding?

**There are 3 major types of chemical bonds: ionic, covalent, and metallic. They all involve *sharing* or *transfer* of electrons in some way. Ionic bonds have electrons transferred from the cation to the anion. Covalent bonds result from shared pairs of electrons between two atoms. Metallic bonding has metal atoms transfer some of their electrons to a "common pool" where they are all shared equally.**

Problem 2.39 When talking about the formula for an ionic compound, why do we typically refer to a formula unit rather than a molecule?

**Molecules are discrete groups of neutral atoms bonded together with covalent bonds. An ionic compound is not made of discrete units but large collections of cations and anions held together with electrostatic attractive forces (ionic bonds). Because ionic compounds**

**are not made of molecules, it is incorrect to refer to them with the term "molecule". The smallest ratio of these ions that gives a neutral formula is called the "formula unit" of the compound.**

Problem 2.41 Explain the difference between ionic and metallic bonding.

**In ionic bonding, electrons are transferred from one atom to another, creating positive cations and negative anions. These are attracted to each other with electrostatic forces, and they form a large array in a particular lattice structure. In metallic bonding, metal atoms release some of their electrons (outer ones) to form a common pool of electrons, from which all the atoms share. The positive metal atom cores are attracted to this "mobile sea of electrons" and are held together in some particular lattice arrangement.**

Problem 2.43 Describe how a covalently bonded molecule is different from compounds that are either ionic or metallic.

**In a molecule, electrons are shared in pairs between two atoms, as opposed to transferring electrons from one atom to another in ionic compounds or transferring electrons to a common pool in metallic solids. Molecular substances exist as discrete particles, but ionic and metallic substances exist as large arrays of particles.**

Problem 2.45 Why are empirical formulas preferred for describing polymer molecules?

**Polymers are made of a very large number of monomers bonded together: hundreds or even thousands. Sometimes it is not even possible to tell exactly how many. It is much more convenient to represent a polymer with its empirical or monomer formula.**

Problem 2.47 Polybutadiene is a synthetic elastomer, or rubber. The corresponding monomer is butadiene, which has a molecular formula $C_4H_6$. What is the empirical formula of butadiene?

**The empirical formula is the smallest whole number ratio of atoms present in the compound. The empirical formula for butadiene, $C_4H_6$, is $C_2H_3$.**

Problem 2.49 How does the periodic table help to make the study of chemistry more systematic?

**The periodic table organizes elements according to the size and structure of the atom. This allows us to identify trends in properties of atoms and to make predictions in their behavior.**

Problem 2.51 How do binary compounds with hydrogen illustrate the concept of periodicity?

**It is observed that hydrogen forms a compound with elements in a given group always in the same ratio. This ratio changes across the periodic table, indicating a change in the structure of the atom from the left to right side of the periodic table. For example:**

**Group 17 – HF, HCl, HBr, HI**

**Group 16 – $H_2O$, $H_2S$, $H_2Se$, $H_2Te$**

**Group 15 – $NH_3$, $PH_3$, $AsH_3$**

Problem 2.53 Name the group to which each of the following elements belongs:

**Groups are the columns of the periodic table. Each one is labeled sequentially from left to right, 1–18.**

(a) K **Group 1**

(b) Mg **Group 2**

(c) Ar **Group 18**

(d) Br **Group 17**

Problem 2.55 ■ Identify the area of the periodic table in which you would expect to find each of the following types of elements. (**a**) a metal, (**b**) a nonmetal, (**c**) a metalloid

**The metals are located in the middle and left side of the table, with the exception of hydrogen, a nonmetal. The nonmetals are found on the right side of the periodic table, and more so the upper right. Metalloids are sandwiched in between the metals and nonmetals along a stair-step line about ¾ of the way from the left side of the periodic table.**

Problem 2.57 What is a metalloid?

**Metalloids are elements that exhibit some properties of both nonmetals and metals. Since they cannot be classified as either metals or nonmetals, they have a separate category and they exist in the region between metals and nonmetals on the periodic table.**

Problem 2.59 ■ Classify the following elements as metals, metalloids, or nonmetals:

**Metals are generally left and in the middle on the periodic table, with nonmetals on the right, and metalloids in between.**

| | | | | | | | | |
|---|---|---|---|---|---|---|---|---|
| (a) | Si | **Metalloid** | (b) | Zn | **Metal** | (c) | B | **Metalloid** |
| (d) | N | **Nonmetal** | (e) | K | **Metal** | (f) | S | **Nonmetal** |

Problem 2.61 The chemistry of the main group elements is generally simpler than that of the transition metals. Why is this so?

**Most of the transition metals can form more than one cation, sometimes five or more. Elements in the main groups usually form only one anion or cation.**

Problem 2.63 What is meant by the phrase *organic* chemistry?

***Organic* chemistry is the branch of chemistry primarily concerned with molecules containing carbon bonded to other carbon atoms. It is sometime referred to as the chemistry of living things because living organisms are made mostly of organic molecules.**

Problem 2.65 What is a *functional group*? How does the concept of a functional group help to make the study of organic chemistry more systematic?

**A *functional group* is a specific group of atoms that is bonded to an organic molecule. This group has certain properties and causes the molecule to behave in a certain way. Organic molecules are classified according to which functional groups are present and the name of the molecule will reflect that functional group.**

Problem 2.67 ■ Not all polymers are formed by simply linking identical monomers together. Polyurethane, for example, can be formed by reacting the two compounds shown below with one another. Write molecular and empirical formulas for each of these two substances.

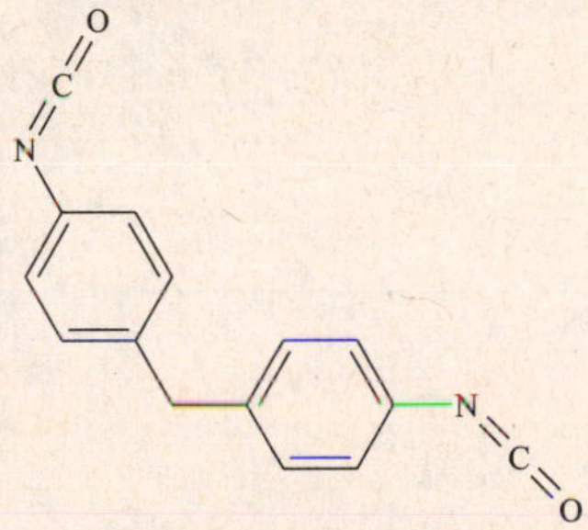

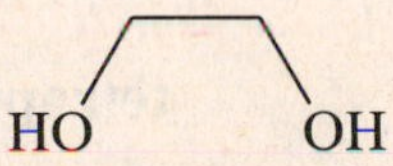

O C N N C O HO OH

**In line structures for organic molecules, each line represents a single bond. The symbols for carbon atoms are not shown, therefore at the terminus of each line, if a symbol is not shown, a C atom is assumed. The symbols of hydrogen atoms bonded to C are also not shown, so if carbon does not have four bonds shown in the structure, we assume the missing bonds are C – H bonds.**

**The empirical and molecular formulas for the molecule on the left are $C_{15}H_{10}N_2O_2$. The molecular formula for the molecule on the right is $C_2H_6O_2$. The empirical formula is $CH_3O$.**

Problem 2.69 The figure below shows the structure gamma-aminobutanoic acid, or GABA. This molecule is a neurotransmitter. Some of the effects of alcohol consumption are due to the interaction between ethanol and GABA. Write the correct molecular formula for this compound.

H2N OH O

**The molecular formula for GABA is $C_4H_9NO_2$.**

Problem 2.71 Why are there different rules for naming covalent and ionic binary compounds?

**In molecules, the atoms combined are neutral and they can bond in many ratios. The only way to distinguish different formulas with a name is by using Greek prefixes. In ionic compounds, the bonded species have a charge and there is only one possible ratio that will give a neutral formula: prefixes are not needed.**

Problem 2.73 ■ Name the following compounds:

**When naming binary covalent compounds, name the first element, leave a space, then name the second element. The second element's name is modified to end in "– ide". For multiples of a type of atom, Greek prefixes are used. Two-syllable prefixes that end in a vowel drop the second vowel when preceding "oxide".**

(a) $N_2O_5$ **dinitrogen pentoxide**

(b) $S_2Cl_2$ **disulfur dichloride**

(c) $NBr_3$ **nitrogen tribromide**

(d) $P_4O_{10}$ **tetraphosphorus decoxide**

Problem 2.75 ■ Write the molecular formula for each of the following covalent compounds:

(a) sulfur hexafluoride **$SF_6$**

(b) bromine pentafluoride **$BrF_5$**

(c) disulfur dichloride **$S_2Cl_2$**

(d) tetrasulfur tetranitride **$S_4N_4$**

Problem 2.77 ■ Name each of the following compounds: **(a)** $MgCl_2$, **(b)** $Fe(NO_3)_2$, **(c)** $Na_2SO_4$, **(d)** $Ca(OH)_2$, **(e)** $FeSO_4$

**When naming ionic compounds, first name the cation, leave a space, and then name the anion. Greek prefixes are never used, even if there is more than one of a type of ion. If a cation has more than one possible charge, a roman numeral in parenthesis indicating the charge directly follows the name of the cation with no space.**

(a) $MgCl_2$ **magnesium chloride**

(b) $Fe(NO_3)_2$ **iron(II) nitrate**

(c) $Na_2SO_4$ **sodium sulfate**

(d) $Ca(OH)_2$ **calcium hydroxide**

(e) $FeSO_4$ **iron(II) sulfate**

Problem 2.79 ■ Name the following compounds:

**When naming ionic compounds with polyatomic ions, the rules are the same. Name the cation, leave a space, and name the anion.**

(a) $PCl_5$ **phosphorus pentachloride**

(b) $Na_2SO_4$ **sodium sulfate**

(c) $Ca_3N_2$ **calcium nitride**

(d) $Fe(NO_3)_3$ **iron(III) nitrate**

(e) $SO_2$ **sulfur dioxide**

(f) $Br_2O_5$ **dibromine pentoxide**

Problem 2.81 How do molecules of low-density polyethylene and high-density polyethylene differ? How do these molecular scale differences explain the differences in the macroscopic properties of these materials?

**Low-density polyethylene (LDPE) is made of polymer molecules that are highly branched. Because of this branching, the molecules cannot pack closely together, much like a pile of twisted branches from a tree. The resulting density is lower because the mass per volume is decreased. High-density polyethylene (HDPE) has polymer molecules that are much straighter with little branching. These can pack more tightly together, increasing the mass per volume, or density. LDPE is weaker, softer, and more flexible; HDPE is stronger and harder.**

Problem 2.83 Use the web to determine the amount of low-density and high-density polyethylene produced annually in the United States. Which uses predominate in the applications of these two materials?

**According to the American Chemistry Council, 6.8 billion pounds of LDPE and 17.6 billion pounds of HDPE were produced in the United States in 2012. The largest use for HDPE is**

**consumer product bottles; milk jugs and quart bottles of motor oil are examples. LDPE is used in plastic liners, garbage and storage bags.**

Problem 2.85 Many transition metals produce more than one ion. For example, iron has ions with charges of 2+ and 3+ that are both common. How could you use the compounds of a transition metal with oxygen to determine the charge of a metal ion? Use iron as your example.

**The key is the fact that almost always oxygen forms a 2– charge when it bonds with metals, especially transition metals like iron. An experiment could be performed where the mass of metal before and after reacting with oxygen is determined. Then the ratios of moles of atoms could be calculated (the formula).**

**In the case of iron the formula of the resulting oxide would be either FeO or $Fe_2O_3$, depending on reaction conditions. Then, assuming that the charge of the oxide ion is 2–, we can calculate the charge of iron that gives us a neutral compound:**

**$Fe + (O^{2-}) = 0$ net charge, $Fe = 2+$ or $Fe^{2+}$**

**$2\ (Fe) + 3\ (O^{2-}) = 0$ net charge, $Fe = 3+$ or $Fe^{3+}$**

Problem 2.87 ■ Strontium has four stable isotopes. Strontium−84 has a very low natural abundance, but $^{86}Sr$, $^{87}Sr$, and $^{88}Sr$ are all reasonably abundant. Knowing that the atomic mass of strontium is 87.62, which of the more abundant isotopes predominates?

**The weighted average mass of Sr is 87.62 amu, which is closest to the mass of strontium-88. Equal abundance of the three isotopes would yield an average around 87 amu. Therefore the most abundant isotope must be strontium−88.**

Problem 2.89 Two common oxides of iron are FeO and $Fe_2O_3$. Based on this information, tell how you would predict two common compounds of chlorine and iron.

**The oxide anion is always a 2– charge. Therefore we can determine what the two charges for iron are in FeO and $Fe_2O_3$.**

**$Fe + (O^{2-}) = 0$ net charge, $Fe = 2+$ or $Fe^{2+}$**

**$2\ (Fe) + 3\ (O^{2-}) = 0$ net charge, $Fe = 3+$ or $Fe^{3+}$**

**Chloride anions are always 1–, so we can predict the formula of compounds containing iron and chlorine based on what combination of cation and anion gives us a neutral formula.**

**$(Fe^{2+}) + 2\ (Cl^-) = 0$ net charge: $FeCl_2$ $(Fe^{3+}) + 3\ (Cl^-) = 0$ net charge: $FeCl_3$**

Problem 2.91 Engineers who design bicycle frames are familiar with the information in Table 1.3, which lists the densities of aluminum (2.699 g/cm$^3$), steel (7.87 g/cm$^3$), and titanium (4.507 g/cm$^3$). How does this information compare with Figure 2.12 and what would it suggest for changes in this figure if more shades were used for the density color-coding? (Iron is the principle component of steel.)

**The density of iron is about 1.75 times greater than that of titanium, yet their shade color is the same in Figure 2.12. The color shades should be darker towards the middle of Period 2, around Groups 7−11, as the density of those metals increases towards the right.**

Problem 2.93 LDPE has a density range of 0.915–0.935 g/cm$^3$, and HDPE has a density in the range of 0.940–0.965 g/cm$^3$. You receive a small disk, 2.0 cm high with a 6.0 cm diameter from a manufacturer of polyethylene, but its label is missing. You measure the mass of the disk and find that it is 53.8 g. Is the material HDPE or LDPE?

**We can calculate the volume of the disk and then use the mass to determine the density.**

**Volume of a cylinder = $\pi \times r^2 \times l$; $V = \pi \times (6.0 \text{ cm} \div 2)^2 \times 2.0 \text{ cm} = 57 \text{ cm}^3$**

**Density = mass/volume; $d = 53.8 \text{ g} \div 57 \text{ cm}^3 = 0.95 \text{ g/cm}^3$**

**This density falls in the range of HDPE; therefore the disk is HDPE.**

## CHAPTER OBJECTIVE QUIZ

This quiz will test your understanding of the basic text chapter objectives and give you additional practice problems. You should work this quiz after completing the end-of-chapter questions. The solutions to these questions are found at the end of this chapter.

1. Which of the following polymers is incorrectly matched with its use?
   a. Polyethylene – shampoo bottles
   b. Poly(vinylchloride) – egg cartons
   c. Polystyrene – foam coffee cups
   d. PVC – water pipe
   e. Poly(vinylidene chloride) – plastic wrap

2. TRUE or FALSE: Isotopes are different forms of an atom with different numbers of protons in the nucleus.

3. Which of the following terms are incorrectly matched with their definitions?
   a. Ion – an atom that has gained or lost one or more electrons.
   b. Functional group – a group of atoms that has unique properties and changes the behavior of the molecule to which it is attached.
   c. Atom – the fundamental unit of matter.
   d. Molecule – a unit of matter consisting of two or more ions bonded to each other.
   e. Polymer – large molecules made of small repeating units of atoms.

4. TRUE or FALSE: The nucleus and the electron cloud occupy about equal volumes in an atom.

5. Identify the number of protons, neurons, and electrons in the following species.

   (a) $^{2}H$  (b) $^{35}Cl$  (c) $^{90}Sr$  (d) $^{235}U$  (e) $^{141}Ba^{2+}$  (f) $^{15}N^{3-}$

6. What monoatomic species has two more protons than electrons, 14 more neutrons than protons, and has mass number equal to 90?

7. Naturally occurring uranium has three isotopes (uranium–235 is the one used for nuclear reactors and bombs):

| Isotope: | $^{234}U$ | $^{235}U$ | $^{238}U$ |
|---|---|---|---|
| Abundance: | 0.0055 % | 0.720 % | 99.2745 % |
| Mass (amu): | 234.0409 | 235.0439 | 238.0508 |

Calculate the atomic mass of uranium with these data.

8. TRUE or FALSE: The empirical formula and the molecular formula of a compound cannot be the same.

9. Answer the following questions based on the chemical formula of these compounds.

   a. How many oxygen atoms are present in one formula unit of $Ca_3(PO_4)_2$?

   b. One molecule of $C_2H_2Cl_2F_2$ contains how many atoms?

   c. How many atoms of each element are present in one formula unit of $KAl(SO_4)_2 \bullet 12H_2O$?

10. Which of the following statements about chemical bonds is false?
    a. All chemical bonds involve sharing or transfer of electrons in some way.
    b. Bonding results in increased energy of the bonded atoms.
    c. Ionic bonds are electrostatic forces of attraction between oppositely charged ions.
    d. Covalent bonds are attractive forces between two atoms when they share a pair of electrons.
    e. In metallic bonds, a mobile sea of electrons is shared by all atoms within the lattice.

11. Match the type of bond (metallic, ionic, or covalent) to each substance listed below.
    a. $S_8$
    b. CuZn alloy
    c. Octane, $C_8H_{18}$
    d. $NH_4Cl$
    e. $FeCO_3$
    f. Pb

12. State the periodic law.

13. Which of the following statements about the periodic table is false?
    a. Dmitri Mendeleev is credited with using the periodic law to establish a table of elements.
    b. Horizontal rows are called periods; there are seven.
    c. Vertical columns are called groups; there are 18.
    d. The alkali metals are the elements in Group 1, including hydrogen.
    e. The noble gases are the elements in Group 18.

14. Match the following elements to a description.

| | |
|---|---|
| a. Nd | i. nonmetal |
| b. I | ii. metalloid |
| c. Pd | iii. halogen |
| d. Sr | iv. lanthanide |
| e. Ge | v. transition metal |
| f. S | vi. alkaline earth metal |

15. Distinguish between the terms *organic* chemistry and *inorganic* chemistry. Identify the following compounds as being either organic or inorganic.

(a) $CH_4$ (b) $CuSO_4$ (c) $PCl_3$ (d) $CHCl_3$ (e) $C_8H_{18}$ (f) FeO

16. Give the correct chemical formula for the following organic molecules.

Styrene

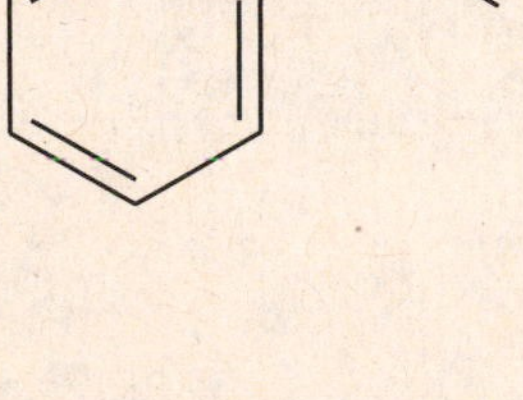

Acrylonitrile

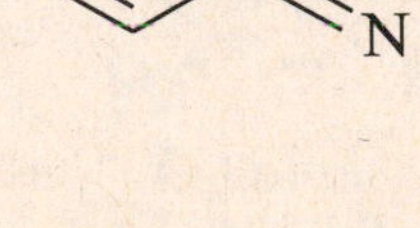

N

17. Identify the type of functional group in the following organic molecules.

a. $(CH_3)_2NH$
b. $CH_3CH_3OH$
c. $H_2C{=}CH_2$
d. Acetanilide
e. Chloroform ($CHCl_3$)

18. Give the correct IUPAC names for the following compounds:

(a) $Cu(NO_3)_2$ (b) $N_2O_7$ (c) CdS (d) $CaCO_3$ (e) $PCl_5$

19. Give the formula for the following compounds.

a. lead (II) sulfate
b. carbon disulfide
c. dibromine pentoxide
d. strontium iodide

# ANSWERS TO THE CHAPTER OBJECTIVE QUIZ

1. b

2. FALSE

3. d

4. FALSE

5.

| | |
|---|---|
| a. $^{2}H$ | 1 proton, 1 electron, 2 – 1 = 1 neutron |
| b. $^{35}Cl$ | 17 protons, 17 electrons, 35 – 17 = 18 neutrons |
| c. $^{90}Sr$ | 38 protons, 38 electrons, 90 – 38 = 52 neutrons |
| d. $^{235}U$ | 92 protons, 92 electrons, 235 – 92 = 143 neutrons |
| e. $^{141}Ba^{2+}$ | 56 protons, 54 electrons, 141 – 56 = 85 neutrons |
| f. $^{15}N^{3-}$ | 7 protons, 10 electrons, 15 – 7 = 8 neutrons |

6. The atomic mass of uranium is the weighted average of the masses of the three isotopes.

$$\frac{0.0055}{100}(234.0409 \text{ amu}) + \frac{0.720}{100}(235.0439 \text{ amu}) + \frac{99.2745}{100}(238.0508 \text{ amu}) =$$

$= 238.03$ amu

7. We'll use the symbols p, n, and e for the number of protons, neutrons and electrons. We have three equations:
The mass number = 90 = p + n;  p = e + 2;  n = p + 14
Substituting n = (e +2) + 14 and then expressing the p and n in the maas number equation in terms of electrons,
90 = (e + 2) + (e + 16);  72 = 2e;  36 = e
So the number of protons = 38, strontium: $^{90}Sr^{2+}$

8. FALSE

9. (a) There are 2 × 4 = 8 oxygen atoms per formula unit of $Ca_3(PO_4)_2$

(b) There are a total of eight atoms in one molecule of $C_2H_2Cl_2F_2$: two C atoms, two H atoms, two Cl atoms, and 2 F atoms.

**(c) This is a hydrate compound; there are twelve water molecules per formula unit of compound. These atoms must be counted as well. There are one K atom, one Al atom, two S atoms, eight O atoms, 24 H atoms, and another 12 O atoms for a total of 48 atoms.**

**10. b**

**11. a. covalent b. metallic c. covalent d. ionic e. ionic f. metallic**

**12. The periodic law states that when the elements are properly arranged, they display regular and periodic variation in their chemical properties.
Proper arrangement refers to placing them in order of increasing numbers of protons and also to how their electrons are structured in repeating patterns.**

**13. d**

**14.**

**a. Nd – iv. lanthanide**

**b. I – iii. halogen**

**c. Pd – v. transition metal**

**d. Sr – vi. alkaline earth metal**

**e. Ge – ii. metalloid**

**f. S – i. nonmetal**

**15. Organic chemistry is the chemistry of compounds containing mainly C – C and C – H bonds. Inorganic chemistry is the chemistry of the main group (groups 1, 2 and 13 – 18) and the transition elements (groups 3 – 12).**

**(a) $CH_4$ --- organic**

**(b) $CuSO_4$ --- inorganic**

**(c) $PCl_3$ --- inorganic**

**(d) $CHCl_3$ --- organic**

**(e) $C_8H_{18}$ --- organic**

**(f) FeO ---- inorganic**

**16. Styrene – $C_8H_8$** **Acrylonitile – $C_3H_3N$**

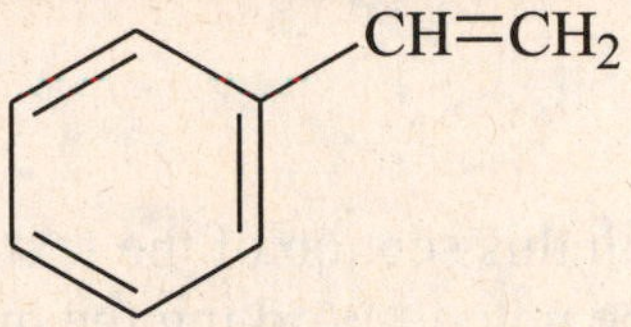

$C_6H_5$–CH=$CH_2$

$H_2C{=}CH{-}C{\equiv}N$:

**17.**

| | | |
|---|---|---|
| a. | $(CH_3)_2NH$ | amine |
| b. | $CH_3CH_3OH$ | alcohol |
| c. | $H_2C{=}CH_2$ | alkene |
| d. | Acetanilide | amide |
| e. | Chloroform ($CHCl_3$) | organic halide |

**18.** (a) $Cu(NO_3)_2$ --- copper (II) nitrate

(b) $N_2O_7$ --- dinitrogen heptoxide

(c) CdS --- cadmium sulfide

(d) $CaCO_3$ --- calcium carbonate

(e) $PCl_5$ --- phosphorus pentachloride

**19.**

a. lead (II) sulfate --- $PbSO_4$

b. carbon disulfide --- $CS_2$

c. dibromine pentoxide --- $Br_2O_5$

d. strontium iodide --- $SrI_2$

# CHAPTER 3

## *Study Goals*

The study goals outline specific concepts to be mastered in this section of the text chapter. Related problems at the end of the text chapter will also be noted. Working the questions noted should aid in mastery of each study goal and also highlight any areas that you may need additional help in.

**Section 3.1 *INSIGHT INTO* Biomass and Biofuel Engineering**

1. Describe the chemical processes used in biomass production and conversion to fossil fuels. *Work Problems 1–5.*

**Section 3.2 Chemical Formulas and Equations**

2. Recognize how a chemical equation must reflect the law of conservation of matter and describe the key features of a chemical equation. *Work Problems 6–10.*
3. Write and balance a chemical equation beginning with either an unbalanced equation or a description in words. *Work Problems 11–20, 77–84.*

**Section 3.3 Aqueous Solutions and Net Ionic Equations**

4. Give definitions for several terms related to solutions, including: solvent, solute, concentration, electrolyte, nonelectrolyte, weak electrolyte, strong electrolyte, acid, and base. *Work Problems 21–25, 32, 34, 93, 94.*
5. Use the solubility guidelines (Table 3.1) to determine if an ionic compound is soluble or insoluble. Given physical data, perform simple solubility calculations. *Work Problems 26–31.*
6. Express chemical reactions using molecular, total ionic, and net ionic equations. *Work Problems 34–36.*

**Section 3.4 Interpreting Equations and the Mole**

7. Understand how Avogadro's number is used to represent matter and interpret chemical equations in terms of moles in addition to molecules (formula units). Write molar ratios relating reactants and products from a chemical equation. *Work Problems 37–40, 85, 86.*
8. Calculate molar mass of a compound using a periodic table. *Work Problems 41–46.*

**Section 3.5 Calculations Using Moles and Molar Mass**

9. Calculate the molar mass of a compound and use it to perform mass/ mole/ molecule conversions. *Work Problems 47–57, 61, 62, 88, 89, 91, 92, 99-109, 115.*
10. Determine the empirical and molecular formulas of a compound given the elemental analysis. *Work Problems 58–60, 87, 90.*
11. Define molarity and perform simple molarity calculations using M = mol/L. *Work Problems 63–66, 70, 95–98, 112, 113.*
12. Perform molarity calculations when a solution is diluted. *Work Problems 67–69, 114.*

**Section 3.6** ***INSIGHT INTO*** **Carbon Sequestration**

1. Describe the concept of "carbon sequestration" and understand how burning fossil fuels redistributes carbon in the environment and may adversely affect the planet's climate. *Work Problems 71–76.*

## *Solutions to the Odd-Numbered Problems*

Problem 3.1 Based on Figure 3.1, determine: (a) the percentage of sunlight that is captured by photosynthesis, (b) the percentage of photosynthesis that occurs in land plants and the percentage that occurs in aquatic plants, and (c) the percentage of the energy captured by all plants that is eventually released when the plants decay.

**The total light energy received from the sun is 86 PW or 8600 TW (recall 1 Petawatt = 1000 terawatts). Light captured by photosynthesis is 90 TW and light captured by land plants is 65 TW with the balance captured by aquatic plants. All energy captured by aquatic plants is released through decay where only 50 TW is released by land plants through decay.**

**(a)** $\dfrac{90\text{ TW}}{8600\text{ TW}} \times 100 = 1.0\%$

**(b) Land plants:** $\dfrac{65\text{ TW}}{90\text{ TW}} \times 100 = 72\%$ **Aquatic plants:** $\dfrac{25\text{ TW}}{90\text{ TW}} \times 100 = 28\%$

**(c)** $\dfrac{75\text{ TW}}{90\text{ TW}} \times 100 = 83\%$

Problem 3.3 Jan Baptist von Helmont is credited with repudiating the idea that trees gain their mass from the soil. In an experiment carried out in 1648, he grew a willow tree in a pot with the amount of soil kept constant. After five years the tree had gained roughly 74 kg. Although von Helmont concluded the weight gain was from water rather than soil, we now know it was from $CO_2$ and $H_2O$ taken in through photosynthesis. Use $C_6H_{12}O_6$ as a representative formula for the carbohydrates produced, and write a simplified equation for the overall process of photosynthesis.

**In photosynthesis, carbon dioxide and water are chemically combined with the help from sun energy (and enzymatic catalysts) to form carbohydrates like glucose or mannose and oxygen. The balanced equation for this reaction is:**

$$6\ CO_2\ (g) + 6\ H_2O\ (\ell) \rightarrow C_6H_{12}O_6 + 6\ O_2\ (g)$$

Problem 3.5 Figure 3.2 shows the production of ethanol as the "economic output" of biofuels. Use the web to research whether this is the only commercially viable product of the process.

**Answers will vary. Biodiesel, butanol, and hydrocarbons are among other fuels that can also be produced in this general process.**

Problem 3.7 Which symbols are used to indicate solids, liquids, gases, and aqueous solutions in chemical equations?

**The symbols are: solid – (s), liquid – (ℓ), gas – (g), and aqueous solution – (aq).**

Problem 3.9 What law of nature underpins the concept of a balanced chemical equation?

**The Law of Conservation of Matter: In ordinary chemical reactions, matter cannot be created nor destroyed. This tells us that a balanced chemical equation must have the same number and type of atom on both sides of the equation.**

Problem 3.11 ■ Balance these equations.

(a) $Al(s) + O_2(g) \rightarrow Al_2O_3(s)$

(b) $N_2(g) + H_2(g) \rightarrow NH_3(g)$

(c) $C_6H_6(\ell) + O_2(g) \rightarrow H_2O(\ell) + CO_2(g)$

**Balancing an equation involves adjusting the coefficients in front of the formulas until there is the same number and type of atoms on both sides.**
**Some hints for balancing equations:**

- **Start balancing an element that appears in only one place on both sides first.**
- **Always balance uncombined elements last.**
- **Try to balance polyatomic ions as a group.**
- **Express the coefficients as whole numbers.**

**(a)** **Balance Al:** $\underline{2}\ Al(s) + O_2(g) \rightarrow Al_2O_3(s)$

**Balance O:** $\underline{2}\ Al(s) + \underline{3/2}\ O_2(g) \rightarrow Al_2O_3(s)$

**Multiply by 2 to express as whole-number coefficients:**

$\underline{4}\ Al(s) + \underline{3}\ O_2(g) \rightarrow \underline{2}\ Al_2O_3(s)$

**(b)** **Balance N:** $N_2(g) + H_2(g) \rightarrow \underline{2}\ NH_3(g)$

**Then Balance H:** $N_2(g) + \underline{3}\ H_2(g) \rightarrow \underline{2}\ NH_3(g)$

**(c)** **Balance C first:** $C_6H_6(\ell) + O_2(g) \rightarrow H_2O(\ell) + \underline{6}\ CO_2(g)$

**Then H:** $C_6H_6(\ell) + O_2(g) \rightarrow \underline{3}\ H_2O(\ell) + \underline{6}\ CO_2(g)$

**Then O:** $C_6H_6(\ell) + \underline{15/2}\ O_2(g) \rightarrow \underline{3}\ H_2O(\ell) + \underline{6}\ CO_2(g)$

**Multiply by 2 to express as whole-number coefficients;**

$\underline{2}\ C_6H_6(\ell) + \underline{15}\ O_2(g) \rightarrow \underline{6}\ H_2O(\ell) + \underline{12}\ CO_2(g)$

Problem 3.13 An explosive whose chemical formula is $C_3H_6N_6O_6$ produces water, carbon dioxide, and nitrogen gas when detonated in oxygen. Write the chemical equation for the detonation reaction of this explosive.

**The basic equation for this explosive detonation is ($O_2$ is a reactant):**

$$C_3H_6N_6O_6 + O_2 \rightarrow H_2O + CO_2 + N_2$$

**See Problem 3.11. Balance C first, then H**

$$\underline{2}\ C_3H_6N_6O_6 + \underline{3}\ O_2 \rightarrow \underline{6}\ H_2O + \underline{6}\ CO_2 + \underline{6}\ N_2$$

Problem 3.15 ■ Ethanol, $C_2H_5OH$, is found in many gasoline blends now in many parts of North America. Write a balanced chemical equation for the combustion of $C_2H_5OH$ to form $CO_2$ and $H_2O$.

**Combustion means reaction of a carbon-based molecule with molecular oxygen, $O_2$. The beginning of the equation is:**

$$C_2H_5OH + O_2 \rightarrow CO_2 + H_2O$$

**See Problem 3.11. We'll balance C first, then H, O last.**

$$C_2H_5OH + \underline{3}\ O_2 \rightarrow \underline{2}\ CO_2 + \underline{3}\ H_2O$$

Problem 3.17 ■ Write balanced chemical equations for the following reactions.
(a) production of ammonia, $NH_3(g)$, by combining $N_2(g)$ and $H_2(g)$
(b) production of methanol, $CH_3OH(\ell)$, by combining $H_2(g)$ and $CO(g)$
(c) production of sulfuric acid by combining sulfur ($S_8$), oxygen, and water

**(a) The unbalanced equation:** $N_2(g) + H_2(g) \rightarrow NH_3(g)$
**Balance N:** $N_2(g) + H_2(g) \rightarrow \underline{2}\ NH_3(g)$
**Then Balance H:** $N_2(g) + \underline{3}\ H_2(g) \rightarrow \underline{2}\ NH_3(g)$

**(b) The unbalanced equation:** $H_2(g) + CO(g) \rightarrow CH_3OH(\ell)$
**C and O are already in balance.**
**Balance H:** $\underline{2}\ H_2(g) + CO(g) \rightarrow CH_3OH(\ell)$

**(c) The unbalanced equation:** $S_8(s) + O_2(g) + H_2O(\ell) \rightarrow H_2SO_4(\ell)$
**Balance S:** $S_8(s) + O_2(g) + H_2O(\ell) \rightarrow \underline{8}\ H_2SO_4(\ell)$
**Balance H:** $S_8(s) + O_2(g) + \underline{8}\ H_2O(\ell) \rightarrow \underline{8}\ H_2SO_4(\ell)$
**Balance O:** $S_8(s) + \underline{12}\ O_2(g) + \underline{8}\ H_2O(\ell) \rightarrow \underline{8}\ H_2SO_4(\ell)$

Problem 3.19 Silicon nitride, $Si_3N_4$, is used as a reinforcing fiber in construction materials. It can be synthesized from silicon tetrachloride and ammonia. The other product is ammonium chloride. Write the balanced chemical reaction for this process.

| | |
|---|---|
| **We begin with the unbalanced equation:** | $SiCl_4 + NH_3 \rightarrow Si_3N_4 + NH_4Cl$ |
| **Balance the Si:** | $\underline{3}\ SiCl_4 + NH_3 \rightarrow Si_3N_4 + NH_4Cl$ |
| **Balance the Cl:** | $\underline{3}\ SiCl_4 + NH_3 \rightarrow Si_3N_4 + \underline{12}\ NH_4Cl$ |
| **Finally the N and H:** | $\underline{3}\ SiCl_4 + \underline{16}\ NH_3 \rightarrow Si_3N_4 + \underline{12}\ NH_4Cl$ |

Problem 3.21 Define the terms solution, solute, and solvent.

**A *solution* is a homogeneous mixture of two or more substances. A *solute* is the minor (lesser amount) component(s) of the solution, and the *solvent* is the major (greater amount) component.**

Problem 3.23 What is a concentrated solution? A dilute solution?

***Concentrated* solutions have relatively high amounts of solute in the solution; *dilute* solutions contain relatively low amounts of solute.**

Problem 3.25 ■ What is an electrolyte? How can you differentiate experimentally between a weak electrolyte and a strong electrolyte? Give an example of each.

**An *electrolyte* is a substance that dissolves in water and produces ions in aqueous solution. This solution then conducts electricity. A *strong* electrolyte produces much higher concentrations of ions than a *weak* electrolyte and therefore conducts electricity much better. For equal concentrations, a solution of a strong electrolyte such as NaCl will have a much higher conductivity than a solution of a weak electrolyte like acetic acid.**

Problem 3.27 ■ The following compounds are water-soluble. What ions are produced by each compound in aqueous solution? **(a)** KOH, **(b)** $K_2SO_4$, **(c)** $LiNO_3$, **(d)** $(NH_4)_2SO_4$

**(a) $K^+$ and $OH^-$ ions**
**(b) $K^+$ and $SO_4^{2-}$ ions**
**(c) $Li^+$ and $NO_3^-$ ions**
**(d) $NH_4^+$ and $SO_4^{2-}$ ions**

Problem 3.29 The solubility of NaCl in water is 35.7 g NaCl/100 g $H_2O$. Suppose that you have 500.0 g of NaCl. What is the minimum volume of water you would need to dissolve it all? (Assume that the density of water is 1.0 g/mL.)

**From the solubility we calculate grams of water required, then use the density to find volume.**

$$500\text{ g NaCl} \times \frac{100\text{ g H}_2\text{O}}{35.7\text{ g NaCl}} \times \frac{1\text{ mL}}{1.0\text{ g H}_2\text{O}} = 1400\text{ mL H}_2\text{O}$$

Problem 3.31 ■ Classify each of these as an acid or a base. Which are strong and which are weak? What ions are produced when each is dissolved in water? (a) KOH, (b) $Mg(OH)_2$, (c) HClO, (d) HBr, (e) LiOH, (f) $H_2SO_3$

**(a) Base, strong, $K^+$ and $OH^-$ ions**
**(b) Base, weak, $Mg^{2+}$ and $OH^-$ ions**
**(c) Acid, weak, $H^+$ and $ClO^-$ ions**
**(d) Acid, strong, $H^+$ and $Br^-$ ions**
**(e) Base, strong, $Li^+$ and $OH^-$ ions**
**(f) Acid, weak, $H^+$ and $HSO_3^-$ ions**

Problem 3.33 What is the difference between a total ionic equation and a net ionic equation?

**The total ionic equation shows all substances that exist completely ionized in the solution as ions. The net ionic equation shows only the reacting species; spectator ions are cancelled out.**

Problem 3.35 ■ Balance the following equations and then write the net ionic equation.

(a) $Zn(s) + HCl(aq) \rightarrow H_2(g) + ZnCl_2(aq)$

**Balanced molecular equation:** $Zn(s) + \underline{2}\ HCl(aq) \rightarrow H_2(g) + ZnCl_2(aq)$

**Balanced total ionic equation:** $Zn(s) + \underline{2}\ H^+(aq) + \underline{2}\ Cl^- \rightarrow H_2(g) + Zn^{2+}(aq) + \underline{2}\ Cl^-(aq)$
**(HCl and $ZnCl_2$ are strong electrolytes)**
**Balanced net ionic equation:** $Zn(s) + \underline{2}\ H^+(aq) \rightarrow H_2(g) + Zn^{2+}(aq)$
**($Cl^-$ is a spectator ion)**

(b) $Mg(OH)_2(s) + HCl(aq) \rightarrow MgCl_2(aq) + H_2O(\ell)$

**Balanced molecular equation:** $Mg(OH)_2(s) + \underline{2}\ HCl(aq) \rightarrow MgCl_2(aq) + \underline{2}\ H_2O(\ell)$

**Balanced total ionic equation:**
$Mg(OH)_2(s) + \underline{2}\ H^+(aq) + \underline{2}\ Cl^-(aq) \rightarrow Mg^{2+}(aq) + \underline{2}\ Cl^-(aq) + \underline{2}\ H_2O(\ell)$
**(HCl and $MgCl_2$ are strong electrolytes)**

**Balanced net ionic equation:** $Mg(OH)_2(s) + \underline{2}\ H^+(aq) \rightarrow Mg^{2+}(aq) + \underline{2}\ H_2O(\ell)$
**($Cl^-$ is a spectator ion)**

(c) $HNO_3(aq) + CaCO_3(s) \rightarrow Ca(NO_3)_2(aq) + H_2O(\ell) + CO_2(g)$

**Balanced molecular equation:** $\underline{2}\ HNO_3(aq) + CaCO_3(s) \rightarrow Ca(NO_3)_2(aq) + H_2O(\ell) + CO_2(g)$

**Balanced total ionic equation:**

**$\underline{2}$ $H^+(aq)$ + $\underline{2}$ $NO_3^-$ (aq) + $CaCO_3(s) \rightarrow Ca^{2+}(aq)$ + $\underline{2}$ $NO_3^-$ (aq) + $H_2O(\ell)$ + $CO_2(g)$**
**($HNO_3$ and $Ca(NO_3)_2$ are strong electrolytes)**

**Balanced net ionic equation:** **$\underline{2}$ $H^+(aq) + CaCO_3(s) \rightarrow Ca^{2+}(aq) + H_2O(\ell) + CO_2(g)$**
**($NO_3^-$ is a spectator ion)**

(d) $(NH_4)_2S(aq) + FeCl_3(aq) \rightarrow NH_4Cl(aq) + Fe_2S_3(s)$

**Balanced molecular equation:** **$\underline{3}$ $(NH_4)_2S(aq)$ + $\underline{2}$ $FeCl_3(aq) \rightarrow$ $\underline{6}$ $NH_4Cl(aq) + Fe_2S_3(s)$**

**Balanced total ionic equation:**

**$\underline{6}$ $NH_4^+(aq)$ + $\underline{3}$ $S^{2-}(aq)$ + $\underline{2}$ $Fe^{3+}(aq)$ + $\underline{6}$ $Cl^-(aq) \rightarrow Fe_2S_3(s)$ + $\underline{6}$ $NH_4^+(aq)$ + $\underline{6}$ $Cl^-(aq)$**
**($(NH_4)_2S$, $NH_4Cl$, and $FeCl_3$ are strong electrolytes)**

**Balanced net ionic equation:** **$\underline{3}$ $S^{2-}(aq)$ + $\underline{2}$ $Fe^{3+}(aq) \rightarrow Fe_2S_3(s)$**
**($NH_4^+$ and $Cl^-$ are spectator ions)**

Problem 3.37 Explain the concept of the mole in your own words.

**The mole is a convenient number for expressing quantities of matter that contain very large numbers of molecules. It is defined as the number of atoms in exactly 12 g of carbon-12. The mole is equal to $6.022 \times 10^{23}$ particles (atoms, molecules, ions, etc.).**

Problem 3.39 If a typical grain of sand occupies a volume of $1.3 \times 10^{-4}$ $cm^3$, what is the volume (in $cm^3$) of one mole of sand (ignoring the space between grains)? What is the volume in liters?

**One mole of anything is $6.022 \times 10^{23}$ particles (Avogadro's number). To find the volume of a mole of sand, we multiply the volume of a single grain times Avogadro's number, then convert to liters.**

$$\frac{1.3 \times 10^{-4}\ \text{cm}^3}{\text{grain}} \times \frac{6.022 \times 10^{23}\ \text{grains}}{1\ \text{mole sand}} = 7.8 \times 10^{19}\ \text{cm}^3\ \text{(per mol sand)}$$

$$7.8 \times 10^{19}\ \text{cm}^3 \times \frac{1\ \text{mL}}{1\ \text{cm}^3} \times \frac{1\ \text{L}}{1000\ \text{mL}} = 7.8 \times 10^{16}\ \text{L (per mol sand)}$$

Problem 3.41 ■ Calculate the molar mass of each of the following compounds. (a) $Fe_2O_3$, iron(III) oxide, (b) $BCl_3$, boron trichloride, (c) $C_6H_8O_6$, ascorbic acid (vitamin C).

**The molar mass is the sum, in grams, of all the masses of all the atoms in the formula. It is equal to the mass of one mole of the formula.**

| | | | |
|---|---|---|---|
| **(a)** | $Fe_2O_3$ | **2 mol Fe:** | **2 × 55.85 g/mol** |
| | | **3 mol O:** | **3 × 16.00 g/mol** |
| | | **Total:** | **159.70 g/mol** |

**(b)** $BCl_3$

| | | |
|---|---|---|
| **1 mol B:** | **1 × 10.81 g/mol** |
| **3 mol Cl:** | **3 × 35.45 g/mol** |
| **Total:** | **117.16 g/mol** |

**(c)** $C_6H_8O_6$

| | |
|---|---|
| **6 mol C:** | **6 × 12.01 g/mol** |
| **8 mol H:** | **8 × 1.008 g/mol** |
| **6 mol O:** | **6 × 15.999 g/mol** |
| **Total:** | **176.12 g/mol** |

Problem 3.43 ■ Calculate the molar mass of each of these compounds and the mass percent of each element. (**a**) PbS, lead(II) sulfide, galena, (**b**) $C_2H_6$, ethane, a hydrocarbon fuel, (**c**) $CH_3COOH$, acetic acid, an important ingredient in vinegar, (**d**) $NH_4NO_3$, ammonium nitrate, a fertilizer

**The molar mass is the sum, in grams, of all the masses of all the atoms in the formula. It is equal to the mass of one mole of the formula. The percentages may not add up to 100% due to rounding.**

**(a)** PbS

| | |
|---|---|
| **1 mol Pb:** | **1 × 207.2 g/mol = 207.2 g Pb** |
| **1 mol S:** | **1 × 32.07 g/mol = 32.07 g S** |
| **Total:** | **239.3 g/mol** |

$$\%Pb = \frac{207.2\text{ g Pb}}{239.3\text{ g PbS}} \times 100 = 86.59\%$$

$$\%S = \frac{32.07\text{ g S}}{239.3\text{ g PbS}} \times 100 = 13.40\%$$

**(b)** $C_2H_6$

| | |
|---|---|
| **2 mol C:** | **2 × 12.01 g/mol = 24.02 g C** |
| **6 mol H:** | **6 × 1.008 g/mol = 6.048 g H** |
| **Total:** | **30.07 g/mol** |

$$\%C = \frac{24.02\text{ g C}}{30.07\text{ g }C_2H_6} \times 100 = 79.88\%$$

$$\%H = \frac{6.048\text{ g H}}{30.07\text{ g }C_2H_6} \times 100 = 20.11\%$$

**(c)** $CH_3COOH$

| | |
|---|---|
| **2 mol C:** | **2 × 12.01 g/mol = 24.02 g C** |
| **2 mol O:** | **2 × 16.00 g/mol = 32.00 g O** |
| **4 mol H:** | **4 × 1.008 g/mol = 4.032 g H** |
| **Total:** | **60.05 g/mol** |

$$\%C = \frac{24.02\text{ g C}}{60.05\text{ g }CH_3COOH} \times 100 = 40.00\%$$

$$\%O = \frac{32.00\text{ g O}}{60.05\text{ g }CH_3COOH} \times 100 = 53.29\%$$

$$\%H = \frac{4.032\text{ g H}}{60.05\text{ g }CH_3COOH} \times 100 = 6.74\%$$

**(d)** $NH_4NO_3$

| | |
|---|---|
| **2 mol N:** | **2 × 14.01 g/mol = 28.02 g N** |
| **4 mol H:** | **4 × 1.008 g/mol = 4.032 g H** |
| **3 mol O:** | **3 × 16.00 g/mol = 48.00 g O** |
| **Total:** | **80.05 g/mol** |

$$\%N = \frac{28.02\text{ g N}}{80.05\text{ g }NH_4NO_3} \times 100 = 35.00\%$$

$$\%H = \frac{4.032\text{ g H}}{80.05\text{ g }NH_4NO_3} \times 100 = 5.037\%$$

$$\%O = \frac{48.00\text{ g O}}{80.05\text{ g }NH_4NO_3} \times 100 = 59.96\%$$

Problem 3.45 Calculate the molar mass of the following compounds:

**See Problem 3.41**

**(a)** magnesium phosphate **Formula: $Mg_3(PO_4)_2$**

| | |
|---|---|
| **3 mol Mg:** | **3 × 24.305 g/mol** |
| **2 mol P:** | **2 × 30.974 g/mol** |
| **8 mol O:** | **8 × 15.999 g/mol** |
| **Total:** | **262.855 g/mol** |

**(b)** sodium sulfide **Formula: $Na_2S$**

| | |
|---|---|
| **2 mol Na:** | **2 × 22.990 g/mol** |
| **1 mol S:** | **1 × 32.066 g/mol** |
| **Total:** | **78.046 g/mol** |

**(c)** dinitrogen tetroxide **Formula: $N_2O_4$**

| | |
|---|---|
| **2 mol N:** | **2 × 14.007 g/mol** |
| **4 mol O:** | **4 × 15.999 g/mol** |
| **Total:** | **92.010 g/mol** |

Problem 3.47 A chemist needs exactly 2 moles of NaOH to make a solution. What mass of NaOH must be used?

**We convert from moles to grams by multiplying by the molar mass of the substance.**

**Mass (g) = moles × molar mass (g/mole)** **or** $\text{mol} \times \frac{\text{g}}{\text{mol}} = \text{g}$

$$2 \text{ moles} \times \frac{40.00\text{ g}}{\text{mol}} = 80.00\text{ g NaOH}$$

Problem 3.49 ■ Calculate the mass in grams of each the following: (**a**) 2.5 mol of aluminum, (**b**) $1.25 \times 10^{-3}$ mol of iron, (**c**) 0.015 mol of calcium, (**d**) 653 mol of neon

**We convert from moles to grams by multiplying by the molar mass of the substance.**

**Grams = moles × molar mass (g/mole)**

**(a)** $2.5 \text{ mol of aluminum} \times \frac{26.98\text{ g Al}}{\text{mol Al}} = 67\text{ g Al}$ **(rounded to two significant figures)**

**(b)** $1.25 \times 10^{-3} \text{ mol of iron} \times \frac{55.85\text{ g Fe}}{\text{mol Fe}} = 0.0698\text{ g Fe}$ **(rounded to three significant figures)**

**(c)** $0.015 \text{ mol of calcium} \times \frac{40.08\text{ g Ca}}{\text{mol Ca}} = 0.60\text{ g Ca}$ **(rounded to two significant figures)**

**(d)** $653 \text{ mol of neon} \times \frac{20.18\text{ g Ne}}{\text{mol Ne}} = 1.32 \times 10^{4}\text{ g Ne}$ **(rounded to three significant figures)**

Problem 3.51 How many moles are present in the given quantities of explosives?

(a) 358.1 g trinitrotoluene (TNT), $C_7H_5N_3O_6$

$$358.1\text{ g TNT} \times \frac{1\text{ mol TNT}}{227.13\text{ g TNT}} = 1.577\text{ mol TNT}$$

(b) 82.6 g nitromethane, $CH_3NO_2$

$$82.6\text{ g } CH_3NO_2 \times \frac{1\text{ mol } CH_3NO_2}{61.04\text{ g } CH_3NO_2} = 1.35\text{ mol } CH_3NO_2$$

(c) 1.68 kg RDX, $C_3H_6N_6O_6$

$$1.68\text{ kg RDX} \times \frac{1000\text{ g}}{1\text{ kg}} \times \frac{1\text{ mol } Na_2SO_4}{221.11\text{ g RDX}} = 7.60\text{ mol RDX}$$

Problem 3.53 Modern instruments can measure a mass as small as 5 nanograms. If one observed 5.0 ng of $CO_2$, how many molecules were measured?

**We will need to convert mass to moles, then moles into molecules using Avogadro's number.**

$$5.0 \text{ ng } CO_2 \times \frac{1\times10^{-9}\text{ g}}{1\text{ ng}} \times \frac{1\text{ mol } CO_2}{44.01\text{ g } CO_2} \times \frac{6.022\times10^{23}\text{ molecules} CO_2}{1\text{ mol } CO_2} = 6.8 \times 10^{13}\text{ molecules}$$

Problem 3.55 How many oxygen atoms are present in 214 g of mannose, $C_6H_{12}O_6$?

**We'll convert grams to moles, then moles to molecules, and finally use the ratio in the chemical formula to calculate O atoms.**

$$214\text{ g mannose} \times \frac{1\text{ mol mannose}}{180.16\text{ g mannose}} \times \frac{6.022\times10^{23}\text{ molecules mannose}}{1\text{ mol mannose}} \times$$

$$\frac{6\text{ O atoms}}{1\text{ molecule mannose}} =$$

$$= 4.29 \times 10^{24}\text{ atoms of oxygen}$$

Problem 3.57 An average person inhales roughly 2.5 g $O_2$ in a minute. How many molecules of oxygen are inhaled in: **(a)** one minute, **(b)** one hour, **(c)** one day by an average person?

**Convert g $O_2$ to moles, then moles to molecules.**

(a) $$2.5\text{ g } O_2\text{ /minute} \times \frac{1\text{ mol } O_2}{32.00\text{ g } O_2} \times \frac{6.022\times10^{23}\text{ molecules} O_2}{1\text{ mol } O_2}$$

$$= 4.7 \times 10^{22}\text{ molecules } O_2\text{ /minute}$$

(b) $$\frac{4.7\times10^{22}\text{ molecules} O_2}{1\text{ min}} \times \frac{60\text{ min}}{1\text{ hour}} = 2.8 \times 10^{24}\text{ molecules } O_2\text{ /hour}$$

(c) $$\frac{2.8\times10^{24}\text{ molecules} O_2}{1\text{ hour}} \times \frac{24\text{ hour}}{1\text{ day}} = 6.8 \times 10^{25}\text{ molecules } O_2\text{ /day}$$

Problem 3.59 ■ Mandelic acid is an organic acid composed of carbon (63.15%), hydrogen (5.30%), and oxygen (31.55%). Its molar mass is 152.14 g/mol. Determine the empirical and molecular formulas of the acid.

**The empirical formula is the simplest formula with the smallest whole-number ratio of atoms present in the compound. To find the empirical formula beginning with percent composition, we first assume 100 g of the compound, meaning each percentage can be**

**expressed as grams. The next step is to find the moles of each element. The final step is to take ratios, because ratios are what a chemical formula shows. We want to divide by whichever element has the fewest moles present when we take ratios, ensuring that the formula we end up with is the simplest one.**

**In 100 g of mandelic acid there are**

**63.15 g C, 5.30 g H, and 31.55 g O.**

**Using the molar masses to find moles:**

$$63.15 \text{ g C} \times \frac{1 \text{ mol C}}{12.011 \text{ g C}} = 5.258 \text{ mol C} \qquad 5.30 \text{ g H} \times \frac{1 \text{ mol H}}{1.0079 \text{ g H}} = 5.258 \text{ mol H}$$

$$31.55 \text{ g O} \times \frac{1 \text{ mol O}}{15.999 \text{ g O}} = 1.972 \text{ mol O}$$

**The fewest moles are for O, so we'll take our ratios by dividing by the moles of O:**

$$5.258 \text{ mol C} \div 1.972 \text{ mol O} = 2.666 \text{ or } 8/3 \qquad 5.258 \text{ mol H} \div 1.972 \text{ mol O} = 2.666 \text{ or } 8/3$$

$$1.972 \text{ mol O} \div 1.972 \text{ mol O} = 1.000$$

**These ratios are the subscripts in our empirical formula but they must be whole numbers. Multiply by three to eliminate the fractions:**
$3 \times [C_{8/3}H_{8/3}O_{1.0}] = C_8H_8O_3$

**This is the empirical formula with a molar mass of 152.1 g/mol. The actual molar mass of mandelic acid was given at 152.14 g/mol, so the empirical and molecular formulas are the same.**

Problem 3.61 ■ The composition of materials such as alloys can be described in terms of mole percentage (mol %), atom percentage (at %), or weight percentage (wt %). Carry out the following conversions among these units. (**a**) 60 wt % Cu and 40 wt % Al to at %, (**b**) 25 mol % NiO and 75 mol % MgO to wt %, (**c**) 40 wt % MgO and 60 wt % FeO to mol %

**(a) 60 wt % Cu and 40 wt % Al represents 60 g Cu and 40 g Al per 100-g mixture. The masses are converted to moles then atoms; then the percentage of each atom type may be expressed.**

$$60 \text{ g Cu} \times \frac{1 \text{ mol Cu}}{63.55 \text{ g Cu}} \times \frac{6.022 \times 10^{23} \text{ atoms Cu}}{1 \text{ mol Cu}} = 5.686 \times 10^{23} \text{ atoms Cu}$$

$$40 \text{ g Al} \times \frac{1 \text{ mol Al}}{26.98 \text{ g Cu}} \times \frac{6.022 \times 10^{23} \text{ atoms Al}}{1 \text{ mol Al}} = 8.928 \times 10^{23} \text{ atoms Al}$$

**Total: $1.461 \times 10^{24}$ atoms**

$$\text{at \% Cu} = \frac{5.686\times10^{23}\ \text{atoms Cu}}{1.461\times10^{24}\ \text{atoms}} \times 100 = 38.92\% \approx 39\% \text{ (two significant figures)}$$

$$\text{at \% Al} = 100\% - 38.92\% = 61.08\% \approx 61\%$$

**(b) In one mole of this mixture there are 0.25 mol NiO and 0.75 mol MgO (25 mol % NiO and 75 mol % MgO).**

$$0.25\ \text{mol NiO} \times \frac{74.69\ \text{g NiO}}{1\ \text{mol NiO}} = 18.67\ \text{g NiO}$$

$$0.75\ \text{mol MgO} \times \frac{40.31\ \text{g MgO}}{1\ \text{mol MgO}} = 30.23\ \text{g MgO}$$

**Total: 48.90 g**

$$\text{wt \% NiO} = \frac{18.67\ \text{g NiO}}{48.90\ \text{g Total}} \times 100 = 38.18\ \text{wt\% NiO} \approx 38\%$$

$$\text{wt \% MgO} = \frac{30.23\ \text{g MgO}}{48.90\ \text{g Total}} \times 100 = 61.82\ \text{wt\% MgO} \approx 62\%$$

**(c) 40 wt % MgO and 60 wt % FeO represents 40 g MgO and 60 g FeO per 100 g mixture.**

$$40\ \text{g MgO} \times \frac{1\ \text{mol MgO}}{40.31\ \text{g MgO}} = 0.9923\ \text{mol MgO}$$

$$60\ \text{g FeO} \times \frac{1\ \text{mol FeO}}{71.85\ \text{g FeO}} = 0.8351\ \text{mol FeO}$$

**Total: 1.8274 mol**

$$\text{mol\% MgO} = \frac{0.9923\ \text{mol MgO}}{1.827\ \text{mol Total}} \times 100 = 54.30\ \text{mol\% MgO} \approx 54\%$$

$$\text{mol\% FeO} = \frac{0.8351\ \text{mol FeO}}{1.827\ \text{mol Total}} \times 100 = 45.70\ \text{mol\% FeO} \approx 46\%$$

Problem 3.63 Calculate the molarity of each of the following solutions:

**Molarity is defined moles per liter of solution,** $\mathbf{M} = \dfrac{\text{moles solute}}{\text{liters solution}}$**.**

**(a)** 1.45 mol HCl in 250. mL of solution $\quad 250.\ \text{mL} = 0.250\ \text{L},\quad \dfrac{1.45\ \text{mol HCl}}{0.250\ \text{L}} = 5.80\ \text{M}$

**(b)** 14.3 mol NaOH in 3.4 L of solution $\quad \dfrac{14.3\ \text{mol NaOH}}{3.4\ \text{L}} = 4.2\ \text{M}$

**(c)** 0.341 mol KCl in 100.0 mL solution **100.0 mL = 0.1000 L,** $\frac{\mathbf{0.341\ mol\ KCl}}{\mathbf{0.1000\ L}}$ **= 3.41 M**

**(d)** $2.5 \times 10^4$ mol $NaNO_3$ in 350 L of solution $\frac{\mathbf{2.5 \times 10^4\ mol\ NaNO_3}}{\mathbf{350\ L}}$ **= 71 M**

Problem 3.65 How many moles of solute are present in each of these solutions?

**Molarity is defined moles per liter of solution,** $\mathbf{M = \frac{moles\ solute}{liters\ solution}}$.

**Therefore, moles solute = liters of solution × molarity (L × M = moles).**

**(a)** 48.0 mL of 3.4 M $H_2SO_4$ **48.0 mL = 0.0480 L**

$\mathbf{0.0480\ L \times \frac{3.4\ mol\ H_2SO_4}{L} = 0.16\ moles}$

**(b)** 1.43 mL of 5.8 M $KNO_3$ **1.43 mL = 0.00143 L**

$\mathbf{0.00143\ L \times \frac{5.8\ mol\ KNO_3}{L} = 0.0083\ moles}$

**(c)** 3.21 L of 0.034 M $NH_3$ $\mathbf{3.21\ L \times \frac{0.034\ mol\ NH_3}{L} = 0.11\ moles}$

**(d)** $1.9 \times 10^3$ L of $1.4 \times 10^{-5}$ M NaF

$\mathbf{1.9 \times 10^3\ L \times \frac{1.4 \times 10^{-5}\ mol\ NaF}{L} = 0.0266\ moles}$

Problem 3.67 Determine the final molarity for the following solutions:

**Dilution problems are most easily solved with a formula called the dilution rule:** $V_1 \times M_1 = V_2 \times M_2$**. This formula is based on the concept that in a dilution, more solvent is added but the number of moles of solute doesn't change. The product of volume in liters and molarity is moles. It doesn't really matter what units are used for volume, since they are in a ratio.**

**(a)** 24.5 mL of 3.0 M solution diluted to 100.0 mL

**Let** $\mathbf{M_2}$ **= the final molarity of the diluted solution.**

**(24.5 mL)(3.0 M) = (100.0 mL)(**$\mathbf{M_2}$**)** **rearranging,** $\mathbf{M_2 = \frac{(24.5\ mL)(3.0\ M)}{(100.0\ mL)} = 0.74\ M}$

**(b)** 15.3 mL of 4.22 M solution diluted to 100.0 mL

**(15.3 mL)(4.22 M) = (1000 mL)(M₂)** **rearranging,** $\mathbf{M_2 = \frac{(15.3\ mL)(4.22\ M)}{(1000\ mL)} = 0.0646\ M}$

(c) 1.45 mL of 0.034 M solution diluted to 10 mL

**(1.45 mL)(0.034 M) = (10 mL)($M_2$) rearranging,** $\mathbf{M_2 = \frac{(1.45\ mL)(0.034\ M)}{(10\ mL)} = 0.0049\ M}$

(d) 2.35 L of 12.5 M solution diluted to 100 L

**(2.35 L)(12.5 M) = (100 L)($M_2$)** **rearranging,** $\mathbf{M_2 = \frac{(2.35\ L)(12.5\ M)}{(100\ L)} = 0.294\ M}$

Problem 3.69 ■ Commercially available concentrated sulfuric acid is 18.0 M $H_2SO_4$. Calculate the volume of concentrated sulfuric acid required to prepare 2.50 L of 1.50 M $H_2SO_4$ solution.

**See Problem 3.67.**
**Let $V_1$ represent the initial volume of concentrated sulfuric acid.**

**($V_1$ L)(18.0 M) = (2.50 L)(1.50 M)** **rearranging,** $\mathbf{V_1\ L = \frac{(2.50\ L)(1.50\ M)}{(18.0\ M)} = 0.208\ L,}$

**or 208 mL of 18.0 M $H_2SO_4$**

Problem 3.71 Carbon dioxide is just one of many greenhouse gases in the atmosphere. What property makes a gas a greenhouse gas?

**An atmospheric greenhouse gas has the ability to absorb infrared radiation that is produced by the sun's light shining on the Earth's surface. This radiation would otherwise escape the atmosphere so the trapped energy may cause climate change.**

Problem 3.73 What is meant by the term *carbon reservoir*? What are the two largest carbon reservoirs for our planet?

**Refer to Figure 3.10**
**A carbon reservoir is an area of the environment where a very large amount of carbon is located.**
**The oceanic pool and fossil fuels are the two largest carbon reservoirs for our planet.**

Problem 3.75 Explain how the burning of fossil fuels, which contributes only 7.0 Pg $yr^{-1}$ to the atmospheric store of carbon, can upset the balance of carbon in the atmosphere if photosynthesis fixes 120 Pg $yr^{-1}$.

**Photosynthesis fixes 120 Pg/yr but respiration from plants and soils offsets that amount. So although burning fossil fuels contributes only 7.0 Pg/yr, it is enough to cause a steady increase in the atmospheric pool.**

Problem 3.77 Nitric acid ($HNO_3$) can be produced by the reaction of nitrogen dioxide ($NO_2$) and water. Nitric oxide (NO) is also formed as a product. Write a balanced chemical equation for this reaction.

**The unbalanced equation is:** $NO_2 + H_2O(\ell) \rightarrow HNO_3(\ell) + NO(g)$

**Balance H first:** $NO_2(g) + H_2O(\ell) \rightarrow \underline{2}\ HNO_3(\ell) + NO(g)$

**Next, both O and N can be balanced by using a coefficient of three for $NO_2$:**

$\underline{3}\ NO_2(g) + H_2O(\ell) \rightarrow \underline{2}\ HNO_3(\ell) + NO(g)$

Problem 3.79 Pyridine has the molecular formula $C_5H_5N$. When pyridine reacts with $O_2$, the products are $CO_2$, $H_2O$, and $N_2$. Write a balanced equation for this reaction.

**The unbalanced equation is:** $C_5H_5N + O_2 \rightarrow CO_2 + H_2O + N_2$

**We'll balance C first, then H, O and N last.** $C_5H_5N + O_2 \rightarrow \underline{5}\ CO_2 + H_2O + N_2$

$C_5H_5N + O_2 \rightarrow \underline{5}\ CO_2 + \underline{5/2}\ H_2O + N_2$

$C_5H_5N + \underline{25/4}\ O_2 \rightarrow \underline{5}\ CO_2 + \underline{5/2}\ H_2O + \underline{1/2}\ N_2$

**Multiplying the entire equation by 4 gives us whole-number coefficients.**

$\underline{4}\ C_5H_5N + \underline{25}\ O_2 \rightarrow \underline{20}\ CO_2 + \underline{10}\ H_2O + \underline{2}\ N_2$

Problem 3.81 Hydrogen cyanide (HCN) is extremely toxic, but it is used in the production of several important plastics. In the most common method for producing HCN, ammonia ($NH_3$) and methane ($CH_4$) react with oxygen to give HCN and water. Write a balanced chemical equation.

**The unbalanced equation is:** $NH_3 + CH_4 + O_2 \rightarrow HCN + H_2O$

**C and N are balanced; we'll balance H next,**

$\underline{1}\ NH_3 + \underline{1}\ CH_4 + O_2 \rightarrow \underline{1}\ HCN + \underline{3}\ H_2O$

**Finally, we'll balance O:** $\underline{1}\ NH_3 + \underline{1}\ CH_4 + \underline{3/2}\ O_2 \rightarrow \underline{1}\ HCN + \underline{3}\ H_2O$

**In whole numbers:** $\underline{2}\ NH_3 + \underline{2}\ CH_4 + \underline{3}\ O_2 \rightarrow \underline{2}\ HCN + \underline{6}\ H_2O$

Problem 3.83 Adipic acid is used in the production of nylon, so it is manufactured in large quantities. The most common method for the preparation of adipic acid is the reaction of cylcohexane with oxygen. Balance the skeleton equation below.

$$\text{cyclohexane} + O_2 \longrightarrow \text{HO(O=)C-CH}_2\text{CH}_2\text{CH}_2\text{CH}_2\text{-C(=O)OH} + H_2O$$

**We'll balance H with $H_2O$ (C is already balanced) and then balance O. The condensed reaction makes it easier to see what needs to be balanced.**

$$2\ \text{cyclohexane} + 5\,O_2 \longrightarrow 2\ \text{HO(O=)C-CH}_2\text{CH}_2\text{CH}_2\text{CH}_2\text{-C(=O)OH} + 2\,H_2O$$

$$2\,C_6H_{12} + 5\,O_2 \rightarrow 2\,C_6H_{10}O_4 + 2\,H_2O$$

Problem 3.85 ■ Answer the following questions. Note that none require difficult calculations.

**(a)** How many molecules are present in 1 mole of octane ($C_8H_{18}$)?

**Avogadro's number, $6.022 \times 10^{23}$ molecules.**

**(b)** How many moles of fluorine atoms are present in 4 moles of $C_2F_6$?

**There are 6 moles F per one mole of $C_2F_6$, therefore, $4 \times 6 = 24$ moles of fluorine atoms are present.**

**(c)** What is the approximate mass (to the nearest gram) of $3 \times 10^{23}$ carbon atoms?

**This is about ½ mole of carbon, therefore ½ × 12 g = 6 g carbon.**

Problem 3.87 Cumene is a hydrocarbon, meaning it contains only carbon and hydrogen. If this compound is 89.94% C by mass and its molar mass is 120.2 g/mol, what is its molecular formula?

**This is an empirical formula determination, see Problem 3.59.**

**In 100 g of cumene,**

$$89.94\text{ g C} \times \frac{1\text{ mol C}}{12.011\text{ g C}} = 7.488\text{ mol C} \qquad 10.06\text{ g H} \times \frac{1\text{ mol H}}{1.0079\text{ g H}} = 9.981\text{ mol H}$$

**The fewest moles are for C, so we'll take our ratios by dividing by the moles of C:**

**7.488 mol C ÷ 7.488 mol C = 1.0** **9.981 mol H ÷ 7.488 mol N = 1.333**

**The empirical formula is $C_1H_{4/3}$ or 3 [ $C_1H_{4/3}$ ] = $C_3H_4$**

**This formula yields a molar mass of 40.06 g/mol.**

**Therefore, the actual formula is** $\frac{120.2 \text{ g/mol}}{40.06 \text{ g/mol}} = 3 \times [\,C_3H_4\,] = C_9H_{12}$

Problem 3.89 A low-grade form of iron is called taconite and the iron in the ore is in the form $Fe_3O_4$. If a 2.0-ton sample of taconite pellets yields 1075 pounds of iron when it is refined, what is the mass percentage of $Fe_3O_4$ in taconite?

**We'll first calculate how much $Fe_3O_4$ would be required to produce 1075 pounds of Fe.**

$$1075 \text{ lbs Fe} \times \frac{454 \text{ g}}{1 \text{ lb}} \times \frac{1 \text{ mol Fe}}{55.847 \text{ g Fe}} \times \frac{1 \text{ mol Fe}_3\text{O}_4}{3 \text{ mol Fe}} \times \frac{231.54 \text{ g Fe}_3\text{O}_4}{1 \text{ mol Fe}_3\text{O}_4} \times \frac{1 \text{ lb}}{454 \text{ g}} =$$

**= 1486 lbs $Fe_3O_4$**

**The percentage of $Fe_3O_4$ in 2.0 tons of taconite is** $\frac{1486 \text{ lb Fe}_3\text{O}_4}{4000 \text{ lbs taconite}} \times 100 = 37.1\%$

Problem 3.91 Iron-platinum alloys may be useful as high-density recording materials because of their magnetic properties. These alloys have been made with a wide range of composition, from 34.0 at % Pt to 81.8 at % Pt. Express this range in mol %.

**This would be the same range, 34.0 mol % Pt to 81.8 mol % Pt. The ratio of atoms in this alloy is the same whether the composition is viewed from an atomic level or a molar one. Multiplying the atoms present by Avogadro's number doesn't change the percentages.**

Problem 3.93 ■ Which (if any) of the following compounds are electrolytes? (**a**) glucose, $C_6H_{12}O_6$, (**b**) ethanol, $C_2H_5OH$, (**c**) magnesium sulfide, MgS, (**d**) sulfur hexafluoride, $SF_6$

**(c) magnesium sulfide, MgS is an electrolyte. The other three are non-electrolytes.**

Problem 3.95 ■ What is the mass in grams of solute in 250. mL of a 0.0125 M solution of $KMnO_4$?

**Molarity is defined as moles per liter of solution,** $M = \frac{\text{moles solute}}{\text{liters solution}}$.

**Therefore, moles solute = liters of solution × molarity (L × M = moles), and once moles are found they can be converted to grams with the molar mass.**

**250. mL of 0.0125 M $KMnO_4$** **250. mL = 0.250 L**

$$0.250\text{ L} \times \frac{0.0125\text{ mol KMnO}_4}{\text{L}} \times \frac{158.04\text{ g KMnO}_4}{1\text{ mol KMnO}_4} = 0.494\text{ g KMnO}_4$$

Problem 3.97 ■ Nitric acid is often sold and transported as a concentrated 16 M aqueous solution. How many gallons of such a solution would be needed to contain the entire $1.41 \times 10^7$ pounds of $HNO_3$ produced in the United States in 2003?

**From the mass of $HNO_3$, the moles can be determined, and then the volume can be found using the solution's molarity: L = mol/M.**

$$1.41 \times 10^7\text{ lbs HNO}_3 \times \frac{454\text{ g}}{1\text{ lb}} \times \frac{1\text{ mol HNO}_3}{63.012\text{ g HNO}_3} \times \frac{\text{L}}{16\text{ mol HNO}_3} \times \frac{1\text{ gal}}{3.785\text{ L}} =$$

$$= 1.7 \times 10^6\text{ gallons}$$

Problem 3.99 ■ As computer processor speeds increase, it is necessary for engineers to increase the number of circuit elements packed into a given area. Individual circuit elements are often connected using very small copper "wires" deposited directly onto the surface of the chip. In current generation processors, these copper interconnects are about 65 nm wide. How many copper atoms would be in a 1-mm length of such an interconnect, assuming a square cross section. (The density of copper is 8.96 g/cm$^3$.)

**1 nm = $10^{-7}$ cm, 1 mm = 0.1 cm**

**Volume of copper = $(65 \times 10^{-7}\text{ cm})(65 \times 10^{-7}\text{ cm})(0.1\text{ cm}) = 4.225 \times 10^{-12}\text{ cm}^3$**

**Mass of copper = volume × density = $4.225 \times 10^{-12}\text{ cm}^3 \times 8.96\text{ g/cm}^3 = 3.786 \times 10^{-11}\text{ g}$**

$$\text{Copper atoms} = 3.786 \times 10^{-11}\text{ g} \times \frac{1\text{ mol Cu}}{63.55\text{ g Cu}} \times \frac{6.022 \times 10^{23}\text{ atoms Cu}}{1\text{ mol Cu}} = 3.6 \times 10^{11}\text{ atoms Cu}$$

Problem 3.101 Materials engineers often create new alloys in an effort to improve the properties of an existing material. ZnO based semiconductors show promise in applications like light-emitting diodes, but their performance can be enhanced by the addition of small amounts of cadmium. One material that has been studied can be represented by the formula $Zn_{0.843}Cd_{0.157}O$. (These materials are solid solutions, and so they can have variable compositions. The non-integer coefficients do not imply fractional atoms.) Express the composition of this alloy in terms of (**a**) at %, (**b**) mol %, and (**c**) wt %.

**(a) One mole of alloy contains**

$$0.843\text{ mol Zn} \times \frac{6.022 \times 10^{23}\text{ atoms Zn}}{1\text{ mol Zn}} = 5.077 \times 10^{23}\text{ atoms Zn}$$

$$\text{at \% Zn} = \frac{5.077 \times 10^{23}\text{ atoms Zn}}{1.204 \times 10^{24}\text{ total atoms}} \times 100 = 42.2\%$$

**$0.157 \text{ mol Cd} \times \dfrac{6.022\times10^{23}\text{ atoms Cd}}{1\text{ mol Cd}} = 9.455\times10^{22}\text{ atoms Cd}$**

**$\text{at \% Cd} = \dfrac{9.455\times10^{22}\text{ atoms Cd}}{1.204\times10^{24}\text{ total atoms}}\times 100 =$ 7.85%**

**$1.00\text{ mol O}\times\dfrac{6.022\times10^{23}\text{ atoms O}}{1\text{ mol O}} = 6.022\times10^{23}\text{ atoms O}$**

**$\text{at \% O} = \dfrac{6.022\times10^{23}\text{ atoms O}}{1.204\times10^{24}\text{ total atoms}}\times 100 = 50.0\%$**

**Total atoms in the alloy: $1.204\times10^{24}$**

**(b) One mole of alloy contains**

**0.843 mol Zn $\quad \text{mol \% Zn} = \dfrac{0.843\text{ mol Zn}}{2.00\text{ mol total}}\times100 = 42.2\%$**

**0.157 mol Cd $\quad \text{mol \% Cd} = \dfrac{0.157\text{ mol Cd}}{2.00\text{ total}}\times100 = 7.85\%$**

**1.00 mol O $\quad \text{mol \% O} = \dfrac{1.00\text{ mol O}}{2.00\text{ total}}\times100 = 50.0\%$**

**Total: 2.00 moles in the alloy**

**Note: The mol % and at % will be the same; the ratio of atoms in the alloy is the same whether or not it is multiplied by Avogadro's number.**

**(c) One mole of alloy contains**

**$0.843\text{ mol Zn}\times\dfrac{65.39\text{ g Zn}}{1\text{ mol Zn}} = 55.12\text{ g Zn} \quad \text{wt \% Zn} = \dfrac{55.12\text{ g Zn}}{88.77\text{ g total}}\times100 = 62.1\%$**

**$0.157\text{ mol Cd}\times\dfrac{112.41\text{ g Cd}}{1\text{ mol Cd}} = 17.65\text{ g Cd} \quad \text{wt \% Cd} = \dfrac{17.65\text{ g Cd}}{88.77\text{ g total}}\times100 = 19.9\%$**

**$1.00\text{ mol O}\times\dfrac{16.00\text{ g O}}{1\text{ mol O}} = 16.00\text{ g O} \quad \text{wt \% O} = \dfrac{16.00\text{ g O}}{88.77\text{ g total}}\times100 = 18.0\ \%$**

**Total grams in the alloy: 88.77 g**

Problem 3.103 The chlorophyll molecule responsible for photosynthesis in plants has 2.72 % Mg by mass. There is only one Mg atom per chlorophyll molecule. How can you determine the molar mass of chlorophyll with this information?

**Assume that you have 100-g of chlorophyll. There would be 2.72 g of magnesium present (2.72% Mg). This corresponds to 0.112 moles of Mg: $2.72\text{ g Mg}\times\dfrac{1\text{ mol Mg}}{24.31\text{ g Mg}} = 0.112$ moles**

**Since there is a one-to-one ratio of Mg atoms and chlorophyll, there is 0.112 moles of chlorophyll in the 100 g sample. The molar mass of chlorophyll is the mass divided by the moles: $\frac{100\text{ g}}{0.112\text{ moles}} = 894$ g/mole.**

Problem 3.105 $MgCl_2$ is often found as an impurity in table salt (NaCl). If a 0.05200-g sample of table salt is found to contain 61.10% Cl by mass, describe how you could determine the percentage $MgCl_2$ in the sample?

**Let "$X$" = the fraction of the sample that is NaCl and let "$1 - X$" = the fraction of the sample that is $MgCl_2$. By calculation, we know that NaCl is 60.66% Cl:**

$$\left(\frac{35.45\text{ g/mol Cl}}{58.44\text{ g/mol NaCl}} \times 100\right) \text{ and } MgCl_2 \text{ is } 74.47\% \text{ Cl: } \left(\frac{2 \times 35.45\text{ g/mol Cl}}{95.21\text{ g/mol } MgCl_2} \times 100\right).$$

**Therefore,** $60.66\%\text{ Cl }(X) + 74.47\%\text{ Cl }(1 - X) = 61.10\%\text{ Cl and } X = 0.9681.$

**So the sample is 96.81% NaCl and 3.19% $MgCl_2$.**

Problem 3.107 The average person exhales 1.0 kg of carbon dioxide in a day. Describe how you would estimate the number of $CO_2$ molecules exhaled per breath for this "average" person.

**First you would have to estimate how many times a day that a person takes a breath. Then calculate the mass of $CO_2$ per breath, then moles, and finally molecules.**

**Suppose a person takes a breath once every three seconds. Then there would be**

$$\frac{1\text{ breath}}{3\text{ seconds}} \times \frac{3600\text{ seconds}}{\text{hour}} \times \frac{24\text{ hours}}{\text{day}} = 28800 \text{ breaths per day.}$$

$$\frac{1\text{ day}}{28800\text{ breaths}} \times \frac{1000\text{ g } CO_2}{\text{day}} \times \frac{1\text{ mole } CO_2}{44.01\text{ g } CO_2} \times \frac{6.022 \times 10^{23}\ CO_2\text{ molecules}}{\text{mole } CO_2} = 4.8 \times 10^{20}$$

**$CO_2$ molecules per breath.**

Problem 3.109 For the following oxides of iron, FeO, $Fe_2O_3$, and $Fe_3O_4$, describe how you would determine which has the greatest percentage by mass of oxygen. Would you need to look up any information to solve this problem?

**To calculate the percent oxygen for each molecule, one must find the mass of oxygen in the compound and divide by that compound's total mass. You could solve the problem by looking up the mass and performing the operations, but $Fe_2O_3$ has the highest ratio of oxygen to iron atoms. Because we know that oxygen has a much lower atomic mass, so it can be reasoned that the higher the O:Fe ratio, the higher the % oxygen.**

Problem 3.111 For the compounds that commonly undergo precipitation reactions, what type of chemical bonding is expected?

**A precipitation reaction results in the formation of a salt, which is an ionic compound. Therefore the bonding would be ionic.**

Problem 3.113 If you have 21.1 g of iron(II) nitrate that is dissolved to give 1.54 L of solution, what is the molarity of the solution? What is the molarity of the nitrate ions?

$$21.1\ \text{g Fe(NO}_3)_2 \times \frac{1\ \text{mol Fe(NO}_3)_2}{179.87\ \text{g Fe(NO}_3)_2} \times \frac{1}{1.54\ \text{L}} = 0.0762\ \text{mol/L}$$

**There are two moles of nitrate ions for every mole of iron(II) nitrate, so the molarity of the nitrate ions would be 2 × 0.0762 M = 0.152 M**

Problem 3.115 Most periodic tables provide molar masses for the elements with four or five significant figures. How accurately would you have to measure the mass of a sample of roughly 100 g to make a calculation of the number of moles of the chemical to have its significant figures limited by the molar mass calculation rather than the mass measurement?

**The mass measurement would have to include at least six significant figures.**

## CHAPTER OBJECTIVE QUIZ

This quiz will test your understanding of the basic text chapter objectives and give you additional practice problems. You should work this quiz after completing the end-of-chapter questions. The solutions to these questions are found at the end of this chapter.

1. Which of the following characteristics applied to biofuels is TRUE?
   a. Biofuels utilize energy from the sun to produce biomass and require no other source of energy.
   b. Biofuels are always a cheaper alternative to traditional fuels.
   c. Our economy could quickly and seamlessly transition to biofuels if the demand was there.
   d. Photosynthesis is the process where plants utilize energy from the sun to produce carbohydrates.

2. TRUE or FALSE: A balanced chemical equation must obey the law of conservation of matter.

3. Which of the following statements about balanced chemical equations is false?
   a. Charge is not necessarily conserved.
   b. There must be the same number and type of atom present in the reactants and products.
   c. Mass must be conserved.
   d. Energy must be conserved.
   e. The coefficients are adjusted to satisfy the law of conservation of matter.

4. Balance the following chemical reactions:

   (a) $ZnS(s) + O_2(g) \rightarrow ZnO(s) + SO_2(g)$

   (b) $CuFeS_2(s) + CuCl_2(aq) \rightarrow CuCl(aq) + FeCl_2(aq) + S(s)$

   (c) $N_2O(g) + Na(s) + NH_3(\ell) \rightarrow NaN_3(s) + NaOH(aq) + N_2(g)$

5. Chlorine dioxide can be made by the reaction of elemental chlorine and sodium chlorite. Write the balanced chemical equation for this reaction. (The other product is sodium chloride).

6. Write and balance an equation showing nitrogen dioxide reacting with water to produce nitric acid and nitrogen oxide.

7. TRUE or FALSE: The mole, Avogadro's number, represents $6.022 \times 10^{23}$ particles of matter: atoms, molecules or ions.

8. Consider the following equation: $2\ NO(g) + O_2(g) \rightarrow 2\ NO_2(g)$

   (a) How many molecules of $NO_2$ form if 14 molecules of $O_2$ react?
   (b) How many total molecules of NO and $O_2$ react to form 6 moles of $NO_2$?
   (c) What is the molar ratio for $O_2$ and $NO_2$?
   (d) 32.5 moles of $O_2$ reacting requires how many moles of NO?

9. What is the molar mass of dimethyldichlorosilane, $(CH_3)_2SiCl_2$? How many moles are present in 32.21 grams? How many molecules are present in 0.158 g $(CH_3)_2SiCl_2$?

10. How many grams of carbon are present in 15.53 g of acetic acid, $CH_3COOH$?

11. What is the elemental analysis of $K_2Cr_2O_7$?

12. The elemental analysis of methanol is 37.5 %C, 12.6 %H, and 49.9 %O. What is the empirical formula of methanol?

13. TRUE or FALSE: Molarity is defined as moles of solute per liter of solvent.

14. How many grams of solute are contained in 255 mL of 0.158 M $FeCl_3$?

15. How many milliliters of solution would contain 45.0 grams sucrose ($C_{12}H_{22}O_{11}$) if the molarity were 4.09 M?

16. How many milliliters of 3.00 M $H_2SO_4$ are required to produce 1.50 L of 0.250 M solution?

17. Which of the following statements about electrolytes is incorrect?
    a. Electrolytes dissolve in water to produce ions.
    b. Nonelectrolytes do not dissolve in water.
    c. Electrolyte solutions conduct electricity.
    d. Acids, bases, and salts are electrolytes.
    e. Nonelectrolyte solutions do not conduct electricity.

18. Predict the species expected to be present in significant amounts when the following substances are added to water:
    (a) $Ba(OH)_2$ (b) $C_{12}H_{22}O_{11}$ (sucrose) (c) $KClO_3$
    (d) $H_2SO_4$ (e) PbS (f) HCN

19. What is the final molarity of sodium ions if 50.0 mL of 0.25 M $Na_2SO_4$ is combined with 50.0 mL of 1.00 M NaCl? Assume the total solution volume is exactly 100.0 mL.

20. Classify the following ionic compounds as soluble or insoluble.

    (a) NaCN (b) AgCl (c) $(NH_4)_2S$ (d) $MgCl_2$
    (e) $Na_3PO_4$ (f) $Sn(NO_3)_2$ (g) $CaCO_3$

21. Classify each of the following as a strong acid, weak acid, strong base, or weak base.
    (a) $NH_3$ (b) HCl (c) $H_2SO_4$ (d) KOH
    (e) $CH_3COOH$ (f) $Sr(OH)_2$ (g) $HNO_2$

22. A solution of sodium sulfide is added to a solution of $Pb(NO_3)_2$. Lead (II) sulfide precipitates. Write the balanced molecular, total ionic, and net ionic equations.

23. TRUE or FALSE: Injection of $CO_2$ deep into the ocean is a possibility to sequester carbon somewhere besides the atmospheric reservoir.

## ANSWERS TO THE CHAPTER OBJECTIVE QUIZ

1. **d**

2. **TRUE**

3. **a**

4. **(a) Zinc and sulfur are already in balance, the $O_2$ needs to be balanced:**

   **$ZnS(s) + \underline{3/2}\ O_2(g) \rightarrow ZnO(s) + SO_2(g)$**

   **To express the equation in terms of the smallest whole-number coefficients, multiply by 2:**

   **$2\ [\ ZnS(s) + \underline{3/2}\ O_2(g) \rightarrow ZnO(s) + SO_2(g)\ ] = \underline{2}\ ZnS(s) + \underline{3}\ O_2(g) \rightarrow \underline{2}\ ZnO(s) + \underline{2}\ SO_2(g)$**

   **(b) We want to balance Fe first (which is balanced), then Cu:**

   **$\underline{1}\ CuFeS_2(s) + \underline{1}\ CuCl_2(aq) \rightarrow \underline{2}\ CuCl(aq) + \underline{1}\ FeCl_2(aq) + S(s)$**

   **Next, balance Cl: we have two Cl on the left and three on the right; the least common multiplier is six, new need six on both sides. We'll change the coefficient of $CuCl_2$ to three, and the coefficient of CuCl to four, to get six Cl on both sides.**

   **$\underline{1}\ CuFeS_2(s) + \underline{3}\ CuCl_2(aq) \rightarrow \underline{4}\ CuCl(aq) + \underline{1}\ FeCl_2(aq) + S(s)$**

   **Finally, we balance S:**

   **$\underline{1}\ CuFeS_2(s) + \underline{3}\ CuCl_2(aq) \rightarrow \underline{4}\ CuCl(aq) + \underline{1}\ FeCl_2(aq) + \underline{2}\ S(s)$**

   **(c) Balance H first, then O, then Na, N last.**

   **$N_2O(g) + Na(s) + NH_3(\ell) \rightarrow NaN_3(s) + \underline{3}\ NaOH(aq) + N_2(g)$**

   **$\underline{3}\ N_2O(g) + Na(s) + NH_3(\ell) \rightarrow NaN_3(s) + \underline{3}\ NaOH(aq) + N_2(g)$**

   **$\underline{3}\ N_2O(g) + \underline{4}\ Na(s) + NH_3(\ell) \rightarrow NaN_3(s) + \underline{3}\ NaOH(aq) + \underline{2}\ N_2(g)$**

5. **The reactants are $NaClO_2$ and $Cl_2$, the products are $ClO_2$ and NaCl. The unbalanced reaction is: $NaClO_2 + Cl_2 \rightarrow ClO_2 + NaCl$**
   **Na and Cl are already balanced; chlorine is balanced with a ½ coefficient:**

$NaClO_2 + \underline{½} Cl_2 \rightarrow ClO_2 + NaCl$

In whole-number coefficients, the equation is:
$\underline{2}\ NaClO_2 + \underline{1}\ Cl_2 \rightarrow \underline{2}\ ClO_2 + \underline{2}\ NaCl$

6. The unbalanced equation is:
$NO_2(g) + H_2O(\ell) \rightarrow HNO_3(aq) + NO(g)$

First balance H:
$NO_2(g) + H_2O(\ell) \rightarrow \underline{2}\ HNO_3(aq) + NO(g)$

Next balance N:
$\underline{3}\ NO_2(g) + H_2O(\ell) \rightarrow \underline{2}\ HNO_3(aq) + NO(g)$

Oxygen is already in balance, the equation is balanced.

7. TRUE

8. (a) $14 \text{ molecules } O_2 \times \dfrac{2 \text{ molecules} NO_2}{1 \text{ molecule} O_2} = 28 \text{ molecules } NO_2$

(b) $6 \text{ molecules } NO_2 \times \dfrac{1 \text{ molecule} O_2}{2 \text{ molecules} NO_2} = 3 \text{ molecules } O_2$

$6 \text{ molecules } NO_2 \times \dfrac{2 \text{ molecules} NO}{2 \text{ molecules} NO_2} = 6 \text{ molecules } NO_2,$ 9 total molecules react

(c) $\dfrac{1 \text{ mole} O_2}{2 \text{ mole} NO_2}$

(d) $32.5 \text{ mol } O_2 \times \dfrac{2 \text{ mole NO}}{1 \text{ mole} O_2} = 65 \text{ moles NO required}$

9. $(CH_3)_2SiCl_2$

| | |
|---|---|
| 2 mol C: | 2 × 12.011 g/mol |
| 6 mol H: | 6 × 1.0079 g/mol |
| 1 mol Si: | 1 × 28.086 g/mol |
| 2 mol Cl: | 2 × 35.453 g/mol |
| Total: | 129.06 g/mol |

$32.21 \text{ g } (CH_3)_2SiCl_2 \times \dfrac{1 \text{ mol } (CH_3)_2SiCl_2}{129.06 \text{ g } (CH_3)_2SiCl_2} = 0.2496 \text{ moles } (CH_3)_2SiCl_2$

$$0.158\text{ g }(CH_3)_2SiCl_2 \times \frac{1\text{ mol }(CH_3)_2SiCl_2}{129.06\text{ g }(CH_3)_2SiCl_2} \times \frac{6.022\times10^{23}\text{ molecules}}{1\text{ mol }(CH_3)_2SiCl_2} =$$

$= 7.37 \times 10^{20}$ **molecules**

10. $$15.53\text{ g }CH_3COOH \times \frac{1\text{ mol }CH_3COOH}{60.052\text{ g }CH_3COOH} \times \frac{2\text{ mol C}}{1\text{ mol }CH_3COOH} \times \frac{12.011\text{ g C}}{1\text{ mol C}} =$$

**6.212 g C**

11. $K_2Cr_2O_7$:

| | |
|---|---|
| 2 mol K: | 2 × 39.098 g/mol = 78.196 g K |
| 2 mol Cr: | 2 × 51.996 g/mol = 103.992 g Cr |
| 7 mol O: | 7 × 15.999 g/mol = 111.993 g O |
| Total: | 294.181 g/mol |

$$\%\ K = \frac{78.196\text{ g K}}{294.181\text{ g/mol}} \times 100 = 26.58\ \%\ K, \qquad \%\ Cr = \frac{103.992\text{ g Cr}}{294.181\text{ g/mol}} \times 100 =$$

$= 35.35\ \%\ Cr$

$$\%\ O = \frac{111.993\text{ g O}}{294.181\text{ g/mol}} \times 100 = 38.07\ \%\ O$$

12. **In 100 g of methanol there are**

**37.5 g C, 12.6 g H, and 49.9 g O.**

**Using the molar masses to find moles:**

$$37.5\text{ g C} \times \frac{1\text{ mol C}}{12.011\text{ g C}} = 3.12\text{ mol C} \qquad 12.6\text{ g H} \times \frac{1\text{ mol H}}{1.0079\text{ g H}} = 12.5\text{ mol H}$$

$$49.9\text{ g O} \times \frac{1\text{ mol O}}{15.999\text{ g O}} = 3.12\text{ mol O}$$

**The fewest moles are for C, so we'll take our ratios by dividing by the moles of C:**

$3.12\text{ mol C} \div 3.12\text{ mol C} = 1.0$ $\qquad$ $12.5\text{ mol H} \div 3.12\text{ mol C} = 4.0$

$3.12\text{ mol O} \div 3.12\text{ mol C} = 1.0$

**These ratios are the subscripts in our empirical formula:**
$C_{1.0}H_{4.0}O_{1.0}$ **or** $CH_4O$ **(This is usually written** $CH_3OH$**)**

13. **FALSE (moles per liter of <u>solution</u>)**

**14.** $0.158\ M\ FeCl_3 = \dfrac{0.158\ \text{mol}\ FeCl_3}{L}$

$$\frac{0.158\ \text{mol}\ FeCl_3}{L} \times 0.255\ L \times \frac{162.20\ g\ FeCl_3}{1\ \text{mol}\ FeCl_3} = 6.54\ g\ FeCl_3$$

**15.** $45.0\ g\ C_{12}H_{22}O_{11} \times \dfrac{1\ \text{mol}\ C_{12}H_{22}O_{11}}{342.29\ g\ C_{12}H_{22}O_{11}} = 0.131\ \text{mol}\ C_{12}H_{22}O_{11}$

$M = \dfrac{\text{moles}}{L}$, therefore, $L = \dfrac{\text{moles}}{M} = \dfrac{0.131\ \text{mol}\ C_{12}H_{22}O_{11}}{4.09\ M} = 0.0321\ L = 32.1\ mL$ solution

**16.** Using the dilution formula, $M_1V_1 = M_2V_2$,

$(3.00\ M)(?\ L) = (0.250\ M)(1.50\ L) \quad L = \dfrac{(0.250\ M)(1.50\ L)}{(3.00\ M)} = 0.125\ L = 125\ mL$ required.

**17.** b

**18.** (a) $Ba^{2+}(aq)$, $OH^-(aq)$; $Ba(OH)_2$ is a strong electrolyte (strong base).

(b) $C_{12}H_{22}O_{11}(aq)$; $C_{12}H_{22}O_{11}$ is soluble but is a nonelectrolyte.

(c) $K^+(aq)$, $ClO_3^-(aq)$, $KClO_3$ is a strong electrolyte (soluble salt).

(d) $H^+(aq)$, $HSO_4^-(aq)$, $H_2SO_4$ is a strong electrolyte (strong acid).

(e) $PbS(s)$, PbS is an insoluble salt.

(f) $HCN(aq)$, HCN is a weak electrolyte (weak acid).

**19.** Recall L × M = moles. We'll calculate the moles of $Na^+$ in each solution, and then divide that total by the solution volume.

$0.25\ M\ Na_2SO_4$: $\quad 0.0500\ L \times \dfrac{2 \times 0.25\ \text{moles}\ Na^+}{L} = 0.025\ \text{moles}\ Na^+$

$1.00\ M\ NaCl$: $\quad 0.0500\ L \times \dfrac{1.00\ \text{moles}\ Na^+}{L} = 0.050\ \text{moles}\ Na^+$

Total: $0.0750\ \text{moles}\ Na^+$

Molarity = $0.0750\ \text{moles}\ Na^+ \div 0.100\ L = 0.75\ M$

20. (a) Soluble. All $Na^+$ salts are soluble.

    (b) Insoluble. AgCl is one of the exceptions to the rule that $Cl^-$ salts are generally soluble.

    (c) Soluble. All $NH_4^+$ salts are soluble.

    (d) Soluble. Chloride salts are generally soluble.

    (e) Soluble. All $Na^+$ salts are soluble.

    (f) Soluble. All $NO_3^-$ salts are soluble.

    (g) Insoluble. Carbonates are generally insoluble.

21. (a) $NH_3$, weak base (b) HCl, strong acid (c) $H_2SO_4$, strong acid

    (d) KOH, Strong base (e) $CH_3COOH$, weak acid (f) $Sr(OH)_2$, strong base

    (g) $HNO_2$, weak acid

22. The reactants are $Na_2S(aq)$ and $Pb(NO_3)_2(aq)$. The products are PbS(s) and $NaNO_3(aq)$

    The balanced molecular equation is:

    $Na_2S(aq) + Pb(NO_3)_2(aq) \rightarrow PbS(s) + 2\ NaNO_3(aq)$

    In the total ionic equation, the soluble salts are written as separate ions but the insoluble salt remains as a neutral formula:

    $2\ Na^+(aq) + S^{2-}(aq) + Pb^{2+}(aq) + 2\ NO_3^-(aq) \rightarrow PbS(s) + 2\ Na^+(aq) + NO_3^-(aq)$

    For the net ionic equation, the spectator ions, $Na^+$ and $NO_3^-$, are cancelled out:

    $S^{2-}(aq) + Pb^{2+}(aq) \rightarrow PbS(s)$

23. TRUE

# CHAPTER 4

## *Study Goals*

The study goals outline specific concepts to be mastered in this section of the text chapter. Related problems at the end of the text chapter will also be noted. Working the questions noted should aid in mastery of each study goal and also highlight any areas that you may need additional help in.

**Section 4.1 *INSIGHT INTO* Gasoline and Other Fuels**

1. Give definitions for several terms related to gasoline and fuels: hydrocarbons, alkanes, fossil fuels, complete combustion, and octane rating. *Work Problems 1–6.*

**Section 4.2 Fundamentals of Stoichiometry**

2. Define stoichoimetry and write sets of mole ratios from a balanced equation. *Work Problem 7.*
3. Solve stoichoimetry problems relating two substances in a chemical equation beginning with moles and ending with moles or molecules. *Work Problems 8–13*
4. Solve stoichoimetry problems relating two substances in a chemical equation beginning with mass given and ending with mass desired. *Work Problems 14–21, 66, 70, 72, 75, 84– 87,90, 91, 93, 95, 96, 98–105.*

**Section 4.3 Limiting Reactants**

5. Determine the limiting reactant in a non-stoichiometric mixture of reactants and calculate the theoretical yield using the limiting reactant. *Work Problems 22–35, 76 – 81, 94, 97.*

**Section 4.4 Theoretical and Percentage Yields**

6. Define percent yield and solve various problems relating percent yield and theoretical yield. *Work Problems 36–50.*

**Section 4.5 Solution Stoichiometry**

7. Use molarity to solve stoichoimetry problems relating substances in aqueous solution. *Work Problems 51, 52, 54–61, 69, 71, 73, 74, 82, 83, 88, 89, 92, 106.*
8. Describe the titration procedure. *Work Problem 53.*

**Section 4.6 *INSIGHT INTO* Alternative Fuels and Fuel Additives**

9. Give several reasons gasoline additives are used and list several common additives. *Work Problems 62–67.*

## *Solutions to the Odd-Numbered Problems*

Problem 4.1 List at least two factors that make it difficult to describe the combustion of gasoline accurately.

**(1) Gasoline is not a single compound; it is a mixture of dozens of different molecules, generally hydrocarbons between 5 – 12 carbons in size. (2) Combustion of gasoline is not complete – meaning that the products are not always carbon dioxide and water.**

Problem 4.3 Explain the difference between complete and incomplete combustion.

***Complete* combustion refers to a combustion reaction between a carbon–containing compound and oxygen where all the carbon atoms are converted to carbon dioxide. In *incomplete* combustion, some carbon atoms are converted to carbon monoxide or even just carbon, because of an insufficient amount of oxygen.**

Problem 4.5 Methane, ethane, and propane are also hydrocarbons, but they are not major components of gasoline. What prevents them from being part of this mixture?

**Methane, ethane, and propane contain only one, two, and three carbon atoms, respectively. They are gases at room temperature so it would not be practical or safe to add them to gasoline. They would evaporate rapidly from an open container of gasoline. These gases would also make gasoline a much more explosive mixture.**

Problem 4.7 For the following reactions, write the ratios that can be established among molar amounts of the various compounds.

**The ratios that can be established are given by the "stoichiometric" or molar coefficients in front of the formula. The molar ratios give the exact quantitative relationships between any two reactants or products. Each ratio can be inverted depending on the order of the calculations being performed.**

(a) $2\ H_2 + O_2 \rightarrow 2\ H_2O$ $\quad \frac{2\ \text{mol}\ H_2}{1\ \text{mol}\ O_2}$ or $\frac{1\ \text{mol}\ O_2}{2\ \text{mol}\ H_2}$ ; $\frac{2\ \text{mol}\ H_2}{2\ \text{mol}\ H_2O}$ or $\frac{2\ \text{mol}\ H_2O}{2\ \text{mol}\ H_2}$ ; $\frac{2\ \text{mol}\ H_2O}{1\ \text{mol}\ O_2}$ or $\frac{1\ \text{mol}\ O_2}{2\ \text{mol}\ H_2O}$

(b) $2\ H_2O_2 \rightarrow 2\ H_2O + O_2$ $\quad \frac{2\ \text{mol}\ H_2O_2}{2\ \text{mol}\ H_2O}$ or $\frac{2\ \text{mol}\ H_2O}{2\ \text{mol}\ H_2O_2}$ ; $\frac{2\ \text{mol}\ H_2O_2}{1\ \text{mol}\ O_2}$ or $\frac{1\ \text{mol}\ O_2}{2\ \text{mol}\ H_2O_2}$ ; $\frac{1\ \text{mol}\ O_2}{2\ \text{mol}\ H_2O}$ or $\frac{2\ \text{mol}\ H_2O}{1\ \text{mol}\ O_2}$

(c) $P_4 + 5\ O_2 \rightarrow P_4O_{10}$ $\frac{1\ \text{mol}\ P_4}{5\ \text{mol}\ O_2}$ or $\frac{5\ \text{mol}\ O_2}{1\ \text{mol}\ P_4}$ ; $\frac{1\ \text{mol}\ P_4}{1\ \text{mol}\ P_4O_{10}}$ or $\frac{1\ \text{mol}\ P_4O_{10}}{1\ \text{mol}\ P_4}$ ;

$\frac{5\ \text{mol}\ O_2}{1\ \text{mol}\ P_4O_{10}}$ or $\frac{1\ \text{mol}\ P_4O_{10}}{5\ \text{mol}\ O_2}$

(d) $2\ KClO_3 \rightarrow 2KCl + 3\ O_2$ $\frac{2\ \text{mol}\ KClO_3}{2\ \text{mol}\ KCl}$ or $\frac{2\ \text{mol}\ KCl}{2\ \text{mol}\ KClO_3}$ ; $\frac{2\ \text{mol}\ KClO_3}{3\ \text{mol}\ O_2}$ or

$\frac{3\ \text{mol}\ O_2}{2\ \text{mol}\ KClO_3}$ ; $\frac{2\ \text{mol}\ KCl}{3\ \text{mol}\ O_2}$ or $\frac{3\ \text{mol}\ O_2}{2\ \text{mol}\ KCl}$

Problem 4.9 ■ Sulfur, $S_8$, combines with oxygen at elevated temperatures to form sulfur dioxide. (a) Write a balanced chemical equation for this reaction. (b) If 200 oxygen molecules are used up in this reaction, how many sulfur molecules react? (c) How many sulfur dioxide molecules are formed in part (b)?

**(a) $S_8(s) + \underline{8}\ O_2(g) \rightarrow \underline{8}\ SO_2(g)$**
**(b) There is an 8:1 ratio of oxygen and sulfur molecules reacting, so 200 ÷ 8 = 25 sulfur molecules.**
**(c) There is a 1:1 ratio of oxygen and sulfur dioxide molecules, so 200 $SO_2$ molecules.**

Problem 4.11 MTBE, $C_5H_{12}O$, is one of the additives that replaced tetraethyl-lead to increase performance in gasoline. (See Example Problem 4.6 and Section 4.6.) How many moles of $O_2$ are needed for the complete combustion of 1.50 mol of MTBE?

**We first need to write a balanced equation showing MTBE, $C_5H_{12}O$, reacting with oxygen to produce carbon dioxide and water (these products are produced if complete combustion occurs). Then, we can find the molar ratio relating moles MTBE reacting to moles of oxygen reacting.**

**$2\ C_5H_{12}O + 15\ O_2 \rightarrow 10\ CO_2 + 12\ H_2O$**

**Therefore the ratio we need is $\frac{15\ \text{mol}\ O_2}{2\ \text{mol MTBE}}$.**

**1.5 moles MTBE × $\frac{15\ \text{mol}\ O_2}{2\ \text{mol MTBE}}$ = 11 moles $O_2$ consumed (rounded to two significant digits).**

Problem 4.13 ■ For the following reactions determine the value of *x*.

**We need to use the appropriate molar ratios to relate the two substances of interest.**

(a) $4\ C + S_8 \rightarrow 4\ CS_2$

3.2 mol $S_8$ yield $x$ mol $CS_2$

$$3.2 \text{ mol } S_8 \times \frac{4 \text{ mol } CS_2}{1 \text{ mol } S_8} = x \text{ mol } CS_2, \quad x = 13 \text{ mol } CS_2$$

(b) $CS_2 + 3\ O_2 \rightarrow CO_2 + 2\ SO_2$

1.8 mol $CS_2$ yields $x$ mol $SO_2$

$$1.8 \text{ mol } CS_2 \times \frac{2 \text{ mol } SO_2}{1 \text{ mol } CS_2} = x \text{ mol } SO_2, \quad x = 3.6 \text{ mol } SO_2$$

(c) $N_2H_4 + 3\ O_2 \rightarrow 2\ NO_2 + 2\ H_2O$

7.3 mol $O_2$ yields $x$ mol $NO_2$

$$7.3 \text{ mol } O_2 \times \frac{2 \text{ mol } NO_2}{3 \text{ mol } O_2} = x \text{ mol } NO_2, \quad x = 4.9 \text{ mol } SO_2$$

(d) $SiH_4 + 2\ O_2 \rightarrow SiO_2 + 2\ H_2O$

$1.3 \times 10^{-3}$ mol $SiH_4$ yields $x$ mol $H_2O$

$$1.3 \times 10^{-3} \text{ mol } SiH_4 \times \frac{2 \text{ mol } H_2O}{1 \text{ mol } SiH_4} = x \text{ mol } H_2O, \quad x = 2.6 \times 10^{-3} \text{ mol } H_2O$$

Problem 4.15 ■ What mass of the unknown compound is formed in the following reactions? (Assume that the reactants for which amounts are not given are in excess.)

**We need to use the appropriate molar ratios to relate the two substances of interest. This time we also need to do mass-mole conversions. The general scheme of these calculations will be:**

| **Grams** | **→** | **Moles A** | **→** | **Moles B** | **→** | **Grams B** |
|---|---|---|---|---|---|---|
| | **divide by molar mass A** | | **multiply by molar ratio: Moles B/Moles A** | | **multiply by molar mass B** | |

**These are called stoichoimetry calculations.**

(a) $C_2H_4 + H_2 \rightarrow C_2H_6$

3.4 g $C_2H_4$ reacts to produce $x$ g $C_2H_6$

$$3.4 \text{ g } C_2H_4 \times \frac{1 \text{ mol } C_2H_4}{28.05 \text{ g } C_2H_4} \times \frac{1 \text{ mol } C_2H_6}{1 \text{ mol } C_2H_4} \times \frac{30.07 \text{ g } C_2H_6}{1 \text{ mol } C_2H_6} = x \text{ g } C_2H_6, \quad x = 3.6 \text{ g } C_2H_6$$

(b) $CS_2 + 3Cl_2 \rightarrow CCl_4 + S_2Cl_2$

5.78 g $Cl_2$ reacts to produce $x$ g $S_2Cl_2$

$$5.78\text{ g }Cl_2 \times \frac{1\text{ mol }Cl_2}{70.91\text{ g }Cl_2} \times \frac{1\text{ mol }S_2Cl_2}{3\text{ mol }Cl_2} \times \frac{135.0\text{ g }S_2Cl_2}{1\text{ mol }S_2Cl_2} = x\text{ g }S_2Cl_2, \quad x = 3.67\text{ g }S_2Cl_2$$

(c) $PCl_3 + 3\,H_2O \rightarrow H_3PO_3 + 3\,HCl$

3.1 mg $PCl_3$ reacts to produce $x$ mg HCl

**This one we also have to convert mg to g.**

$$3.1\text{ mg }PCl_3 \times \frac{1\text{ g}}{1000\text{ mg}} \times \frac{1\text{ mol }PCl_3}{137.3\text{ g }PCl_3} \times \frac{3\text{ mol HCl}}{1\text{ mol }PCl_3} \times \frac{36.46\text{ g HCl}}{1\text{ mol HCl}} \times \frac{1000\text{ mg}}{1\text{ g}} = x\text{ mg}$$

**HCl,**
**$x = 2.5$ mg HCl**

**Note that the conversion from g to mg and then back again does not change our result. It's a good idea to include it in your calculations anyway.**

(d) $B_2H_6 + 3\,O_2 \rightarrow B_2O_3 + 3\,H_2O$

4.6 kg $B_2H_6$ reacts to produce $x$ kg $B_2O_3$

**Here we have to convert kg to g. Again it won't really affect our results.**

$$4.6\text{ kg }B_2H_6 \times \frac{1000\text{ g}}{1\text{ kg}} \times \frac{1\text{ mol }B_2H_6}{27.67\text{ g }B_2H_6} \times \frac{1\text{ mol }B_2O_3}{1\text{ mol }B_2H_6} \times \frac{69.62\text{ g }B_2O_3}{1\text{ mol }B_2O_3} \times \frac{1\text{ kg}}{1000\text{ g}} = x\text{ kg }B_2O_3,$$

**$x = 12$ kg $B_2O_3$**

Problem 4.17 Phosgene is a highly toxic gas that has been used as a chemical weapon at times in the past. It is now used in the manufacture of polycarbonates, which are used to make compact discs and plastic eyeglass lenses. Phosgene is produced by the reaction, $CO + Cl_2 \rightarrow COCl_2$. Given an excess of carbon monoxide, what mass of chlorine gas must be reacted to form 4.5 g of phosgene?

**Refer to our strategy in Problem 4.15. We'll convert g phosgene to moles, relate moles phosgene formed to moles chlorine reacted, and then convert moles chlorine to g.**

$$4.5\text{ g }COCl_2 \times \frac{1\text{ mol }COCl_2}{98.92\text{ g }COCl_2} \times \frac{1\text{ mol }Cl_2}{1\text{ mol }COCl_2} \times \frac{70.91\text{ g }Cl_2}{1\text{ mol }Cl_2} = x\text{ g }Cl_2\text{ required to react,}$$

**$x = 3.2$ g $Cl_2$**

Problem 4.19 How many metric tons of carbon are required to reduce 7.83 metric tons of $Fe_2O_3$ according to the following reaction?

$$2\ Fe_2O_3 + 3\ C \rightarrow 3\ CO_2 + 4\ Fe$$

How many metric tons of iron are produced?

**A metric ton is 1000 kg. These are more stoichiometry calculations and we will use the same basic strategy as described in Problem 4.15. First we relate amount of $Fe_2O_3$ reacted to amount carbon reacting.**

$$7.83 \text{ metric tons } Fe_2O_3 \times \frac{1000\ kg}{1\ \text{metric ton}} \times \frac{1000\ g}{1\ kg} \times \frac{1\ mol\ Fe_2O_3}{159.7\ g\ Fe_2O_3} \times \frac{3\ mol\ C}{2\ mol\ Fe_2O_3} \times$$

$$\frac{12.01\ g\ C}{1\ mol\ C} \times \frac{1\ kg}{1000\ g} \times \frac{1\ \text{metric ton}}{1000\ kg} = 0.883 \text{ metric tons C reacted.}$$

**The second part of the problem is very similar. We need to do another stoichoimetry calculation but now we need to relate the $Fe_2O_3$ reacted to the Fe produced.**

$$7.83 \text{ metric tons } Fe_2O_3 \times \frac{1000\ kg}{1\ \text{metric ton}} \times \frac{1000\ g}{1\ kg} \times \frac{1\ mol\ Fe_2O_3}{159.7\ g\ Fe_2O_3} \times \frac{4\ mol\ Fe}{2\ mol\ Fe_2O_3} \times$$

$$\frac{55.85\ g\ Fe}{1\ mol\ Fe} \times \frac{1\ kg}{1000\ g} \times \frac{1\ \text{metric ton}}{1000\ kg} = 5.48 \text{ metric tons Fe produced.}$$

Problem 4.21 Ammonium nitrate, $NH_4NO_3$, will decompose explosively to form $N_2$, $O_2$, and $H_2O$, a fact that has been exploited in terrorist bombings. What mass of nitrogen is formed by the decomposition of 2.6 kg of ammonium nitrate?

**Refer to our strategy in Problem 4.15. We first have to write a balanced equation representing the decomposition of ammonium nitrate. Then we do a stoichoimetry calculation relating mass of ammonium nitrate reacted to mass of nitrogen formed.**

$$2\ NH_4NO_3\ (s) \rightarrow 2\ N_2\ (g) + O_2\ (g) + 4\ H_2O\ (g)$$

$$2.6\ kg\ NH_4NO_3 \times \frac{1000\ g}{1\ kg} \times \frac{1\ mol\ NH_4NO_3}{80.04\ g\ NH_4NO_3} \times \frac{2\ mol\ N_2}{2\ mol\ NH_4NO_3} \times \frac{28.01\ g\ N_2}{1\ mol\ N_2} =$$

$$= 910\ g\ N_2 \text{ formed}$$

Problem 4.23 If 8.4 moles of disilane, $Si_2H_6$, are combined with 15.1 moles of $O_2$, which is the limiting reactant?

$$2\ Si_2H_6 + 7\ O_2 \rightarrow 4\ SiO_2 + 6\ H_2O$$

**To determine a limiting reactant, we must do a stoichiometry calculation. We pick one reactant (it doesn't matter which) and calculate how much of the *other* reactant is required for complete reaction. Then we compare that result with how much we *actually* have to determine which the limiting or excess reactant is.**

$$8.4 \text{ moles } Si_2H_6 \times \frac{7 \text{ mol } O_2}{2 \text{ mol } Si_2H_6} = 29.4 \text{ moles } O_2 \text{ required to react with \textit{all} the } Si_2H_6$$

**Since more $O_2$ is required than we actually have, $O_2$ is the limiting reactant.**

Problem 4.25 In the reaction of arsenic with bromine, $AsBr_5$ will form only when excess bromine is present. Write a balanced chemical equation for this reaction. Determine the minimum number of moles of bromine that are needed if 9.6 moles of arsenic are present.

**The two reactants are arsenic, As, and bromine, $Br_2$. The balanced equation is:**

$$2\,As + 5\,Br_2 \rightarrow 2\,AsBr_5$$

**We need to do a stoichiometry calculation to the minimum number of moles $Br_2$ required to react with 9.6 moles As.**

$$9.6 \text{ moles As} \times \frac{5 \text{ mol } Br_2}{2 \text{ mol As}} = 24 \text{ moles } Br_2 \text{ required to react with \textit{all} the As}$$

**There must be more than 24 moles $Br_2$.**

Problem 4.27 When octane is combusted with inadequate oxygen, carbon monoxide may form. If 100 g of octane are burned in 200 g of $O_2$, are conditions conducive to forming carbon monoxide?

**To answer this we first need to write a balanced equation showing complete combustion ($CO_2$ and $H_2O$ are products) of octane, $C_8H_{18}$.**

$$2\,C_8H_{18} + 25\,O_2 \rightarrow 16\,CO_2 + 18\,H_2O$$

**We then do a stoichiometry calculation to determine if there is excess $O_2$ present.**

$$100 \text{ g } C_8H_{18} \times \frac{1 \text{ mol } C_8H_{18}}{114.2 \text{ g } C_8H_{18}} \times \frac{25 \text{ mol } O_2}{2 \text{ mol } C_8H_{18}} \times \frac{32.00 \text{ g } O_2}{1 \text{ mol } O_2} = 350 \text{ g } O_2 \text{ required to react}$$

**Therefore, there is insufficient $O_2$ present for complete combustion, and carbon monoxide will probably form. Conditions are conductive to the formation of CO.**

Problem 4.29 Copper reacts with sulfuric acid according to the following equation:

$$2\,H_2SO_4 + Cu \rightarrow CuSO_4 + 2\,H_2O + SO_2$$

How many grams of sulfur dioxide are created by this reaction if 14.2 g of copper react with 18.0 g sulfuric acid?

**This is a two-step problem. First, the limiting reactant must be determined. We find this by doing a stoichiometry calculation to determine how many grams of one reactant are required to completely react with the other.**

$$14.2\text{ g Cu} \times \frac{1\text{ mol Cu}}{63.55\text{ g Cu}} \times \frac{2\text{ mol }H_2SO_4}{1\text{ mol Cu}} \times \frac{98.08\text{ g }H_2SO_4}{1\text{ mol }H_2SO_4} = 43.83\text{ g }H_2SO_4\text{ required to}$$

**react with all the Cu. There are only 18.0 g $H_2SO_4$ present, so $H_2SO_4$ is the limiting reactant.**
**The second step is to calculate the amount of $SO_2$ produced when the limiting reactant completely reactants.**

$$18.0\text{ g }H_2SO_4 \times \frac{1\text{ mol }H_2SO_4}{98.08\text{ g }H_2SO_4} \times \frac{1\text{ mol }SO_2}{2\text{ mol }H_2SO_4} \times \frac{64.06\text{ g }SO_2}{1\text{ mol }SO_2} = 5.88\text{ g }SO_2\text{ produced.}$$

<u>Problem 4.31</u> When $Al(OH)_3$ reacts with sulfuric acid, the following reaction occurs:

$$2\,Al(OH)_3 + 3\,H_2SO_4 \rightarrow Al_2(SO_4)_3 + 6\,H_2O$$

If $1.7 \times 10^3$ g of $Al(OH)_3$ are combined with 680 g of $H_2SO_4$, how much aluminum sulfate can form?

**See Problem 4.29. First, the limiting reactant must be determined.**

$$1700\text{ g }Al(OH)_3 \times \frac{1\text{ mol }Al(OH)_3}{78.00\text{ g }Al(OH)_3} \times \frac{3\text{ mol }H_2SO_4}{2\text{ mol }Al(OH)_3} \times \frac{98.08\text{ g }H_2SO_4}{1\text{ mol }H_2SO_4} = 3200\text{ g }H_2SO_4$$

**are required to react. Therefore, $H_2SO_4$ is the limiting reactant because there is only 680 g present.**

**Second, calculate the amount of aluminum sulfate formed when the limiting reactant completely reacts.**

$$680\text{ g }H_2SO_4 \times \frac{1\text{ mol }H_2SO_4}{98.08\text{ g }H_2SO_4} \times \frac{1\text{ mol }Al_2(SO_4)_3}{3\text{ mol }H_2SO_4} \times \frac{342.1\text{ g }Al_2(SO_4)_3}{1\text{ mol }Al_2(SO_4)_3} = 790\text{ g }Al_2(SO_4)_3$$

**produced.**

Problem 4.33 How much $HNO_3$ can be formed in the following reaction if 3.6 kg of $NO_2$ gas are bubbled through 2.5 kg of water?

$$3\ NO_2(g) + H_2O(\ell) \rightarrow 2\ HNO_3(aq) + NO(g)$$

**This is another limiting reactant problem. We will first find the limiting reactant by doing a stoichiometry calculation to see how much water is required to react with 3.6 kg of $NO_2$.**

**3.6 kg $NO_2$ = 3600 g $NO_2$**

$$3600\ g\ NO_2 \times \frac{1\ mol\ NO_2}{46.00\ g\ NO_2} \times \frac{1\ mol\ H_2O}{3\ mol\ NO_2} \times \frac{18.01\ g\ H_2O}{1\ mol\ H_2O} = 470\ g\ H_2O \text{ are required.}$$

**There is more water present (2.5 kg) than required, so water is the excess reactant and $NO_2$ is the limiting reactant.**

**Second we will do another stoichiometry calculation to determine how much nitric acid ($HNO_3$) is formed when all of the limiting reactant (3600 g $NO_2$) reacts.**

$$3600\ g\ NO_2 \times \frac{1\ mol\ NO_2}{46.00\ g\ NO_2} \times \frac{2\ mol\ HNO_3}{3\ mol\ NO_2} \times \frac{63.01\ g\ HNO_3}{1\ mol\ HNO_3} = 3300\ g \text{ or } 3.3\ kg \text{ of } HNO_3$$

**are produced.**

Problem 4.35 Silicon carbide, an abrasive, is made by the reaction of silicon dioxide with graphite (solid carbon).

$$SiO_2 + C \xrightarrow{heat} SiC + CO \qquad \text{(balanced?)}$$

We mix 150.0 g of $SiO_2$ and 101.5 g of C. If the reaction proceeds as far as possible, which reactant is left over? How much of this reactant remains?

**The balanced equation is $SiO_2 + \underline{3}\ C \xrightarrow{heat} SiC + \underline{2}\ CO$.**
**This is another limiting reactant problem. We will first find the limiting reactant by doing a stoichiometry calculation to see how much C is required to react with 150.0 g of $SiO_2$.**

$$150.0\ g\ SiO_2 \times \frac{1\ mol\ SiO_2}{60.08\ g\ SiO_2} \times \frac{3\ mol\ C}{1\ mol\ SiO_2} \times \frac{12.01\ g\ C}{1\ mol\ C} = 89.96\ g\ C \text{ are required.}$$

**There are 101.5 g C present, but our calculation shows that only 89.96 g will be required to react with all the $SiO_2$. Therefore the C is the excess reactant. So next we will calculate how much C will be left over.**

**101.5 g – 89.96 g = 11.54 g C remains**

Problem 4.37 What types of events happen in chemical reactions that lead to percentage yields of less than 100%?

**A variety of things can cause less than 100% yield. Side reactions can occur that produce something other than the desired product or sometimes the reaction does not go to completion. Sometimes physical problems like reactants evaporating or being spilled occur, or possibly some product is stuck in a beaker or on filter paper.**

Problem 4.39 The theoretical yield and the actual yield for various reactions are given below. Determine the corresponding percentage yields.

**The theoretical yield is the maximum amount of product that can be formed assuming all the limiting reactant reacts. The actual yield is the product actually formed in a real experiment. Percentage yield is a way of determining how close to the maximum your experiment is.**

$$\% \text{ Yield} = \frac{\text{Actual yield (grams)}}{\text{Theoretical yield (grams)}} \times 100$$

**(The units for yield don't really matter as long as they are the same.)**

Reaction 1 **Theoretical yield = 35.0 g, Actual yield = 12.8 g,**

$$\%\text{Yield} = \frac{12.8 \text{ g}}{35.0 \text{ g}} \times 100 = 36.6\%$$

Reaction 2 **Theoretical yield = 9.3 mg, Actual yield = 120 mg,**

$$\%\text{Yield} = \frac{120 \text{ mg}}{9.3 \text{ mg}} \times 100 = 1290\%$$ **This is obviously not actually possible.**

Reaction 3 **Theoretical yield = 3.7 metric tons** $\times \frac{1000 \text{ kg}}{1 \text{ metric ton}} = 3700 \text{ kg}$,

**Actual yield = 1250 kg,** $\%\text{Yield} = \frac{1250 \text{ kg}}{3700 \text{ kg}} \times 100 = 34\%$

Reaction 4 **Theoretical yield = 40 g, Actual yield = 41 g,**

$$\%\text{Yield} = \frac{41 \text{ g}}{40 \text{ g}} \times 100 = 103\%$$

Problem 4.41 ■ Methanol, $CH_3OH$, is used in racing cars because it is a clean-burning fuel. It can be made by this reaction:

$$CO(g) + 2\,H_2(g) \rightarrow CH_3OH(\ell)$$

What is the percent yield if $5.0 \times 10^3$ g $H_2$ reacts with excess CO to form $3.5 \times 10^3$ g

$CH_3OH$?

**Given the balanced chemical equation, the mass of the limiting reactant and the actual yield, determine the percent yield.**
**First, we need to calculate the theoretical yield. We get this by determining the maximum amount of product that could have been made from the given quantities of reactant: Take the mass of the limiting reactant and convert it to moles. Then use stoichiometry to find the moles of product. Then convert to grams using molar mass. Take the actual yield and divide by the calculated theoretical yield and multiply by 100% to get percent yield.**
**The limiting reactant is $H_2$. From its mass, find the maximum grams of $CH_3OH$ that could be made. The mole ratio comes from the balanced equation.**

$$5.0 \times 10^3 \text{ g } H_2 \times \frac{1 \text{ mol } H_2}{2.02 \text{ g } H_2} \times \frac{1 \text{ mol } CH_3OH}{2 \text{ mol } H_2} \times \frac{32.04 \text{ g } CH_3OH}{1 \text{ mol } CH_3OH} = 4.0 \times 10^4 \text{ g}$$

**$CH_3OH$**

**The given mass of $CH_3OH$ is the actual yield. Use these two masses to calculate the percent yield.**

$$\%\text{Yield} = \frac{3.5 \times 10^3 \text{ g}}{4.0 \times 10^4 \text{ g}} \times 100 = 8.8\% \text{ yield}$$

Problem 4.43 ■ The percent yield of the following reaction is consistently 87%.

$$CH_4(g) + 4\ S(g) \rightarrow CS_2(g) + 2\ H_2S(g)$$

How many grams of sulfur would be needed to obtain 80.0 g of $CS_2$?

**The actual yield = 0.87 × theoretical yield of $CS_2$ = 80.0 g $CS_2$**
**The theoretical yield = 80.0 g $CS_2$ ÷ 0.87 = 92.0 g $CS_2$**

**Now we calculate how much sulfur must react to form 92.0 g $CS_2$.**

$$92.0 \text{ g } CS_2 \times \frac{1 \text{ mol } CS_2}{76.15 \text{ g } CS_2} \times \frac{4 \text{ mol S}}{1 \text{ mol } CS_2} \times \frac{32.07 \text{ g S}}{1 \text{ mol S}} = 1.5 \times 10^2 \text{ g S}$$

Problem 4.45 Magnesium nitride forms in a side reaction when magnesium metal burns in air. This reaction may also be carried out in pure nitrogen.

$$3\ Mg(s) + N_2(g) \rightarrow Mg_3N_2(s)$$

If 18.4 g of $Mg_3N_2$ forms from the reaction of 20.0 g of magnesium with excess nitrogen, what is the percentage yield?

**This is very similar to Problem 4.43. Magnesium is the limiting reactant and the actual yield is 18.4 g magnesium nitride.**

$$20.0\text{ g Mg} \times \frac{1\text{ mol Mg}}{24.305\text{ g Mg}} \times \frac{1\text{ mol }Mg_3N_2}{3\text{ mol Mg}} \times \frac{100.93\text{ g }Mg_3N_2}{1\text{ mol }Mg_3N_2} = 27.7\text{ g }Mg_3N_2$$

**are produced.**

**This is the theoretical yield.**

**Now we'll calculate the % Yield:**

$$\%\text{ Yield} = \frac{\text{Actual yield}}{\text{Theoretical yield}} \times 100 \quad = \frac{18.4\text{ g }Mg_3N_2}{27.7\text{ g }Mg_3N_2} \times 100 = 66.4\ \%\text{ Yield}$$

Problem 4.47 If 21 g of $H_2S$ is mixed with 38 g of $O_2$ and 31 g of $SO_2$ form, what is the percentage yield?

$$2\ H_2S + 3\ O_2 \rightarrow 2\ SO_2 + 2\ H_2O$$

**This problem is both a limiting reactant problem and a percent yield calculation. We first will do a stoichoimetry calculation to determine how much $O_2$ is required to react with 21 g of $H_2S$.**

$$21\text{ g }H_2S \times \frac{1\text{ mol }H_2S}{34.08\text{ g }H_2S} \times \frac{3\text{ mol }O_2}{2\text{ mol }H_2S} \times \frac{32.00\text{ g }O_2}{1\text{ mol }O_2} = 29.6\text{ g }O_2\text{ are required.}$$

**There is more oxygen (38 g) than is required; therefore $H_2S$ is the limiting reactant.**

**Next, we'll calculate how much $SO_2$ is formed when 21 g $H_2S$ reacts.**

$$21\text{ g }H_2S \times \frac{1\text{ mol }H_2S}{34.08\text{ g }H_2S} \times \frac{2\text{ mol }SO_2}{2\text{ mol }H_2S} \times \frac{64.06\text{ g }SO_2}{1\text{ mol }SO_2} = 39\text{ g }SO_2\text{ is produced.}$$

**This is the theoretical yield.**

**Finally, we'll calculate the percent yield. The actual yield of $SO_2$ is 31 g.**

$$\%\text{ Yield} = \frac{\text{Actual yield}}{\text{Theoretical yield}} \times 100 \quad = \frac{31\text{ g }SO_2}{39\text{ g }SO_2} \times 100 = 79\ \%\text{ Yield}$$

Problem 4.49 Silicon carbide is an abrasive used in the manufacture of grinding wheels. A company is investigating whether they can be more efficient in the construction of such wheels by making their own SiC via the reaction

$$SiO_2 + 3\ C \rightarrow SiC + 2\ CO$$

(a) If this reaction consistently has a yield of 85%, what is the minimum amount of both silicon dioxide and carbon needed to produce 3400 kg of SiC for the manufacture of cutting wheels?

**To find the amount of $SiO_2$ and C needed, the theoretical yield must first be found. Then stoichiometry calculations will show how much reactants are needed.**

$$85\% = \frac{3400\text{ kg SiC}}{\text{Theoretical yield}}\text{; Theoretical yield} = 3400\text{ kg} \div 0.85 = 4000\text{ kg SiC}$$

$$4000 \times 10^3\text{ g SiC} \times \frac{1\text{ mol SiC}}{40.10\text{ g SiC}} \times \frac{1\text{ mol SiO}_2}{1\text{ mol SiC}} \times \frac{60.09\text{ g SiO}_2}{1\text{ mol SiO}_2} \times \frac{1\text{ kg}}{1000\text{ g}} = 6000\text{ kg SiO}_2$$

$$4000 \times 10^3\text{ g SiC} \times \frac{1\text{ mol SiC}}{40.10\text{ g SiC}} \times \frac{3\text{ mol C}}{1\text{ mol SiC}} \times \frac{12.01\text{ g C}}{1\text{ mol SiO}_2} \times \frac{1\text{ kg}}{1000\text{ g}} = 3600\text{ kg C}$$

(b) The silicon dioxide is to be obtained from sand and the carbon is derived from coal. If the available sand is 95% $SiO_2$ by weight and the coal is 73% C by weight, what mass of coal is needed for each metric ton of sand used?

**One metric ton = 1000 kg. The ratio of C to $SiO_2$ required is** $\frac{3600\text{ kg C}}{6000\text{ kg SiO}_2} = 0.60$

**One metric ton of sand contains 0.95 × (1000 kg) = 950 kg $SiO_2$**
**So 0.60 × (950 kg) = 570 kg C is needed.**
**The coal is 73% carbon so the mass of coal needed is 570 kg ÷ 0.73 = 780 kg coal.**

**For every metric ton of sand, 780 kg of coal is required.**

Problem 4.51 Small quantities of hydrogen gas can be prepared by the following reaction:

$$Zn(s) + H_2SO_4(aq) \rightarrow ZnSO_4(aq) + H_2(g)$$

How many grams of $H_2$ can be prepared from 25.0 mL of 6.00 M $H_2SO_4$ and excess zinc?

**This is a stoichiometry problem but instead of relating grams of reactant to grams of product, we are given a volume of a certain concentration of solution. Our approach will be very similar, only we'll do a molarity calculation to find moles of sulfuric acid (the limiting reactant).**

$$25.0\text{ mL H}_2\text{SO}_4 \times \frac{1\text{ L}}{1000\text{ mL}} \times \frac{6.00\text{ mol H}_2\text{SO}_4}{1\text{ L H}_2\text{SO}_4} \times \frac{1\text{ mol H}_2}{1\text{ mol H}_2\text{SO}_4} \times \frac{2.016\text{ g H}_2}{1\text{ mol H}_2} = 0.302\text{ g H}_2$$

**are produced.**

Problem 4.53 What is the role of an indicator in a titration?

**The role of an indicator is to provide a color change near the *stoichiometric point* of the titration. This is where you have added stoichiometric amounts of both reactants and therefore this is where you want to stop the titration. The color change of the indicator is called the *endpoint*.**

Problem 4.55 ■ What volume, in milliliters, of 0.512 M NaOH is required to react completely with 25.0 mL 0.234 M $H_2SO_4$?

**We will solve this stoichoimetry problem much like Problem 4.49, but here we must first write the balanced equation of this acid/base reaction. Note that sulfuric acid has two hydrogen ions but sodium hydroxide only has one hydroxide ion to donate.**

**$H_2SO_4 + 2\ NaOH \rightarrow 2\ H_2O + Na_2SO_4$ The molar ratio is 1:2.**
**We must invert the molarity of the NaOH in the end of the calculation.**

$$25.0\text{ mL } H_2SO_4 \times \frac{1\text{ L}}{1000\text{ mL}} \times \frac{0.234\text{ mol } H_2SO_4}{1\text{ L } H_2SO_4} \times \frac{2\text{ mol NaOH}}{1\text{ mol } H_2SO_4} \times \frac{1\text{ L NaOH}}{0.512\text{ mol NaOH}} \times \frac{1000\text{ mL}}{1\text{ L}} = 22.9\text{ mL NaOH are required for neutralization.}$$

Problem 4.57 ■ Hydrazine, $N_2H_4$, a weak base, can react with an acid such as sulfuric acid.

$$2\ N_2H_4(aq) + H_2SO_4(aq) \rightarrow 2\ N_2H_5^+(aq) + SO_4^{2-}(aq)$$

What mass of hydrazine can react with 250. mL 0.225 M $H_2SO_4$?

**We will solve this stoichoimetry problem much like Problem 4.49. This is a typical titration problem, which means a stoichoimetry problem involving solutions. From the volume and molarity of $H_2SO_4$, we'll calculate moles, and then relate moles $H_2SO_4$ reacted to moles hydrazine reacted. Finally, we'll convert moles hydrazine to grams.**

$$250.\text{ mL } H_2SO_4 \times \frac{1\text{ L}}{1000\text{ mL}} \times \frac{0.225\text{ mol } H_2SO_4}{1\text{ L } H_2SO_4} \times \frac{2\text{ mol } N_2H_4}{1\text{ mol } H_2SO_4} \times \frac{32.05\text{ g } N_2H_4}{1\text{ mol } N_2H_4} = 3.61\text{ g } N_2H_4$$

Problem 4.59 Many mining operations produce tailings that can be oxidized to form acids, including sulfuric acid. One chemical that reacts in this way is chalcopyrite, $CuFeS_2$, whose net oxidation can be described by the equation:

$$4\ CuFeS_2(s) + 17\ O_2(g) + 10\ H_2O(\ell) \rightarrow 4\ Fe(OH)_3(s) + 4\ CuSO_4(aq) + 4\ H_2SO_4(aq)$$

In a laboratory experiment to study weathering, a mining engineer put a 2.3-kg sample of

$CuFeS_2$(s) near an empty 40-L container. Simulated weathering experiments were conducted and the runoff was collected, leading to 35.8 L of solution. A 25.00-mL aliquot of this solution was analyzed with 0.0100 M NaOH(aq). The titration required 38.1 mL to reach the equivalence point.
(a) What was the concentration of sulfuric acid?

**The titration of NaOH and $H_2SO_4$ is described by the equation:**
**$H_2SO_4 + 2\ NaOH \rightarrow 2\ H_2O + Na_2SO_4$**
**The solution stoichiometry calculation for the acid concentration is:**

$$\mathbf{0.0381\,L\ NaOH \times \frac{0.0100\ mol\ NaOH}{1\ L\ NaOH} \times \frac{1\ mol\ H_2SO_4}{2\ mol\ NaOH} \times \frac{1}{0.02500\ L}}$$

**$= 0.00762$ M $H_2SO_4$**

(b) Assuming the only acid in the container comes from the reacted chalcopyrite, what mass of the chalcopyrite has reacted?

**The total moles acid formed:** $\mathbf{\frac{0.00762\ M\ H_2SO_4}{L} \times 35.8\ L = 0.273\ moles\ H_2SO_4}$

$$\mathbf{0.273\ moles\ H_2SO_4\ formed \times \frac{4\ mol\ CuFeS_2}{4\ mol\ H_2SO_4} \times \frac{183.54\ g\ CuFeS_2}{1\ mol\ CuFeS_2} = 50.1\ g\ CuFeS_2}$$

(c) What percentage of the chalcopyrite reacted in this experiment?

$$\mathbf{Percent = \frac{50.1\ g\ CuFeS_2}{2300\ g\ sample\ CuFeS_2} \times 100 = 2.2\ \%}$$

Problem 4.61 Aluminum dissolves in HCl according to the equation written below, whereas copper does not react with HCl.

$$2\ Al(s) + 6\ HCl(aq) \rightarrow 2\ AlCl_3(aq) + 3\ H_2(g)$$

A 35.0-g sample of a copper-aluminum alloy is dropped into 750 mL of 3.00 M HCl, and the reaction above proceeds as far as possible. If the alloy contains 77.1% Al by mass, what mass of hydrogen gas would be produced?

**The alloy contains 0.771 × (35.0 g) = 26.985 g Al**
**It must be determined whether HCl or Al is the limiting reactant. This can be done by finding the amount of product that can form from each reactant**

$$\mathbf{26.985\ g\ Al \times \frac{1\ mol\ Al}{26.98\ g\ Al} \times \frac{3\ mol\ H_2}{2\ mol\ Al} \times \frac{2.016\ g\ H_2}{1\ mol\ H_2} = 3.02\ g\ H_2}$$

$$\mathbf{0.750\ L\ NaOH \times \frac{3.00\ mol\ HCl}{1\ L\ HCl} \times \frac{3\ mol\ H_2}{6\ mol\ HCl} \times \frac{2.016\ g\ H_2}{1\ mol\ H_2} = 2.27\ g\ H_2}$$

**HCl is the limiting reactant (it produces the smaller yield of $H_2$). The reaction would produce 2.27 g $H_2$.**

Problem 4.63 What is actually measured by the octane ratings of different grades of gasoline?

**The octane rating measures how much the fuel mixture can be compressed before it ignites (without a spark). This property is desirable because premature burning of the gasoline in an engine throws off the timing of the cylinders (knocking). Pure octane is very compressible and was chosen arbitrarily as a standard for comparison. Octane is given a "rating" of 100. Some gasoline mixtures are more or less compressible than octane; their ratings are higher or lower than 100. Generally, the higher the octane rating, the better the fuel burns in an engine.**

Problem 4.65 Perform a web search to learn what restrictions (if any) on the use of MTBE in gasoline might exist in your local area.

**Answers will vary according to location. Restrictions on the use of MTBE (methyl tertiary butyl ether) are beginning to be common because of evidence that it might be a carcinogen.**

Problem 4.67 Using the web, find out how lead "poisons" the catalyst in a catalytic converter.

**Some of the catalysts in a catalytic converter are metals: platinum, palladium, and rhodium. These catalysts greatly increase the speed of reactions to reduce harmful gaseous emissions in automobile exhaust. Lead will react with these metals, essentially covering them up and not allowing them to act as catalysts. Therefore, use of tetraethyl lead as a gasoline additive would destroy a catalytic converter.**

Problem 4.69 ■ You have 0.954 g of an unknown acid, $H_2A$, which reacts with NaOH according to the balanced equation

$$H_2A(aq) + 2\ NaOH(aq) \rightarrow Na_2A(aq) + 2\ H_2O(\ell)$$

If 36.04 mL of 0.509 M NaOH is required to titrate the acid to the equivalence point, what is the molar mass of the acid?

**This is another titration problem. This time we will find the moles of base reacted, then moles of the acid. Then we'll divide the grams by the moles to find the molar mass.**

$$\mathbf{36.04\ mL \times \frac{1\ L}{1000\ mL} \times \frac{0.509\ mol\ NaOH}{1\ L\ NaOH} \times \frac{1\ mol\ H_2A}{2\ mol\ NaOH} = 0.00917\ mol\ H_2A}$$

$$\mathbf{\frac{0.954\ g\ H_2A}{0.00917\ mol} = 104\ g/mol}$$

Problem 4.71 ■ Phosphoric acid ($H_3PO_4$) is important in the production of both fertilizers and detergents. It is distributed commercially as a solution with a concentration of about 14.8 M. Approximately $2.1 \times 10^9$ gallons of this concentrated phosphoric acid solution are produced annually in this country alone. Assuming that all of this $H_3PO_4$ is produced in the reaction below, what mass of the mineral fluoroapatite ($Ca_5(PO_4)_3F$) would be required each year?

$$Ca_5(PO_4)_3F + 5\ H_2SO_4 \rightarrow 3\ H_3PO_4 + 5\ CaSO_4 + HF$$

**We need to do a stoichiometry calculation relating phosphoric acid to fluoroapatite. First, we will use the molarity and volume (converting gallons to liters) to determine moles of $H_3PO_4$. Second, we'll calculate the grams of $Ca_5(PO_4)_3F$ formed from those moles of phosphoric acid.**

$$2.1 \times 10^9 \text{ gal } H_3PO_4 \times \frac{1\text{ L}}{0.2640\text{ gal}} \times \frac{14.8\text{ mol } H_3PO_4}{1\text{ L } H_3PO_4} \times \frac{1\text{ mol } Ca_5(PO_4)_3F}{3\text{ mol } H_3PO_4} \times$$

$$\frac{504.3\text{ g } Ca_5(PO_4)_3F}{1\text{ mol } Ca_5(PO_4)_3F} = 2.0 \times 10^{13}\text{ g } Ca_5(PO_4)_3F$$

Problem 4.73 Blood alcohol levels are usually reported in mass percent: A level of 0.10 means that 0.10 g of alcohol ($C_2H_5OH$) was present in 100 g of blood. This is just one way of expressing the concentration of alcohol dissolved in the blood. Other concentration units could be used. Find the molarity of alcohol in the blood of a person with a blood alcohol level of 0.12. (The density of blood is 1.2 g/mL.)

**A blood alcohol level of 0.12 means the person has 0.12 g alcohol per 100 g blood. To convert this to molarity, we will convert grams alcohol to moles and convert grams blood to liters.**

$$\frac{0.12\text{ g } C_2H_5OH}{100\text{ g blood}} \times \frac{1\text{ mol } C_2H_5OH}{46.07\text{ g } C_2H_5OH} \times \frac{1.2\text{ g blood}}{\text{mL blood}} \times \frac{1000\text{ mL}}{1\text{ L}} = 0.031\text{ M } C_2H_5OH$$

Problem 4.75 Ammonium sulfate ($(NH_4)_2SO_4$) is a common fertilizer that can be produced by the reaction of ammonia ($NH_3$) with sulfuric acid ($H_2SO_4$). Each year about $2 \times 10^9$ kg of ammonium sulfate is produced worldwide. How many kilograms of ammonia would be needed to generate this quantity of $(NH_4)_2SO_4$? (Assume that the reaction goes to completion and no other products formed.)

**First, we'll write the balanced equation for this reaction.**

$$2\ NH_3 + H_2SO_4 \rightarrow (NH_4)_2SO_4$$

**Second, we'll do a stoichiometry calculation to determine how much ammonia is needed to produce $2 \times 10^9$ kg of $(NH_4)_2SO_4$.**

$$2 \times 10^9 \text{ kg } (NH_4)_2SO_4 \times \frac{1000 \text{ g}}{1 \text{ kg}} \times \frac{1 \text{ mol } (NH_4)_2SO_4}{132.1 \text{ g } (NH_4)_2SO_4} \times \frac{2 \text{ mol } NH_3}{1 \text{ mol } (NH_4)_2SO_4} \times \frac{17.03 \text{ g } NH_3}{1 \text{ mol } NH_3}$$

$$\times \frac{1 \text{ kg}}{1000 \text{ g}} = 5 \times 10^8 \text{ kg } NH_3 \text{ required.}$$

Problem 4.77 The pictures below show a molecular scale view of a chemical reaction between $H_2$ and CO to produce methanol, $CH_3OH$. The box on the left represents the reactants at the instant of mixing, and the box on the right shows what is left once the reaction has gone to completion. Was there a limiting reagent in this reaction? If so, what was it? Write a balanced chemical equation for this reaction. As usual, your equation should use the smallest possible whole number coefficients for all substances.

**According to the picture, there is leftover hydrogen (a homonuclear diatomic molecule) after completion of the reaction. Carbon monoxide (CO) must be the limiting reactant.**

$$2\ H_2 + CO \rightarrow CH_3OH$$

Problem 4.79 In the cold vulcanization of rubber, disulfur dichloride ($S_2Cl_2$) is used as a source of sulfur atoms, and those sulfur atoms form "bridges," or cross-links, between polymer chains. $S_2Cl_2$ can be produced by reacting molten sulfur ($S_8(\ell)$) with chlorine ($Cl_2(g)$). What is the maximum mass of $S_2Cl_2$ that can be produced by reacting 32.0 g of sulfur with 71.0 g of chlorine?

**We'll start by writing the balanced equation.**

$$S_8(\ell) + 4\ Cl_2(g) \rightarrow 4\ S_2Cl_2(s)$$

**The question is asking for the theoretical yield. We first must find the limiting reactant, though. See Problem 4.29.**

$$32.0 \text{ g } S_8 \times \frac{1 \text{ mol } S_8}{256.5 \text{ g } S_8} \times \frac{4 \text{ mol } Cl_2}{1 \text{ mol } S_8} \times \frac{70.91 \text{ g } Cl_2}{1 \text{ mol } Cl_2} = 35.4 \text{ g } Cl_2 \text{ required to react}$$

**with all the $S_8$. There is 71.0 g $Cl_2$ present, more than enough, so $Cl_2$ is the excess reactant and $S_8$ is the limiting reactant.**

$$32.0 \text{ g } S_8 \times \frac{1 \text{ mol } S_8}{256.5 \text{ g } S_8} \times \frac{4 \text{ mol } S_2Cl_2}{1 \text{ mol } S_8} \times \frac{135.0 \text{ g } S_2Cl_2}{1 \text{ mol } S_2Cl_2} = 67.4 \text{ g } S_2Cl_2 \text{ are produced.}$$

Problem 4.81 ■ Cryolite ($Na_3AlF_6$) is used in the commercial production of aluminum from ore. Cryolite itself is produced by the following reaction:

$$6\ NaOH + Al_2O_3 + 12\ HF \rightarrow 2\ Na_3AlF_6 + 9\ H_2O$$

A mixture containing 800 kg of NaOH, 300 kg of $Al_2O_3$, and 600 kg of HF is heated to 950°C and reacts to completion. What is the maximum mass of $Na_3AlF_6$ formed?

**In this limiting reactant problem, we have three reactants to consider for the limiting reactant.**

$$800 \text{ kg NaOH} \times \frac{1000\text{ g}}{1\text{ kg}} \times \frac{1\text{ mol NaOH}}{40.00\text{ g NaOH}} \times \frac{1\text{ mol Al}_2\text{O}_3}{6\text{ mol NaOH}} \times \frac{102.0\text{ g Al}_2\text{O}_3}{1\text{ mol Al}_2\text{O}_3} \times \frac{1\text{ kg}}{1000\text{ g}} =$$

**340 kg $Al_2O_3$ required. This is more than we actually have but we must first check if the HF will limit the reaction.**

$$300 \text{ kg Al}_2\text{O}_3 \times \frac{1000\text{ g}}{1\text{ kg}} \times \frac{1\text{ mol Al}_2\text{O}_3}{102.0\text{ g Al}_2\text{O}_3} \times \frac{12\text{ mol HF}}{1\text{ mol Al}_2\text{O}_3} \times \frac{20.01\text{ g HF}}{1\text{ mol HF}} \times \frac{1\text{ kg}}{1000\text{ g}} =$$

**700 kg HF required. This is more than we actually have, therefore the HF will be used up before the $Al_2O_3$. HF is the limiting reactant.**

**When the problem asks for the maximum mass of product formed, they are asking for the theoretical yield.**

$$600 \text{ kg HF} \times \frac{1000\text{ g}}{1\text{ kg}} \times \frac{1\text{ mol HF}}{20.01\text{ g HF}} \times \frac{2\text{ mol Na}_3\text{AlF}_6}{12\text{ mol HF}} \times \frac{209.9\text{ g Na}_3\text{AlF}_6}{1\text{ mol Na}_3\text{AlF}_6} \times \frac{1\text{ kg}}{1000\text{ g}} =$$

**1049 kg $Na_3AlF_6$ formed ≈ $1.05 \times 10^3$ kg.**

Problem 4.83 ■ Calcium carbonate (limestone, $CaCO_3$) dissolves in hydrochloric acid, producing water and carbon dioxide as shown in the following *unbalanced* net ionic equation.

$$CaCO_3(s) + H_3O^+(aq) \rightarrow H_2O(\ell) + CO_2(g) + Ca^{2+}(aq)$$

Suppose 5.0 g of $CaCO_3$ is added to 700. mL of 0.10 M HCl. What is the maximum mass of $CO_2$ that could be formed? What would be the final concentration of $Ca^{2+}$ ions? (Assume that the final solution volume is still 700 mL.)

**Balancing the equation is the first task.**

$$CaCO_3(s) + 2\ H_3O^+(aq) \rightarrow 3\ H_2O(\ell) + CO_2(g) + Ca^{2+}(aq)$$

**We can calculate the moles of HCl (which gives the moles of $H_3O^+$) and the moles of $CaCO_3$ and then do a stoichiometry calculation to find the limiting reactant. From the limiting reactant, the mass of $CO_2$ that could be formed can be found.**

$$0.700\ \text{L} \times \frac{0.10\ \text{mol HCl}}{\text{L HCl}} = 0.070\ \text{mol HCl} = 0.070\ \text{moles H}^+ = 0.070\ \text{moles H}_3\text{O}^+$$

$$5.0\ \text{g CaCO}_3 \times \frac{1\ \text{mol CaCO}_3}{100.1\ \text{g CaCO}_3} \times \frac{2\ \text{mol H}_3\text{O}^+}{1\ \text{mol CaCO}_3} = 0.10\ \text{mol H}_3\text{O}^+\ \text{required.}$$

**More moles of $H_3O^+$ are required than are actually present, so $H_3O^+$ (HCl) is the limiting reactant.**
**The maximum mass of $CO_2$ that can be formed is the theoretical yield, which is calculated using the limiting reactant.**

$$0.070\ \text{moles H}_3\text{O}^+ \times \frac{1\ \text{mol CO}_2}{2\ \text{mol H}_3\text{O}^+} \times \frac{44.01\ \text{g CO}_2}{1\ \text{mol CO}_2} = 1.5\ \text{g CO}_2\ \text{formed}$$

**We can calculate the moles of calcium ion produced as well, and then divide by liters of solution to get molarity of $Ca^{2+}$.**

$$0.070\ \text{moles H}_3\text{O}^+ \times \frac{1\ \text{mol Ca}^{2+}}{2\ \text{mol H}_3\text{O}^+} \div 0.700\ \text{L} = 0.050\ \text{M Ca}^{2+}$$

Problem 4.85 ■ Calcium sulfate is the essential component of plaster and sheet rock. Waste calcium sulfate can be converted into quicklime, CaO, by reaction with carbon at high temperatures. The following two reactions represent a sequence of reactions that might take place:

$$CaSO_4(s) + 4\ C(s) \rightarrow CaS(\ell) + 4\ CO(g)$$
$$CaS(\ell) + 3\ CaSO_4(s) \rightarrow 4\ CaO(s) + 4\ SO_2(g)$$

What weight of sulfur dioxide (in grams) could be obtained from 1.250 kg of calcium sulfate?

**To find the molar ratio between $SO_2$ and $CaSO_4$, the two sequential reactions must be combined.**
**The overall reaction is:** $4\ C(s) + 4\ CaSO_4(s) \rightarrow 4\ CaO(s) + 4\ SO_2(g) + 4\ CO(g)$

$$1.250\ \text{kg CaSO}_4 \times \frac{1000\ \text{g}}{1\ \text{kg}} \times \frac{1\ \text{mol CaSO}_4}{136.15\ \text{g CaSO}_4} \times \frac{4\ \text{mol SO}_2}{4\ \text{mol CaSO}_4} \times \frac{64.07\ \text{g SO}_2}{1\ \text{mol SO}_2} = 588.2\ \text{g SO}_2$$

Problem 4.87 A mixture of methane ($CH_4$) and propane ($C_3H_8$) has a total mass of 29.84 g. When the mixture was burned completely in excess oxygen, the $CO_2$ and $H_2O$ products had a combined mass of 142.97 g. Calculate the mass of methane in the original mixture.

**The two combustion reactions are:**
$CH_4 + 2\ O_2 \rightarrow CO_2 + 2\ H_2O$

**$C_3H_8 + 5\ O_2 \rightarrow 3\ CO_2 + 4\ H_2O$**

**Let X = moles $C_3H_8$ and let Y = moles $CH_4$.**

**The total mass of reactants can be calculated using the molar masses:**
**(X)44.09 g/mol $C_3H_8$ + (Y)16.04 g/mol $CH_4$ = 29.84 g total mass reactants.**

**Solving for X:**

$$X = \frac{29.84 - 16.04Y}{44.09}$$

**The first reaction produces Y moles $CO_2$ (1:1 ratio) and 2Y moles $H_2O$ (2:1).**
**The second reaction produces 3X moles $CO_2$ (3:1 ratio) and 4X moles $H_2O$ (4:1).**

**The total mass of product formed:**
**(Y+3X)44.01 g/mol $CO_2$ + (2Y+4X)18.015 g/mol $H_2O$ = 142.97 g total mass products**
**Substituting for X into the above equation yields**

$$80.04Y + 204.09\left(\frac{29.84 - 16.04Y}{44.09}\right) = 142.97$$

**Solving for Y, Y = 0.836 moles $CH_4$.**

**0.836 moles $CH_4$ × 16.04 g/mol $CH_4$ = 13.41 g $CH_4$**

Problem 4.89 A solution contains both $Ca^{2+}$ and $Pb^{2+}$ ions. A 50.0-mL sample of this solution was treated with 0.528 M NaF, and 53.8 mL of the NaF solution was needed to precipitate all of the $Ca^{2+}$ and $Pb^{2+}$ as $CaF_2$(s) and $PbF_2$(s). The precipitate was dried and weighed, and had a total mass of 3.141 g. What was the concentration of $Ca^{2+}$ in the original solution?

**The volume and molarity of the NaF solution will allow us to calculate the total moles of $F^-$ that reacted with $Ca^{2+}$ and $Pb^{2+}$.**

$$0.0538\text{L NaF} \times \frac{0.528\text{ mol NaF}}{1\text{ L}} = 0.0284\text{ moles F}^-$$

**Let X = moles $Ca^{2+}$ present and let Y = moles $Pb^{2+}$ present.**
**There are two moles of $F^-$ for each mole of $CaF_2$ and $PbF_2$ so:**
**0.0284 moles $F^-$ = 2X + 2Y.**

**Using the molar masses, we can set up an equation relating the mass of precipitate recovered and the moles of each compound:**
**3.141 g = X(78.08 g/mol of $CaF_2$) + Y(245.2 g/mol $PbF_2$)**

**After expressing Y in terms of X,**

$$Y = \frac{0.0284 - 2X}{2}$$

**This can be substituted into the equation and we can solve for X,**

$$3.141\ g = X(78.08\ g/mol\ of\ CaF_2) + \left(\frac{0.0284 - 2X}{2}\right)(245.2\ g/mol\ PbF_2)$$

**3.141 = 78.08X + 3.483 – 245.2X; X = 0.002044 moles $Ca^{2+}$**

**The original molarity is 0.002044 moles $Ca^{2+}$ ÷ 0.0500 L = 0.0409 M**

Problem 4.91 You are designing a process for a wastewater treatment facility intended to remove aqueous HCl from the input water. Several bases are being considered for use, and their costs are listed in the table below. If cost is the primary concern, which base would you recommend using? Use stoichiometric arguments to justify your answer.

| Material | $CaCO_3$ | $Ca(OH)_2$ | $NH_3$ | NaOH |
|---|---|---|---|---|
| Cost per metric ton | $270 | $165 | $255 | $315 |

**The cost can be expressed in dollars per mole base, then multiplied by the moles base-to-acid ratio to find the dollars of cost per mole of acid neutralized.**

**The balanced equations are:**
**$CaCO_3 + 2\ HCl \rightarrow H_2O + CO_2 + CaCl_2$**
**$Ca(OH)_2 + 2\ HCl \rightarrow 2\ H_2O + CaCl_2$**
**$NH_3 + HCl \rightarrow NH_4Cl$**
**$NaOH + HCl \rightarrow H_2O + NaCl$**

| Material | $CaCO_3$ | $Ca(OH)_2$ | $NH_3$ | NaOH |
|---|---|---|---|---|
| Cost per metric ton | $270 | $165 | $255 | $315 |
| Cost per gram | $2.70 × 10^{-4}$ | $1.65 × 10^{-4}$ | $2.55 × 10^{-4}$ | $3.15 × 10^{-4}$ |
| Molar mass | 100.09 g/mol | 74.10 g/mol | 17.03 g/mol | 40.00 g/mol |
| Cost per mole | $2.70 × 10^{-2}$ | $1.22 × 10^{-2}$ | $4.34 × 10^{-3}$ | $1.26 × 10^{-2}$ |

| Moles base per mole acid neutralized | 1:2 | 1:2 | 1:1 | 1:1 |
|---|---|---|---|---|
| Cost per mole acid neutralized | $1.35 × 10^{-2}$ | $6.10 × 10^{-3}$ | $4.34 × 10^{-3}$ | $1.26 × 10^{-2}$ |

**The most cost effective way to neutralize the acid is to use $NH_3$; it has the lowest cost per mole of acid neutralized.**

Problem 4.93 Chromium metal is found in ores as an oxide, $Cr_2O_3$. This ore can be converted to metallic Cr by the Goldschmidt process,

$$Cr_2O_3(s) + 2\ Al(s) \rightarrow 2\ Cr(\ell) + Al_2O_3(s)$$

Suppose that you have 4.5 tons of chromium(III) oxide. Describe how you would determine how much aluminum metal you would need to carry out this process. List any information you would have to look up.
**Calculate the molar masses of chromium(III) oxide and aluminum metal. From this information you can determine the moles of chromium(III) oxide present. Use the balanced equation to calculate the moles of Al metal consumed. Convert moles Al metal to kg metal to solve the problem.**

Problem 4.95 When heated in air, some element, X, reacts with oxygen to form an oxide with the chemical formula, $X_2O_3$. If a 0.5386-g sample of this unknown element yields an oxide with a mass of 0.7111 g, describe how you would determine the atomic mass of the element.

**The mass of oxygen in the sample is 0.7111 g – 0.5386 g = 0.1725 g O.**

**The moles of oxygen** $= 0.1725\text{ g O} \times \dfrac{1\text{ mol O}}{16.00\text{ g O}} = 0.01078\text{ moles O}$

**Using the ratio of 2 moles of "X" for each 3 moles of O (from the chemical formula), then there are 2/3 × 0.01078 = 0.007188 moles of "X" present in the sample.**

**The molar mass of "X" is**

$$\frac{0.5386\text{ g "X"}}{0.007188\text{ mol "X"}} = 74.93\text{ g/mol}$$ **(Probably arsenic)**

Problem 4.97 Bismuth selenide, $Bi_2Se_3$, is used in semiconductor research. It can be prepared directly from its elements by the reaction, $2\ Bi + 3\ Se \rightarrow Bi_2Se_3$. Suppose that you need 5 kg of bismuth selenide for a prototype system you are working on. You find a large bottle of bismuth metal but only 560 g of selenium. Describe how you can calculate if you have enough selenium on hand to carry out the reaction or if you need to order more.

**Determine the theoretical yield of bismuth selenide from 560 g of Se. (Find moles Se, then use molar ratio 1 mol $Bi_2Se_3$: 3 mol Se, then convert to mass of $Bi_2Se_3$). If the answer is less than 5 kg, then more Se is needed.**

Problem 4.99 The best way to generate small amounts of oxygen gas in a laboratory setting is by decomposing potassium permanganate.

$$2\ KMnO_4(s) \rightarrow K_2MnO_4 + MnO_2 + O_2$$

Suppose that you are testing a new fuel for automobiles and you want to generate specific amounts of oxygen with this reaction. For one test, you need exactly 500 g of oxygen. Describe how you would determine the mass of potassium permanganate you need to obtain this amount of oxygen.

**Determine the moles of $O_2$ needed. Multiply by a factor of 2 (based on the balanced equation) to find the moles of potassium permanganate you need to decompose. Convert moles to grams to determine the exact mass of potassium permanganate needed.**

Problem 4.101 Existing stockpiles of the refrigerant Freon-12, $CF_2Cl_2$, must be destroyed under the terms of the Montreal Protocol because of their potential for harming the ozone layer. One method for doing this involves reaction with sodium oxalate:

$$CF_2Cl_2 + 2\ Na_2C_2O_4 \rightarrow 2\ NaF + 2\ NaCl + C + 4\ CO_2$$

If you had 150 tons of Freon-12, describe how you would know how much sodium oxalate you would need to make that conversion.

**Convert 150 tons of Freon-12 into moles of Freon-12 by dividing by the molar mass; Double the number of moles to find the moles of sodium oxalate necessary.**

Problem 4.103 Benzopyrene is a hydrocarbon that is known to cause cancer. Combustion analysis (see problem 4.102) of a 2.44-g sample of this compound shows that combustion produces 8.54 g of $CO_2$ and 1.045 g of $H_2O$. What is the empirical formula of benzopyrene?

**We will first calculate the moles of H in the water and moles of C in the carbon dioxide. The chemical formulas give us the molar ratios between atoms and molecules. Then we'll use those amounts to determine the empirical formula of the benzopyrene.**

$$\mathbf{1.045\ g\ H_2O \times \frac{1\ mol\ H_2O}{18.01\ g\ H_2O} \times \frac{2\ mol\ H}{1\ mol\ H_2O} = 0.116\ mol\ H}$$

$$8.54 \text{ g } CO_2 \times \frac{1 \text{ mol } CO_2}{44.01 \text{ g } CO_2} \times \frac{1 \text{ mol C}}{1 \text{ mol } CO_2} = 0.194 \text{ mol C}$$

**Next we take ratios:**

**0.116 mol H ÷ 0.116 mol H = 1.0**

$$0.194 \text{ mol C} \div 0.116 \text{ mol H} = 1.672 \approx 1\tfrac{2}{3} = \frac{5}{3}$$

**Multiply both ratios by 3 to eliminate the fraction.**

$$3[\, C_{\frac{5}{3}} H_1 \,] = C_5H_3$$ **, this is the empirical formula.**

Problem 4.105 Aluminum metal reacts with sulfuric acid to form hydrogen gas and Aluminum sulfate. **(a)** Write a balanced chemical equation for this reaction. **(b)** Suppose that a 0.792-g sample of aluminum that contains impurities is reacted with excess sulfuric acid and 0.0813 g of $H_2$ is collected. Assuming that none of the impurities reacts with sulfuric acid to produce hydrogen, what is the percentage of aluminum in the sample?

**(a)** $2\,Al + 3\,H_2SO_4 \rightarrow 3\,H_2 + Al_2(SO_4)_3$

**(b) From the mass of hydrogen collected, the amount of aluminum in the original sample may be determined.**

$$0.0813 \text{ g } H_2 \times \frac{1 \text{ mol } H_2}{2.016 \text{ g } H_2} \times \frac{2 \text{ mol Al}}{3 \text{ mol } H_2} \times \frac{26.98 \text{ g Al}}{1 \text{ mol Al}} = 0.725 \text{ g Al}$$

$$\% \text{ Al in original sample} = \frac{0.725 \text{ g Al}}{0.792 \text{ g Al}} \times 100 = 91.6\%$$

## CHAPTER OBJECTIVE QUIZ

This quiz will test your understanding of the basic text chapter objectives and give you additional practice problems. You should work this quiz after completing the end-of-chapter questions. The solutions to these questions are found at the end of this chapter.

1. Which of the following statements about gasoline is false?
    a. Gasoline is a complex mixture of molecules.
    b. Most molecules in gasoline are hydrocarbons.
    c. Incomplete combustion of gasoline can form soot (carbon) and CO.
    d. Most molecules in gasoline have between one and six carbon atoms.
    e. Octane ($C_8H_{18}$) can be used as a model for the combustion of gasoline.

2. Balance the following combustion reactions:

   (a) $C_5H_{12}(\ell) + O_2(g) \rightarrow CO_2(g) + H_2O(\ell)$

   (b) $C_3H_8(g) + O_2(g) \rightarrow CO(g) + H_2O(\ell)$

   (c) $C_{12}H_{26}(\ell) + O_2(g) \rightarrow C(s) + H_2O(\ell)$

3. Write at least 4 molar ratios that can be obtained from the following equation.

   $2\ HCl + K_2CO_3 \rightarrow 2KCl + CO_2 + H_2O$

4. How many moles of carbon tetrachloride will form if 1.3 moles of chlorine react?
   $CH_4(g) + 4\ Cl_2(g) \rightarrow CCl_4(\ell) + 4\ HCl(g)$

5. How many grams of carbon dioxide are formed if 13.34 g of methanol reacts with excess $O_2$?

   $2\ CH_3OH(\ell) + 3\ O_2(g) \rightarrow 2\ CO_2(g) + 4\ H_2O(\ell)$

6. How many grams of <u>each</u> product are formed if 57.6 g of carbon reacts completely?

   $2\ Fe_2O_3(s) + 3\ C(s) \rightarrow 3\ CO_2(g) + 4\ Fe(s)$

7. How many grams of aluminum are required to react with 107.3 g $Fe_2O_3$?

   $2\ Al + Fe_2O_3 \rightarrow Al_2O_3 + Fe$

8. How many grams of chlorine are required to produce 250.0 g $CCl_4$?

   $CH_4(g) + 4\ Cl_2(g) \rightarrow CCl_4(\ell) + 4\ HCl(g)$

9. Consider the complete combustion of octane:

   $2\ C_8H_{18}(\ell) + 25\ O_2(g) \rightarrow 16\ CO_2(g) + 18\ H_2O(\ell)$

   If a reaction begins with 25.3 g octane and 104.1 g $O_2$, which is the limiting reactant?

10. Formaldehyde, $CH_2O$, burns to produce carbon dioxide and water. If a reaction is initiated with 10.5 g of <u>each</u> reactant, how many grams of the excess reactant are left over? You must write a balanced equation.

11. Isocyanic acid (HNCO) reacts with $NO_2$ to produce nitrogen, carbon dioxide, and water.

$$8\ HNCO + 6\ NO_2 \rightarrow 7\ N_2 + 8\ CO_2 + 4\ H_2O$$

What is the theoretical yield of each product if a reaction is initiated with 24.0 g of both reactants?

12. Aspirin is produced from salicylic acid and acetic anhydride.

$$C_7H_6O_3 + C_4H_8O_3 \rightarrow C_9H_8O_4 + HC_2H_3O_2$$

Salicylic Acid, Acetic Anhydride, Aspirin

If 29.7 g salicylic acid reacts completely and produces 18.3 g aspirin, what is the percent yield?

13. If 155 g of $AlCl_3$ are formed in an experiment with a 94.3 % yield, how many grams of chlorine must have reacted?

$$2\ Al + 3\ Cl_2 \rightarrow 2\ AlCl_3$$

14. Which of the following statements about titrations is false?
    a. The reactants are in solution.
    b. The titration should stop when stoichiometric amounts of reactants are combined to obtain meaningful data.
    c. A buret is commonly used to deliver one reactant.
    d. It is not important that the indicator changes color when stoichiometric amounts have been combined.
    e. The indicator is not part of the reaction.

15. What volume of 1.57 M HCl would react completely with 22.5 g Mg?

$$Mg(s) + 2\ HCl(aq) \rightarrow H_2(g) + MgCl_2(aq)$$

16. If 75.0 mL of sulfuric acid solution reacts to form 0.15 g of hydrogen, what was the molarity of the acid?

$$Zn(s) + H_2SO_4(aq) \rightarrow ZnSO_4(aq) + H_2(g)$$

17. What is the molarity of $HClO_4$ if a 25.00-mL sample requires exactly 15.79 mL of 0.386 M $Sr(OH)_2$ to react completely?

18. Which of the following statements about gasoline additives is incorrect?
    a. The octane rating refers to the amount of octane ($C_8H_{18}$) in the gasoline blend.
    b. Tetraethyl lead is added to increase the octane rating of gasoline.
    c. A more compressible gas mixture burns better in the car engine.
    d. Methyl tertiary-butyl ether (MTBE) contains oxygen and helps reduce pollutant molecules that form when gasoline burns.
    e. MTBE is a suspected carcinogen.

## ANSWERS TO THE CHAPTER OBJECTIVE QUIZ

1. d

2. (a) C will be balanced first, then H, finally oxygen.

$C_5H_{12}(\ell) + O_2(g) \rightarrow \underline{5}\ CO_2(g) + H_2O(\ell)$
$C_5H_{12}(\ell) + O_2(g) \rightarrow \underline{5}\ CO_2(g) + \underline{6}\ H_2O(\ell)$
$C_5H_{12}(\ell) + \underline{8}\ O_2(g) \rightarrow \underline{5}\ CO_2(g) + \underline{6}\ H_2O(\ell)$

(b) Same procedure as (a).

$C_3H_8(g) + O_2(g) \rightarrow \underline{3}\ CO(g) + H_2O(\ell)$
$C_3H_8(g) + O_2(g) \rightarrow \underline{3}\ CO(g) + \underline{4}\ H_2O(\ell)$
$C_3H_8(g) + \underline{7/2}\ O_2(g) \rightarrow \underline{3}\ CO(g) + \underline{4}\ H_2O(\ell)$

In whole-number coefficients: $\underline{2}\ C_3H_8(g) + \underline{7}\ O_2(g) \rightarrow \underline{6}\ CO(g) + \underline{8}\ H_2O(\ell)$

(c) Same procedure as (a).

$C_{12}H_{26}(\ell) + O_2(g) \rightarrow \underline{12}\ C(s) + H_2O(\ell)$
$C_{12}H_{26}(\ell) + O_2(g) \rightarrow \underline{12}\ C(s) + \underline{13}\ H_2O(\ell)$
$C_{12}H_{26}(\ell) + \underline{13/2}\ O_2(g) \rightarrow \underline{12}\ C(s) + \underline{13}\ H_2O(\ell)$

In whole-number coefficients: $\underline{2}\ C_{12}H_{26}(\ell) + \underline{13}\ O_2(g) \rightarrow \underline{24}\ C(s) + \underline{26}\ H_2O(\ell)$

3. The molar ratios are ratios of the <u>stoichiometric coefficients</u> in the balanced equation.

$$\frac{2\text{ mol HCl}}{1\text{ mol }K_2CO_3} \qquad \frac{2\text{ mol HCl}}{2\text{ mol KCl}} \qquad \frac{2\text{ mol KCl}}{1\text{ mol }K_2CO_3} \qquad \frac{2\text{ mol HCl}}{1\text{ mol }CO_2}$$

$$\frac{1\text{ mol }H_2O}{1\text{ mol }CO_2} \qquad \frac{2\text{ mol HCl}}{1\text{ mol }H_2O}$$

(There are more ratios than just these.)

4. $1.3\text{ mol }Cl_2 \times \frac{1\text{ mol }CCl_4}{4\text{ mol }Cl_2} = 0.33\text{ moles }CCl_4$

5. $13.34\text{ g }CH_3OH \times \frac{1\text{ mol }CH_3OH}{32.042\text{ g }CH_3OH} \times \frac{2\text{ mol }CO_2}{2\text{ mol }CH_3OH} \times \frac{44.009\text{ g }CO_2}{1\text{ mol }CO_2} = 18.32\text{ g }CO_2$

6. The theoretical yield of both products are:

$$57.6\text{ g C} \times \frac{1\text{ mol C}}{12.011\text{ g C}} \times \frac{3\text{ mol }CO_2}{3\text{ mol C}} \times \frac{44.009\text{ g }CO_2}{1\text{ mol }CO_2} = 211\text{ g }CO_2\text{ produced}$$

$$57.6\text{ g C} \times \frac{1\text{ mol C}}{12.011\text{ g C}} \times \frac{4\text{ mol Fe}}{3\text{ mol C}} \times \frac{55.847\text{ g Fe}}{1\text{ mol Fe}} = 357\text{ g Fe produced}$$

7. $$107.3\text{ g }Fe_2O_3 \times \frac{1\text{ mol }Fe_2O_3}{159.69\text{ g }Fe_2O_3} \times \frac{2\text{ mol Al}}{1\text{ mol }Fe_2O_3} \times \frac{26.988\text{ g Al}}{1\text{ mol Al}} = 36.3\text{ g Al required}$$

8. $$250.0\text{ g }CCl_4 \times \frac{1\text{ mol }CCl_4}{153.82\text{ g }CCl_4} \times \frac{4\text{ mol }Cl_2}{1\text{ mol }CCl_4} \times \frac{70.906\text{ g }Cl_2}{1\text{ mol }Cl_2} = 461.0\text{ g }Cl_2\text{ required}$$

9. The limiting reactant will be found by calculating how many grams of oxygen are required to completely react with all 25.3 g octane.

$$25.3\text{ g }C_8H_{18} \times \frac{1\text{ mol }C_8H_{18}}{114.23\text{ g }C_8H_{18}} \times \frac{25\text{ mol }O_2}{2\text{ mol }C_8H_{18}} \times \frac{31.998\text{ g }O_2}{1\text{ mol }O_2} = 88.6\text{ g }O_2\text{ required}$$

to react with all the octane. We have more than this amount, therefore $O_2$ is the excess reactant and octane is the limiting reactant.

10. The balanced equation is: $CH_2O(l) + O_2(g) \rightarrow CO_2(g) + H_2O(\ell)$

The first step is to find the limiting reactant by calculating how many grams of $O_2$ is required to react with 10.5 g of $CH_2O$.

$$10.5\text{ g }CH_2O \times \frac{1\text{ mol }CH_2O}{30.026\text{ g }CH_2O} \times \frac{1\text{ mol }O_2}{1\text{ mol }CH_2O} \times \frac{31.998\text{ g }O_2}{1\text{ mol }O_2} = 11.2\text{ g }O_2\text{ required}$$

to react with all the formaldehyde. This is more than we actually have, so $O_2$ is the limiting reactant and $CH_2O$ is the excess reactant. Next, we have to find how many grams of $CH_2O$ will react with the $O_2$.

$$10.5\text{ g }O_2 \times \frac{1\text{ mol }O_2}{31.998\text{ g }O_2} \times \frac{1\text{ mol }CH_2O}{1\text{ mol }O_2} \times \frac{30.026\text{ g }CH_2O}{1\text{ mol }CH_2O} = 9.85\text{ g }CH_2O\text{ react.}$$

This leaves 10.5 g – 9.85 g = 0.65 g $CH_2O$ in excess.

11. To find the theoretical yields of the products, we need to identify the limiting reactant.

$$24.0 \text{ g HNCO} \times \frac{1 \text{ mol HNCO}}{43.025 \text{ g HNCO}} \times \frac{6 \text{ mol } NO_2}{8 \text{ mol HNCO}} \times \frac{46.005 \text{ g } NO_2}{1 \text{ mol } NO_2} = 19.2 \text{ g } O_2$$

**So, 19.2 g $O_2$ are required to react with all the HNCO. This is less than we actually have, so HNCO is the limiting reactant. We'll calculate the theoretical yields using 24.0 g HNCO.**

$$24.0 \text{ g HNCO} \times \frac{1 \text{ mol HNCO}}{43.025 \text{ g HNCO}} \times \frac{7 \text{ mol } N_2}{8 \text{ mol HNCO}} \times \frac{28.013 \text{ g } N_2}{1 \text{ mol } N_2} = 13.7 \text{ g } N_2$$

$$24.0 \text{ g HNCO} \times \frac{1 \text{ mol HNCO}}{43.025 \text{ g HNCO}} \times \frac{8 \text{ mol } CO_2}{8 \text{ mol HNCO}} \times \frac{44.009 \text{ g } CO_2}{1 \text{ mol } CO_2} = 24.5 \text{ g } CO_2$$

$$24.0 \text{ g HNCO} \times \frac{1 \text{ mol HNCO}}{43.025 \text{ g HNCO}} \times \frac{4 \text{ mol } H_2O}{8 \text{ mol HNCO}} \times \frac{18.015 \text{ g } H_2O}{1 \text{ mol } H_2O} = 5.02 \text{ g } N_2$$

**12. The actual yield is 18.3 g aspirin. We need to find the theoretical yield of aspirin, and then determine the percent yield.**

$$29.7 \text{ g } C_7H_6O_3 \times \frac{1 \text{ mol } C_7H_6O_3}{138.12 \text{ g } C_7H_6O_3} \times \frac{1 \text{ mol } C_9H_8O_4}{1 \text{ mol } C_7H_6O_3} \times \frac{180.16 \text{ g } C_9H_8O_4}{1 \text{ mol } C_9H_8O_4} = 38.7 \text{ g}$$

**$C_9H_8O_4$ formed (theoretically).**

$$\% \text{ Yield} = \frac{\text{Actual yield}}{\text{Theoretical yield}} \times 100 \quad = \frac{18.3 \text{ g aspirin}}{38.7 \text{ g aspirin}} \times 100 = 47.2\% \text{ Yield}$$

**13. From the percent yield, we can find the theoretical yield, and from that; we can calculate the amount of reactant required.**

$$\% \text{ Yield} = \frac{\text{Actual yield}}{\text{Theoretical yield}} \times 100; \qquad \text{Theoretical yield} = \frac{\text{Actual yield}}{\% \text{ Yield}} \times 100$$

$$\text{Theoretical yield} = \frac{155 \text{ g } AlCl_3}{94.3\ \%} \times 100 = 164 \text{ g } AlCl_3$$

$$164 \text{ g } AlCl_3 \times \frac{1 \text{ mol } AlCl_3}{133.34 \text{ g } AlCl_3} \times \frac{3 \text{ mol } Cl_2}{2 \text{ mol } AlCl_3} \times \frac{70.905 \text{ g } Cl_2}{1 \text{ mol } Cl_2} = 131 \text{ g } Cl_2 \text{ are required}$$

**to produce an actual yield of 155 g $AlCl_3$ (assuming a percent yield of 94.3 %).**

**14. d**

**15.** $22.5 \text{ g Mg} \times \frac{1 \text{ mol Mg}}{24.305 \text{ g Mg}} \times \frac{2 \text{ mol HCl}}{1 \text{ mol Mg}} = 1.85$ **moles HCl reacted**

$$\text{Volume HCl} = \frac{\text{mol}}{\frac{\text{mol}}{\text{L}}} = \frac{1.85 \text{ mol HCl}}{1.57 \frac{\text{mol}}{\text{L}} \text{HCl}} = 1.18 \text{ L} = 1180 \text{ mL of } 1.57 \text{ M HCl}$$

**16.** $0.15 \text{ g } H_2 \times \frac{1 \text{ mol } H_2}{2.016 \text{ g } H_2} \times \frac{1 \text{ mol } H_2SO_4}{1 \text{ mol } H_2} = 0.0744$ **moles** $H_2SO_4$

**0.0744 moles** $H_2SO_4 \div 0.0750 \text{ L} = 0.99 \text{ M}$

**17. We need a balanced equation for this acid/base reaction:**

$$HClO_4 + Sr(OH)_2 \rightarrow 2\,H_2O + Sr(ClO_4)_2$$

$$0.01579 \text{ L} \times \frac{0.586 \text{ mol } Sr(OH)_2}{\text{L}} \times \frac{2 \text{ mol } HClO_4}{1\, Sr(OH)_2} \times \frac{1}{0.02500 \text{ L}} = 0.740 \text{ M}$$

**18. a**

# CHAPTER 5

## *Study Goals*

The study goals outline specific concepts to be mastered in this section of the text chapter. Related problems at the end of the text chapter will also be noted. Working the questions noted should aid in mastery of each study goal and also highlight any areas that you may need additional help in.

**Section 5.1 *INSIGHT INTO* Air Pollution**

1. Know the composition of the atmosphere and what common air pollutants may be present. *Work Problems 1–6, 104.*
2. List several observable properties that are common to all gases. *Work Problem 7.*

**Section 5.2 Pressure**

3. Define pressure and explain how a gas generates pressure in the atmosphere and in a closed container. *Work Problems 8–10, 17.*
4. Be familiar with several common pressure units and be able to convert between them. *Work Problems 11–16.*

**Section 5.3 History and Application of the Gas Law**

5. State Boyle's, Charles', and Avogadro's laws and use them to calculate changes in P, V, T, or moles. *Work Problems 18–28, 105.*
6. Use the ideal gas law to relate P, V, T, and n for a sample of an ideal gas. *Work Problems 29–36, 85, 86, 88, 92, 95, 98, 101, 102, 109–111, 113, 114.*

**Section 5.4 Partial Pressure**

7. State Dalton's law of partial pressures and use partial pressure and mole fraction to describe mixtures of gases. *Work Problems 37–50, 94, 97, 108.*

**Section 5.5 Stoichiometry of Reactions Involving Gases**

8. Use the ideal gas law to solve stoichoimetry problems involving gases. *Work Problems 51–67, 87, 89–91, 93, 96, 99, 103, 106, 107, 112.*
9. Be familiar with standard temperature and pressure conditions (STP) and standard molar volume of an ideal gas. *Work Problem 84.*

**Section 5.6 Kinetic-Molecular Theory and Ideal Versus Real Gases**

10. Describe ideal gases using the kinetic molecular theory. *Work Problems 68–70.*
11. Represent the velocities of a sample of gas molecules using Maxwell-Boltzmann distributions and understand why temperature and average kinetic energy are related. *Work Problems 71–75.*
12. Identify several situations where gases do not behave as ideal gases. *Work Problem 76, 79.*

13. Use the van der Waals equation to relate P, V, T, and moles in situations where gases do not behave ideally. *Work Problems 77, 78.*

**Section 5.7** ***INSIGHT INTO*** **Gas Sensors**

14. Be familiar with several types of gas pressure sensors. *Work Problems 80–83.*

## *Solutions to the Odd-Numbered Problems*

Problem 5.1 List two types of chemical compounds that must be present in air for photochemical smog to form. What are the most common sources of these compounds?

**Nitrogen oxides and volatile organic compounds (VOCs) must be present. Nitrogen oxides are produced by automobiles due to the reaction between oxygen and nitrogen in air used for combustion. Industrial plants, particularly power plants, also produce some nitrogen oxides. Automobiles are also responsible for emission of VOCs in exhaust and during fueling, and chemical plants release VOCs.**

Problem 5.3 In the production of urban air pollution shown in Figure 5.1, why does the concentration of NO decrease during daylight hours?

**Nitrogen oxide is produced primarily by automobiles. Concentrations peak around morning rush hour. Since NO is a component of photochemical smog, sunlight during daylight hours causes a reaction that converts NO into nitrogen dioxide ($NO_2$) and other compounds. So NO concentration decreases during the day but the concentration of other harmful pollutants increases.**

Problem 5.5 Asphalt is composed of a mixture of organic chemicals. Does an asphalt parking lot contribute directly to the formation of photochemical smog? Explain your answer.

**While asphalt contains organic chemicals, these are mainly large molecules that are not considered to be *volatile*. Therefore, asphalt is not considered to be a contributor to the photochemical smog problem, which requires VOCs. It is possible for some of these chemicals to leach out of the asphalt into the soil or groundwater, causing problems.**

Problem 5.7 One observable property of gases is the variability of density based on conditions. Use this observation to explain why hot air balloons rise.

**The temperature of the gas in a hot air balloon is higher that that of the surrounding air. Gases have lower densities at higher temperature; therefore the gas in the balloon will float on the cooler, denser surrounding air.**

Problem 5.9 How do gases exert atmospheric pressure?

**Pressure is defined as force per unit area. Atmospheric pressure is created when the Earth's gravitational attraction to the gas particles creates a force acting over the surface of the Earth and all objects on it. This force depends on how many gas particles are present (more mass = more force). This means that as one goes higher in altitude where there are fewer gas particles, the atmospheric pressure decreases.**

Problem 5.11 If you had a liquid whose density was half that of mercury, how tall would you need to build a barometer to measure atmospheric pressure in a location where the record pressure recorded was 750 mm Hg?

**The height of a mercury column produced by atmospheric pressure is directly related to the density of that liquid (and mass). If a liquid with a density one-half that of mercury were used, the column would be twice as high or 2 × 750 mm = 1500 mm. This would give a column with the same mass as the mercury column.**

Problem 5.13 Water has a density that is 13.6 times less than that of mercury. If an undersea vessel descends to 1.5 km, how much pressure does the water exert in atm?

**This depth under the ocean would create a pressure equal to manometer reading of 1500 meters of water. Converting from water to mercury (13.6 times more dense) will give the pressure in mm Hg.**

$$1.5 \text{ km } H_2O \times \frac{1000 \text{ m}}{1 \text{ km}} \times \frac{1000 \text{ mm}}{1 \text{ m}} \times \frac{\frac{1.00 \text{ g}}{\text{mL } H_2O}}{\frac{13.6 \text{ g}}{\text{mL Hg}}} = 1.1 \times 10^5 \text{ mm Hg}$$

**Now we can convert from mm Hg to atm.**

$$1.1 \times 10^5 \text{ mm Hg} \times \frac{1.00 \text{ atm}}{760 \text{ mm Hg}} = 150 \text{ atm of pressure at a depth of 1.5 km under the ocean.}$$

Problem 5.15 ■ Gas pressures can be expressed in units of mm Hg, atm, torr, and kPa. Convert these pressure values.
(a) 722. mm Hg to atm, (b) 1.25 atm to mm Hg, (c) 542. mm Hg to torr, (d) 745. mm Hg to kPa, (e) 708. kPa to atm

$$\text{(a) } 722 \text{ mm Hg} \times \frac{1.00 \text{ atm}}{760 \text{ mm Hg}} = 0.950 \text{ atm}$$

$$\text{(b) } 1.25 \text{ atm} \times \frac{760 \text{ mm Hg}}{1.00 \text{ atm}} = 950. \text{ mm Hg}$$

$$\text{(c) } 542 \text{ mm Hg} \times \frac{760 \text{ torr}}{760 \text{ mm Hg}} = 542 \text{ torr}$$

**(d)** $745.\ \text{mmHg} \times \dfrac{1.00\ \text{atm}}{760\ \text{mm Hg}} \times \dfrac{101.325\ \text{kPa}}{1.00\ \text{atm}} = 99.3\ \text{kPa}$

**(e)** $708.\ \text{kPa} \times \dfrac{1.00\ \text{atm}}{101.325\ \text{kPa}} = 6.99\ \text{atm}$

Problem 5.17 Why do your ears "pop" on occasion when you swim deep underwater?

**The increased pressure under the water due to the mass of the water compresses air in the ear, which can cause a "pop".**

Problem 5.19 ■ A sample of $CO_2$ gas has a pressure of 56.5 mm Hg in a 125-mL flask. The sample is transferred to a new flask, where it has a pressure of 62.3 mm Hg at the same temperature. What is the volume of the new flask?

**Pressure and volume are being varied at constant temperature and moles of gas. This is a problem which is most easily answered using Boyle's Law: $P_1V_1 = P_2V_2$.**

**We are solving for the new volume,** $V_2 = \dfrac{P_1V_1}{P_2} \times \dfrac{(56.5\ \text{mm Hg})(125\ \text{mL})}{(62.3\ \text{mm Hg})} = 113\ \text{mL}$

**The units for pressure and volume don't really matter in this formula, as long as they are consistent.**

Problem 5.21 When you buy a Mylar™ balloon in the winter months in colder places, the shopkeeper will often tell you not to worry about it losing shape when you take it home (outside), because it will return to shape once inside. What behavior of gases is responsible for this advice?

**Volume and absolute temperature are directly related when moles of gas and pressure remain constant. This relationship is known as Charles's Law. When the balloon is taken outside, the temperature of the gas decreases and so the volume of the balloon decreases as well. When the balloon is returned to room temperature, however, the balloon returns to its original size.**

Problem 5.23 What evidence gave rise to the establishment of the absolute temperature scale?

**The scientist, for whom Charles's Law is named, Jacques Charles, first observed the direct relationship between volume and temperature of a gas sample at constant pressure. Plotting this type of data (*V* vs. *T*) would yield a straight line. The line intersects the *X*-axis where the volume of the gas becomes zero. Since the volume can go no lower than this, the temperature at this volume is as low as possible – evidence of an "absolute zero" temperature. A real gas would condense into liquid and then form a solid before reaching absolute zero but an ideal gas would not.**

Problem 5.25 Why is it dangerous to store compressed gas cylinders in places that could become very hot?

**At constant volume, the pressure of a gas is directly related to the absolute temperature. A compressed gas cylinder represents a constant volume situation, so if the gas becomes hot, the pressure could increase drastically. The cylinder might rupture, and the rapidly escaping gas would propel the cylinder like a projectile.**

Problem 5.27 A gas bubble forms inside a vat containing a hot liquid. If the bubble is originally at 68°C and a pressure of 1.6 atm with a volume of 5.8 mL, what will its volume be if the pressure drops to 1.2 atm and the temperature drops to 31°C?

**Pressure, volume and temperature are all changing in this problem. We'll start with the ideal gas law.**

**$PV = nRT$ or $\frac{PV}{nT} = R$, a constant. Since this ratio will stay the same even after the variables are changed, we can write $\frac{P_1V_1}{n_1T_1} = \frac{P_2V_2}{n_2T_2}$. The number of moles stays constant so our gas law becomes $\frac{P_1V_1}{T_1} = \frac{P_2V_2}{T_2}$. This expression is sometimes known as the *Combined Gas Law*. Now we just need to insert our data (temperature units must be Kelvin) and solve for the new volume:**

**$T_1 = 68°C + 273 = 341$ K; $T_2 = 31°C + 273 = 304$ K**
**$P_1 = 1.2$ atm; $P_2 = 1.6$ atm; $V_1 = 5.8$ mL**

$$V_2 = \frac{T_2P_1V_1}{P_2T_1} = \frac{(304\text{ K})(1.6\text{ atm})(5.8\text{ mL})}{(1.2\text{ atm})(341\text{ K})} = 6.9\text{ mL}$$

**We could have also solved the problem by using the ideal gas law twice: first to calculate the moles of gas with the original conditions, then to calculate the volume after pressure and temperature is changed.**

Problem 5.29 ■ A balloon filled with helium has a volume of $1.28 \times 10^3$ L at sea level where the pressure is 0.998 atm and the temperature is 31°C. The balloon is taken to the top of a mountain where the pressure is 0.753 atm and the temperature is –25°C. What is the volume of the balloon at the top of the mountain?

**This is another combined gas law problem: $\frac{P_1V_1}{T_1} = \frac{P_2V_2}{T_2}$**

**$T_1 = 31°C + 273 = 304$ K; $T_2 = -25°C + 273 = 248$ K**
**$P_1 = 0.998$ atm; $P_2 = 0.753$ atm; $V_1 = 1.28 \times 10^3$ L**

$$V_2 = \frac{T_2 P_1 V_1}{P_2 T_1} = \frac{(248\text{ K})(0.998\text{ atm})(1.28\times 10^3\text{ L})}{(0.753\text{ atm})(304\text{ K})} = 1.38\times 10^3\text{ L}$$

Problem 5.31 ■ A newly discovered gas has a density of 2.39 g/L at 23.0°C and 715 mm Hg. What is the molar mass of the gas?

**We have insufficient information to take a straightforward approach to this task, so let's look at what we know: Density is mass per unit volume, so density times volume is mass. If we can calculate the molar volume (the volume per mole) and multiply it by the density, we can get the molar mass (M = grams per mole). Use the ideal gas law to determine the molar volume of the gas.**

$$d\times\frac{V}{n} = \frac{m}{n} = M \qquad \text{molar volume} = \frac{V}{n} = \frac{RT}{P}$$

$$d\times\frac{RT}{P} = M \qquad T = 23.0°\text{C} + 273.15 = 296.2\text{ K}$$

$$715\text{ mm Hg}\times\frac{1\text{ atm}}{760\text{ mm Hg}} = 0.941\text{ atm}$$

$$M = \frac{dRT}{P} = \frac{\left(2.39\ \frac{\text{g}}{\text{L}}\right)\times\left(0.08206\ \frac{\text{L}\cdot\text{atm}}{\text{mol}\cdot\text{K}}\right)\times(296.2\text{ K})}{(0.941\text{ atm})} = 61.7\ \frac{\text{g}}{\text{mol}}$$

Problem 5.33 What are the densities of the following gases at STP? (a) $CF_2Cl_2$ (b) $CO_2$ (c) HCl

**Standard temperature and pressure conditions (STP) are 273 K and 1.00 atm. In this problem, we'll use the molar mass and the ideal gas law to calculate density, in g/L.**

$$\text{Molar mass } (MM) = \frac{\text{g}}{\text{moles}(n)} \quad \text{or } n = \frac{\text{g}}{MM}$$

**Now substituting the expression for n into the ideal gas law, we get:**

$$PV = nRT = \frac{gRT}{MM}. \qquad \text{Rearranging, } \frac{g}{V} = \text{density (g/L)} = \frac{(MM)P}{RT}$$

**(a) For $CF_2Cl_2$, the molar mass is 120.9 g/mol.**

$$\text{density (g/L)} = \frac{(MM)P}{RT} = \frac{(120.9\text{ g/mol})(1.00\text{ atm})}{(0.0821\text{ L}\cdot\text{atm/mol}\cdot\text{K})(273\text{ K})} = 5.39\text{ g/L}$$

**(b) For $CO_2$, the molar mass is 44.01 g/mol.**

$$\text{density (g/L)} = \frac{(MM)P}{RT} = \frac{(44.01\text{ g/mol})(1.00\text{ atm})}{(0.0821\text{ L}\cdot\text{atm/mol}\cdot\text{K})(273\text{ K})} = 1.96\text{ g/L}$$

**(c) For HCl, the molar mass is 36.46 g/mol.**

$$\text{density (g/L)} = \frac{(MM)P}{RT} = \frac{(36.46\text{ g/mol})(1.00\text{ atm})}{(0.0821\text{ L}\cdot\text{atm/mol}\cdot\text{K})(273\text{ K})} = 1.63\text{ g/L}$$

Problem 5.35 ■ A cylinder is filled with toxic COS gas to a pressure of 800 torr at 24°C. According to the manufacturer's specifications, the cylinder may rupture if the pressure exceeds 35 psi (pounds per square inch; 1 atm = 14.7 psi). What is the maximum temperature to which the cylinder could be heated without exceeding the pressure rating?

**The volume of the gas and the moles of gas are held constant in this problem. Starting with the ideal gas law, $PV = nRT$ or $\frac{PV}{nT} = R$, a constant. Since this ratio will stay the same even after the variables are changed, we can write $\frac{P_1V_1}{n_1T_1} = \frac{P_2V_2}{n_2T_2}$. The number of moles and volume stay constant so our gas law becomes $\frac{P_1}{T_1} = \frac{P_2}{T_2}$.**

**We want to find the temperature which pressure = 35 psi. Therefore,**

$$P_1 = 800\text{ torr} \times \frac{1\text{ atm}}{760\text{ torr}} \times \frac{14.7\text{ psi}}{1\text{ atm}} = 15.5\text{ psi}$$

$$P_2 = 35\text{ psi} \qquad T_1 = 24^\circ\text{C} + 273\text{ K} = 297\text{ K}$$

$$T_2 = \frac{P_2T_1}{P_1} = \frac{(35\text{ psi})(297\text{ K})}{(15.5\text{ psi})} = 671\text{ K}.$$

**At temperatures over 671 K, the tank will likely burst.**

Problem 5.37 Define the term *partial pressure*.

**Partial pressure is the pressure exerted by a gas in a mixture of gases, acting as if it were alone in the container.**

Problem 5.39 How does the mole fraction relate to the partial pressure?

**Both mole fraction and partial pressure depend directly on *n*, the moles of gas.**

**Starting with mole fraction, $X_i = \frac{n_i}{n_T}$, we can use the ideal gas law to express $n_i$, moles of a gas, and $n_T$, total moles of gas.**

$$n_i = \frac{P_iV}{RT} \text{ and } n_T = \frac{P_TV}{RT}\text{; so the mole fraction becomes } X_i = \frac{\frac{P_iV}{RT}}{\frac{P_TV}{RT}}.$$

**Canceling *V*, *T*, and *R*, we get:**

$$X_i = \frac{P_i}{P_T} \text{ or } P_i = X_i(P_T)$$

Problem 5.41 What is the total pressure of a 15.0-L container at 28°C that contains 3.5 g of $N_2$, 4.5 g of $O_2$, and 13.0 g of $Cl_2$?

**The total pressure in the container is calculated using the total number of moles present. We will first determine the moles of each gas, and then use the ideal gas law to calculate pressure.**

$$3.5\text{ g }N_2 = \frac{1\text{ mol }N_2}{28.01\text{ g }N_2} = 0.125\text{ mol }N_2 \qquad 4.5\text{ g }O_2 = \frac{1\text{ mol }O_2}{32.00\text{ g }O_2} = 0.141\text{ mol }O_2$$

$$13.0\text{ g }Cl_2 = \frac{1\text{ mol }Cl_2}{70.91\text{ g }Cl_2} = 0.183\text{ mol }Cl_2$$

**Total moles, $n_T$, = 0.125 mol $N_2$ + 0.141 mol $O_2$ + 0.183 mol $Cl_2$ = 0.449 mol total.**

$$T = 28^oC + 273 = 301\text{ K}$$

$$PV = nRT \text{ or } P = \frac{nRT}{V} = \frac{(0.449\text{ mol})(0.08206\text{ L}\cdot\text{atm/mol}\cdot\text{K})(301\text{ K})}{15.0\text{ L}} = 0.74\text{ atm}$$

Problem 5.43 A sample containing only $NO_2$ and $SO_2$ has a total pressure of 120. torr. Measurements show that the partial pressure of $NO_2$ is 43 torr. If the vessel has a volume of 800.0 mL and the temperature is 22.0°C, how many moles of each gas are present?

**From the partial pressure of each gas, we can calculate the moles using the ideal gas law.**

**$P_{NO2}$ = 43 torr, $P_{SO2}$ = 120 torr – 43 torr = 77 torr (Dalton's Law)**

**$T$ = 22.0°C + 273 K = 295 K $\qquad$ $V$ = 800 mL $\quad$ We'll use $R$ = 62,400 mL·torr/mol·K.**

$$PV = nRT \qquad \text{or } n = \frac{PV}{RT}$$

$$n_{NO2} = \frac{P_{NO_2}V}{RT} = \frac{(43 \text{ torr})(800 \text{ mL})}{(62{,}400 \text{ mL}\cdot\text{torr/mol}\cdot\text{K})(295 \text{ K})} = 0.0019 \text{ mol } NO_2$$

$$n_{SO2} = \frac{P_{SO_2}V}{RT} = \frac{(77 \text{ torr})(800 \text{ mL})}{(62{,}400 \text{ mL}\cdot\text{torr/mol}\cdot\text{K})(295 \text{ K})} = 0.0033 \text{ mol } SO_2$$

Problem 5.45 ■ A sample of a smokestack emission was collected into a 1.25-L tank at 752 mm Hg and analyzed. The analysis showed 92% $CO_2$, 3.6% NO, 1.2% $SO_2$, and 4.1% $H_2O$ by mass. What is the partial pressure exerted by each gas?

**Assume a 100-g sample.**

**moles $CO_2$ = (92 g)$\left(\frac{1 \text{ mol } CO_2}{44.0 \text{ g } CO_2}\right)$ = 2.1 ; moles NO = (3.6 g)$\left(\frac{1 \text{ mol NO}}{30.0 \text{ g NO}}\right)$ = 0.12**

**moles $SO_2$ = (1.2 g)$\left(\frac{1 \text{ mol } SO_2}{64.1 \text{ g } SO_2}\right)$ = 0.019 ; moles $H_2O$ = (4.1 g)$\left(\frac{1 \text{ mol } H_2O}{18.0 \text{ g } H_2O}\right)$ = 0.23**

**total number of moles = 2.1 + 0.12 + 0.019 + 0.23 = 2.5**

**$P_{CO2}$ = (2.1/2.5)(752 mm Hg) = $6.3 \times 10^2$ mm Hg; $P_{NO}$ = (0.12/2.5)(752 mm Hg) = 36 mm Hg**
**$P_{SO2}$ = (0.019/2.5)(752 mm Hg) = 5.7 mm Hg; $P_{H2O}$ = (0.23/2.5)(752 mm Hg) = 69 mm Hg**

Problem 5.47 In an experiment, a mixture of gases occupies a volume of 3.0 L at a temperature of 22.5°C. The mixture contains 14.0 g of water, 11.5 g of oxygen, and 37.3 g of nitrogen. Calculate the total pressure and the partial pressure of each gas.

**By calculating the number of moles of each gas and adding them to get the total moles, we can calculate the total pressure using the ideal gas law. The mole fraction of each gas can be determined and then each partial pressure.**

$$14.0 \text{ g } H_2O = \frac{1 \text{ mol } H_2O}{18.01 \text{ g } H_2O} = 0.777 \text{ mol } H_2O \qquad 11.5 \text{ g } O_2 = \frac{1 \text{ mol } O_2}{32.00 \text{ g } O_2} = 0.359 \text{ mol } O_2$$

$$37.3 \text{ g } N_2 = \frac{1 \text{ mol } N_2}{28.01 \text{ g } N_2} = 1.34 \text{ mol } N_2$$

**Total moles, $n_T$, = 0.777 mol $H_2O$ + 0.359 mol $O_2$ + 1.34 mol $N_2$ = 2.48 mol total.**

**$T = 22.5°C + 273 = 295.5$ K**

$$PV = nRT \text{ or } P = \frac{nRT}{V} = \frac{(2.48 \text{ mol})(0.08206 \text{ L} \cdot \text{atm/mol} \cdot \text{K})(295.5 \text{ K})}{3.0 \text{ L}} = 20.1 \text{ atm}$$

**Next we'll calculate the mole fractions.**

$$X_{H2O} = \frac{n_{H_2O}}{n_T} = \frac{(0.777 \text{ mol } H_2O)}{(2.48 \text{ mol Total})} = 0.313 \qquad X_{O2} = \frac{n_{O_2}}{n_T} = \frac{(0.359 \text{ mol } O_2)}{(2.48 \text{ mol Total})} = 0.145$$

$$X_{Cl2} = \frac{n_{Cl_2}}{n_T} = \frac{(1.34 \text{ mol } H_2O)}{(2.48 \text{ mol Total})} = 0.540$$

**Finally, we'll calculate the partial pressures using $P_i = X_i(P_T)$.**
**$P_{H2O} = 0.313\ (20.1 \text{ atm}) = 6.29$ atm** **$P_{O2} = 0.145\ (20.1 \text{ atm}) = 2.91$ atm**

**$P_{Cl2} = 0.540\ (20.1 \text{ atm}) = 10.9$ atm**

Problem 5.49 Use the web to determine the range of partial pressures that oxygen sensors must measure in the exhaust manifold of automobile engines. Why is it important for an engineer to know the amount of oxygen in an exhaust stream?

**Answers will vary. The amount of oxygen in the exhaust stream tells the engineer if the correct ratio of fuel-to-oxygen is being used in the engine. In a modern engine, the oxygen sensor will cause an automatic adjustment in the fuel-to air mix in the engine. This helps ensure complete combustion in the engine.**

Problem 5.51 HCl(g) reacts with ammonia gas, $NH_3$(g), to form solid ammonium chloride. If a sample of ammonia occupying 250 mL at 21°C and a pressure of 140 torr is allowed to react with excess HCl, what mass of $NH_4Cl$ will form?

**The equation for this reaction is HCl(g) + $NH_3$(g) → $NH_4Cl$(s).**

**This is a stoichoimetry problem; only we need to use the ideal gas law to find the moles of $NH_3$. Then we can relate moles $NH_3$ reacted to moles $NH_4Cl$ formed.**

**$T = 21°C + 273 = 294$ K and $R = 62{,}400$ mL·torr/mol·K.**

$$PV = nRT \text{ or } n = \frac{PV}{RT} = \frac{(140 \text{ torr})(250 \text{ mL})}{(62{,}400 \text{ mL} \cdot \text{torr/mol} \cdot \text{K})(294 \text{ K})} = 0.0019 \text{ moles } NH_3 \text{ reacted.}$$

$$0.0019 \text{ moles } NH_3 \text{ reacted} \times \frac{1 \text{ mol } NH_4Cl}{1 \text{ mol } NH_3} \times \frac{53.49 \text{ g } NH_4Cl}{1 \text{ mol } NH_4Cl} = 0.102 \text{ g } NH_4Cl \text{ formed.}$$

Problem 5.53 If you need 400.0 mL of hydrogen gas to conduct an experiment at 20.5°C and 748 torr, how many grams of Zn should be reacted with excess HCl to obtain this much gas?

**This is another gas stoichoimetry problem. Using the ideal gas law, the moles of $H_2$ can be found. The moles of $H_2$ formed can be related to moles of Zn reacted.**
**When zinc reacts with hydrochloric acid, the products are hydrogen gas and zinc chloride.**

$$Zn(s) + 2\,HCl(aq) \rightarrow H_2(g) + ZnCl_2(aq)$$

$T = 20.5\,^\circ C + 273 = 293.5\text{ K}$ and $R = 62{,}400\text{ mL·torr/mol·K}$.

$$PV = nRT \text{ or } n = \frac{PV}{RT} = \frac{(748\text{ torr})(400\text{ mL})}{(62{,}400\text{ mL}\cdot\text{torr/mol}\cdot\text{K})(293.5\text{ K})} = 0.0163\text{ moles } H_2 \text{ formed.}$$

$$0.0163\text{ moles } H_2 \text{ formed} \times \frac{1\text{ mol Zn}}{1\text{ mol } H_2} \times \frac{65.39\text{ g Zn}}{1\text{ mol Zn}} = 1.07\text{ g Zn reacted.}$$

Problem 5.55 What volume of oxygen at 24°C and 0.88 atm is needed to completely react via combustion with 45 g of methane gas?

**First, we need to write a combustion reaction for methane ($CH_4$). We'll assume complete combustion so the products are $CO_2$ and $H_2O$.**

$$CH_4(g) + 2\,O_2(g) \rightarrow CO_2(g) + 2\,H_2O(\ell)$$

**We can determine the moles of methane by dividing by the molar mass, relate moles methane to moles $O_2$, and then use the ideal gas law to calculate the volume of $O_2$.**

$$45\text{ g } CH_4 \times \frac{1\text{ mol } CH_4}{16.04\text{ g } CH_4} \times \frac{2\text{ mol } O_2}{1\text{ mol } CH_4} = 5.6\text{ mol } O_2 \text{ required.}$$

**The value of the gas constant to use is $R = 0.08206$ L·atm/mol·K.**

**Temperature must be expressed in Kelvin: $T = 24^\circ C + 273 = 297$ K.**

$$PV = nRT \text{ or } V = \frac{nRT}{P} = \frac{(5.6\text{ mol})(0.08206\text{ L}\cdot\text{atm/mol}\cdot\text{K})(297\text{ K})}{(0.88\text{ atm})} = 160\text{ L } O_2 \text{ required.}$$

**(two significant figures)**

Problem 5.57 $N_2O_5$ is an unstable gas that decomposes according to the following reaction:

$$2\,N_2O_5(g) \rightarrow 4\,NO_2(g) + O_2(g)$$

What would be the total pressure of gases present if a 10.0-L container at 22.0°C begins with 0.400 atm of $N_2O_5$ and the gas completely decomposes?

**First, we'll calculate how many moles of $N_2O_5$ reacted.**

**$T = 22.0\ ^{o}C + 273 = 295\ K$ and $R = 0.08206$ L·atm/mol·K.**

$$PV = nRT \text{ or } n = \frac{PV}{RT} = \frac{(0.40\ \text{atm})(10.0\ \text{L})}{(0.08206\ \text{L}\cdot\text{atm/mol}\cdot\text{K})(295\ \text{K})} = 0.165 \text{ moles } N_2O_5 \text{ reacted.}$$

**Second, we'll calculate the moles of each product formed.**

$$0.165 \text{ moles } N_2O_5 \text{ reacted} \times \frac{4\ \text{mol } NO_2}{2\ \text{mol } N_2O_5} = 0.330 \text{ mol } NO_2 \text{ formed.}$$

$$0.165 \text{ moles } N_2O_5 \text{ reacted} \times \frac{1\ \text{mol } O_2}{2\ \text{mol } N_2O_5} = 0.083 \text{ mol } O_2 \text{ formed.}$$

**Total moles of gas formed = 0.330 mol + 0.083 mol = 0.413 mol.**

**Finally, we'll use the ideal gas law to calculate the pressure in the container after the reaction is complete.**

$$PV = nRT \text{ or } P = \frac{nRT}{V} = \frac{(0.413\ \text{mol})(0.08206\ \text{L}\cdot\text{atm/mol}\cdot\text{K})(295\ \text{K})}{(10.0\ \text{L})} = 1.00\ \text{atm}$$

Problem 5.59 Ammonia is not the only possible fertilizer. Others include urea, which can be produced by the reaction, $CO_2(g) + 2\ NH_3(g) \rightarrow CO(NH_2)_2(s) + H_2O(g)$. A scientist has 75 g of dry ice to provide the carbon dioxide. If 4.50 L of ammonia at 15°C and a pressure of 1.4 atm is added, which reactant is limiting? What mass of urea will form?

**Using the ideal gas law, we can calculate the moles of ammonia present:**

**$T = 15^{o}C + 273 = 288\ K$ and $R = 0.08206$ L·atm/mol·K.**

$$PV = nRT \text{ or } n = \frac{PV}{RT} = \frac{(1.4\ \text{atm})(4.50\ \text{L})}{(0.08206\ \text{L}\cdot\text{atm/mol}\cdot\text{K})(288\ \text{K})} = 0.266 \text{ moles } NH_3$$

$$\text{The moles of } CO_2 \text{ are: } 75\ \text{g } CO_2 \times \frac{1\ \text{mol } CO_2}{44.01\ \text{g } CO_2} = 1.70 \text{ moles } CO_2$$

**Recall we need to do a stoichoimetry calculation to determine a limiting reactant. We'll calculate the moles of $CO_2$ required reacting with all the moles of ammonia.**

**$0.266 \text{ moles } NH_3 \times \dfrac{1 \text{ mol } CO_2}{2 \text{ mol } NH_3} = 0.133 \text{ moles of } CO_2$ required.**

**This is less than we have (1.70 moles), so <u>ammonia</u> is the limiting reactant.**

**Finally, we'll calculate the mass of urea that would form if 0.266 moles of ammonia react.**

$$0.266 \text{ moles } NH_3 \times \frac{1 \text{ mol } CO(NH_2)_2}{2 \text{ mol } NH_3} \times \frac{60.06 \text{ g } CO(NH_2)_2}{1 \text{ mol } CO(NH_2)_2} = 8.0 \text{ g } CO(NH_2)_2 \text{ forms.}$$

<u>Problem 5.61</u> ■ What volume of hydrogen gas, in liters, is produced by the reaction of 3.43 g of iron metal with 40.0 mL of 2.43 M HCl? The gas is collected at 2.25 atm of pressure and 23°C. The other product is $FeCl_2(s)$.

**Balanced equation: $Fe(s) + 2\ HCl(aq) \rightarrow H_2(g) + FeCl_2(s)$**

**The limiting reactant must be determined by calculating the yield of hydrogen from each reactant.**

**Strategy: g Fe $\rightarrow$ mol Fe $\rightarrow$ mol $H_2$**

$$3.43 \text{ g Fe} \times \frac{1 \text{ mol Fe}}{55.85 \text{ g Fe}} \times \frac{1 \text{ mol } H_2}{1 \text{ mol Fe}} = 0.0614 \text{ moles } H_2$$

**Strategy: L HCl $\rightarrow$ mol HCl $\rightarrow$ mol $H_2$**

$$0.0400 \text{ L} \times \frac{2.43 \text{ mol HCl}}{1 \text{ L HCl}} \times \frac{1 \text{ mol } H_2}{2 \text{ mol HCl}} = 0.0486 \text{ mol } H_2$$

**HCl is the limiting reactant since it yields less $H_2$.**

**Known quantities: $n = 0.0486$ mol $H_2$; $P = 2.25$ atm; $T = 23^{\circ}C + 273 = 296$ K**

**Solving $PV = nRT$ for $V$ gives**

$$V = \frac{nRT}{P} = \frac{(0.0486 \text{ mol})(0.08206 \text{ L} \cdot \text{atm/mol} \cdot \text{K})(296 \text{ K})}{(2.25 \text{ atm})} = 0.525 \text{ L}$$

<u>Problem 5.63</u> ■ During a collision, automobile air bags are inflated by the $N_2$ gas formed by the explosive decomposition of sodium azide, $NaN_3$.

$$2\ NaN_3 \rightarrow 2\ Na + 3\ N_2$$

What mass of sodium azide would be needed to inflate a 30.0-L bag to a pressure of 1.40 atm at 25°C?

**This is another gas stoichoimetry problem. The moles of nitrogen will be calculated using the ideal gas law. The moles of $N_2$ formed can be related to the moles of sodium azide reacted and then grams $NaN_3$.**

**$T = 25^{o}C + 273 = 298$ K and $R = 0.08206$ L·atm/mol·K.**

$$PV = nRT \text{ or } n = \frac{PV}{RT} = \frac{(1.40\ \text{atm})(30.0\ \text{L})}{(0.08206\ \text{L}\cdot\text{atm/mol}\cdot\text{K})(298\ \text{K})} = 1.72 \text{ moles } N_2 \text{ formed.}$$

$$1.72 \text{ moles } N_2 \text{ formed} \times \frac{2\ \text{mol NaN}_3}{3\ \text{mol N}_2} \times \frac{65.01\ \text{g NaN}_3}{1\ \text{mol NaN}_3} = 74.4 \text{ g NaN}_3 \text{ that must react to}$$

**produce 30.0 L of $N_2$ at $25^{o}C$ and 1.40 atm.**

Problem 5.65 ■ As one step in its purification, nickel metal reacts with carbon monoxide to form a compound called nickel tetracarbonyl, $Ni(CO)_4$, which is a gas at temperatures above about 316 K. A 2.00-L flask is filled with CO gas to a pressure of 748 torr at 350.0 K, and then 5.00 g of Ni is added. If the reaction described above occurs and goes to completion at constant temperature, what will the final pressure in the flask be?

**The equation for this reaction is $Ni(s) + 4\ CO(g) \rightarrow Ni(CO)_4(g)$.**

**This is a limiting reactant problem similar to Problem 5.59. After the initial moles of Ni are found, we need to use the ideal gas law to determine the moles of CO initially present, and then perform a stoichiometry calculation to determine if CO or Ni is the limiting reactant. Finally, we must calculate the total number of moles of gas present after the reaction is complete to determine the final pressure.**

$$5.00 \text{ g Ni} \times \frac{1\ \text{mol Ni}}{58.69\ \text{g Ni}} = 0.0852 \text{ moles Ni}$$

**$T = 350$ K $V = 2.00$ L $= 2000$ mL and $R = 62{,}400$ mL·torr/mol·K.**

$$PV = nRT \text{ or } n = \frac{PV}{RT} = \frac{(748\ \text{torr})(2000\ \text{mL})}{(62{,}400\ \text{mL}\cdot\text{torr/mol}\cdot\text{K})(350\ \text{K})} = 0.0685 \text{ moles CO.}$$

$$0.0685 \text{ moles CO} \times \frac{1\ \text{mol Ni}}{4\ \text{mol CO}} = 0.0171 \text{ moles Ni required to react with all the CO. This is}$$

**less than is actually present, so Ni is in excess, CO is the limiting reactant.**

**The moles of $Ni(CO)_4$ present at the end of the reaction can be calculated using the moles of CO. These will be the only moles of gas present (all the CO reacts).**

$$0.0685 \text{ moles CO} \times \frac{1 \text{ mol Ni(CO)}_4}{4 \text{ mol CO}} = 0.0171 \text{ moles Ni(CO)}_4 \text{ formed.}$$

**The final pressure can now be calculated.**

$$PV = nRT \text{ or } P = \frac{nRT}{V} = \frac{(0.0171 \text{ mol})(62{,}400 \text{ mL}\cdot\text{torr/mol}\cdot\text{K})(350 \text{ K})}{(2000 \text{ mL})} = 187 \text{ torr}$$

$$\text{or } 188 \text{ torr} \times \frac{1 \text{ atm}}{760 \text{ torr}} = 0.246 \text{ atm.}$$

Problem 5.67 ■ Clouds of hydrogen molecules have been detected deep in interstellar space. It is estimated that these clouds contain about $1 \times 10^{10}$ hydrogen molecules per $m^3$, and have a temperature of just 25 K. Using these data, find the approximate pressure in such a cloud.

**A molecular density is given. We need to convert this from molecules per cubic meter to moles per liter, and then use the ideal gas law to calculate the pressure.**

$$\frac{1\times10^{10} \text{ molecules H}_2}{\text{m}^3} \times \frac{1 \text{ mol}}{6.022\times10^{23} \text{ molecules}} \times \frac{0.001 \text{ m}^3}{1 \text{ L}} = \frac{1.7\times10^{-17} \text{ mol}}{\text{L}}$$

**Using the ideal gas law,**

$$PV = nRT \text{ or } P = \frac{n}{V}RT = \frac{1.7\times10^{-17} \text{ mol}}{\text{L}}(0.08206 \text{ L·atm/mol·K})(25 \text{ K}) = 3.5 \times 10^{-17} \text{ atm}$$

Problem 5.69 Under what conditions do the postulates of the gas kinetic theory break down?

**Gas kinetic theory breaks down under condition of high pressure and low temperature. This means that gases do not behave ideally under these conditions.**

**One of the postulates states that ideal gases themselves occupy no volume. This of course cannot be true but is especially inaccurate in high-pressure situations where there are a large number of molecules in a relatively small volume of space.**

**Another postulate states that ideal gases have no attractive or repulsive forces acting between molecules. Actually there are such forces (gases would not condense to liquids otherwise), and these become very significant at lower temperature where molecules are moving more slowly.**

Problem 5.71 ■ Place these gases in order of increasing average molecular speed at 25°C: Kr, $CH_4$, $N_2$, $CH_2Cl_2$.

**Kinetic energy is proportional to temperature. With all of the samples at the same temperature, their average kinetic energies are the same. Kinetic energy is related to mass and velocity: $E_{kin} = \frac{1}{2}mv^2$. Velocity is related to kinetic energy and mass: $v^2 = \frac{2E}{m}$. Therefore, molecules with smaller mass have the faster molecular speed.**

**To rank the molecules with increasing speed, rank them from the largest molar mass to the smallest.**

**Estimate the molar masses: Kr molar mass = 83.8 g/mol, $CH_4$ molar mass = 16.0 g/mol, $N_2$ molar mass = 28.0 g/mol, $CH_2Cl_2$ molar mass = 84.9 g/mol.**

**slowest speed $CH_2Cl_2 < Kr < N_2 < CH_4$ fastest speed**

Problem 5.73 The figure below shows the distribution of speeds for two samples of $N_2$ gas. One sample is at 300 K, and the other is at 1000 K. Which is which?

**The velocity distribution shifts more to higher speeds at higher temperatures; so the light blue colored distribution (broader one) represents the higher temperature, 1000 K.**

Problem 5.75 Define the term *mean free path*.

**Mean free path is the average distance a gas particle travels between collisions with other gas particles.**

Problem 5.77 Calculate the pressure of 15.0 g of methane gas in a 1.50-L vessel at 45.0°C using (a) the ideal gas law and (b) the van der Waals equation using constants in Table 5.2.

**(a) $T = 45.0°\text{C} + 273 = 318\text{ K}$ moles $CH_4 = 15.0\text{ g} \times \frac{1\text{ mol }CH_4}{16.04\text{ g }CH_4} = 0.935$ moles $CH_4$**

**$R = 0.08206$ L·atm/mol·K**

$$PV = nRT \text{ or } P = \frac{nRT}{V} = \frac{(0.935\text{ mol})(0.08206\text{ L}\cdot\text{atm/mol}\cdot\text{K})(318\text{ K})}{(1.50\text{ L})} = 16.3\text{ atm}$$

**(b) The van der Waals equation is:**

$$\left(P + \frac{an^2}{V^2}\right)(V - nb) = nRT.$$

**Rearranging to solve for $P$,** $P = \frac{nRT}{(V - nb)} - \frac{an^2}{V^2}$.

**From Table 5.2, the van der Waals constants for methane are**

$$a = 2.253\ \frac{\text{L}^2\cdot\text{atm}}{\text{mol}^2} \qquad b = 0.04278\ \frac{\text{L}}{\text{mol}}$$

$$P = \frac{(0.935\text{ mol})(0.08206\text{ L}\cdot\text{atm/mol}\cdot\text{K})(318\text{ K})}{(1.50\text{ L} - (0.935\text{ mol})(0.04278\text{ L/mol}))} - \frac{(2.253\dfrac{\text{L}^2\cdot\text{atm}}{\text{mol}^2})(0.935\text{ mol})^2}{(1.50\text{ L})^2}$$

**$P$ = 16.7 atm – 0.875 atm = 15.8 atm**

Problem 5.79 ■ Consider a sample of $N_2$ gas under conditions in which it obeys the ideal gas law exactly. Which of these statements are true?
(a) A sample of Ne(g) under the same conditions must obey the ideal gas law exactly.
(b) The speed at which one particular $N_2$ molecule is moving changes from time to time.
(c) Some $N_2$ molecules are moving more slowly than some of the molecules in a sample of $O_2$(g) under the same conditions.
(d) Some $N_2$ molecules are moving more slowly than some of the molecules in a sample of Ne(g) under the same conditions.
(e) When two $N_2$ molecules collide, it is possible that both may be moving faster after the collision than they were before.

**(a) $N_2$ is a larger molecule than Ne is an atom and therefore $N_2$ is not as much like an ideal gas. If these conditions allow $N_2$ to behave as an ideal gas, they would certainly be sufficient for Ne to also behave like an ideal gas. Hence, this statement: "A sample of Ne(g) under the same conditions must obey the ideal gas law exactly." is true.**

**(b) All collisions are elastic; however, energy can be transferred during an elastic collision from one molecule to another. A faster molecule hitting a slower molecule might make the slow molecule go faster if the first one ends up slower. In addition, each time the molecule hits the wall, there is a split second when its speed is zero as it bounces off the wall and goes flying away in the opposite direction. Hence, the statement: "The speed at which one particular $N_2$ molecule is moving changes from time to time." is true.**

**(c) The average speed of the $N_2$ molecules will be faster than the average speed of the $O_2$ molecules, but ideal gas particles move at varying speeds, so some molecules in each sample will be moving very slowly and others very quickly. Hence, the statement: "Some $N_2$ molecules are moving more slowly than some of the molecules in a sample of $O_2$ (g) under the same conditions." is true.**

**(d) The average speed of the $N_2$ molecules will be slower than the average speed of the Ne molecules; hence, the statement: "Some $N_2$ molecules are moving more slowly than some of the molecules in a sample of Ne(g) under the same conditions." is true.**

**(e) All collisions are elastic, so collisions must conserve energy. There is no way that both molecules could be going faster, since that implies that energy has increased. Hence, the statement: "When two $N_2$ molecules collide, it is**

**possible that both may be moving faster after the collision than they were before." is false.**

Problem 5.81 What is actually measured in an ionization gauge pressure sensor? How is this actual measurement related to pressure?

**An electric current is what is actually measured. The number of electrons in the flow of electricity is directly related to the number of gas molecules present. By converting the current measured into a number of gas molecules present, the pressure of the gas can be calculated and displayed by the sensor.**

Problem 5.83 Do a web search to learn about at least one other type of pressure sensor not described in the chapter. Describe how the gauge operates in a way that your roommate could understand.

**Answers will vary according to the websites found.**

Problem 5.85 A gas mixture contains 10.0% $CH_4$ and 90.0% Ar by moles. Find the density of this gas at 250°C and 2.5 atm.

**We can express the percent of each gas as a mole fraction, $X_{CH4} = 0.10$, $X_{Ar} = 0.90$.**

**Using these, we'll calculate a weighted-average molar mass (MM) for the gas mixture:**

**$X_{CH4}$ (molar mass of $CH_4$) + $X_{Ar}$ (molar mass of Ar) = Weighted-Average Molar Mass**
**0.10(16.04 g/mol) + 0.90( 39.95 g/mol) = 37.56 g/mol**

**We can use the weighted-average molar mass (*MM*) in the ideal gas law to solve for the ratio of grams: volume, which equals the density, ($\frac{g}{V}$):**

$$PV = nRT, \qquad n = g \div MM \qquad \rightarrow \qquad PV = \frac{g}{MM}RT$$

**Now solving for the ratio:** $\frac{g}{V} = \frac{P(MM)}{RT} = \frac{(2.5\text{ atm})(37.56\text{ g/mol})}{(0.08206\text{ L}\cdot\text{atm/mol}\cdot\text{K})(523\text{ K})} = 2.19$ **g/L.**

Problem 5.87 ■ A number of compounds containing the heavier noble gases, and especially xenon, have been prepared. One of these is xenon hexafluoride ($XeF_6$), which can be prepared by heating a mixture of xenon and fluorine gases. $XeF_6$ is a white crystalline solid at room temperature and melts at about 325 K. A mixture of about 0.0600 g of Xe and 0.0304 g of $F_2$ is sealed into a 100-mL bulb. (The bulb contains no air or other gases.) The bulb is heated, and the reaction above goes to completion. Then the sealed bulb is cooled back to 20°C. What will be the final pressure in the bulb, expressed in torr?

**The reaction is** $Xe(g) + 3 F_2(g) \rightarrow XeF_6(s)$
**After the reaction is over and the bulb is cooled to 20°C, the product, $XeF_6$, will be a solid. To calculate the pressure, we need to perform a stoichoimetry calculation to determine the limiting reactant. Then we must determine how many moles of the excess reactant are left over. With this number we'll calculate the pressure in the container.**

$$0.0600 \text{ g Xe} \times \frac{1 \text{ mol Xe}}{131.3 \text{ g Xe}} \times \frac{3 \text{ mol } F_2}{1 \text{ mol Xe}} \times \frac{38.00 \text{ g } F_2}{1 \text{ mol } F_2} = 0.0521 \text{ g } F_2 \text{ required to react with all}$$

**the Xe. This is more than is present, so $F_2$ is the limiting reactant, Xe the excess reactant.**

**The moles of Xe originally present:**

$$0.0600 \text{ g Xe} \times \frac{1 \text{ mol Xe}}{131.3 \text{ g Xe}} = 4.57 \times 10^{-4} \text{ moles Xe}$$

**Next, we calculate the moles of Xe reacted.**

$$0.0304 \text{ g } F_2 \times \frac{1 \text{ mol } F_2}{38.00 \text{ g } F_2} \times \frac{1 \text{ mol Xe}}{3 \text{ mol } F_2} = 2.67 \times 10^{-4} \text{ mol Xe}$$

**Now, we need to find the moles of Xe leftover.**

$$4.57 \times 10^{-4} \text{ mol Xe} - 2.67 \times 10^{-4} \text{ mol Xe} = 1.90 \times 10^{-4} \text{ mol Xe}$$

**Finally, we use the ideal gas law to calculate the pressure.**

$T = 20°C + 273 = 293$ K $\qquad R = 62{,}400$ mL·torr/mol·K

$$PV = nRT \text{ or } P = \frac{nRT}{V} = \frac{(1.90 \times 10^{-4} \text{ mol})(62{,}400 \text{ mL} \cdot \text{torr/mol} \cdot \text{K})(293 \text{ K})}{(100 \text{ mL})} = 34.7 \text{ torr}$$

Problem 5.89 ■ A 0.2500-g sample of an Al-Zn alloy reacts with HCl to form hydrogen gas:

$$2 \text{ Al(s)} + 6 \text{ H}^+\text{(aq)} \rightarrow \text{Al}^{3+}\text{(aq)} + 3 \text{ H}_2\text{(g)}$$
$$\text{Zn(s)} + 2 \text{ H}^+\text{(aq)} \rightarrow \text{Zn}^{2+}\text{(aq)} + \text{H}_2\text{(g)}$$

The hydrogen produced has a volume of 0.147 L at 25°C and 755 mm Hg. What is the percentage of zinc in the alloy?

**From the data in the problem, the moles of hydrogen produced may be found.**

$$n = \frac{PV}{RT} = \frac{(\frac{755}{760}\text{ atm})(0.147\text{ L})}{(0.08206\text{ L}\cdot\text{atm/mol}\cdot\text{K})(298\text{ K})} = 0.00597\text{ moles }H_2$$

**These moles of $H_2$ were formed by both Al and zinc, according to the molar ratios given in the balanced equations.**
**Let $X$ = mass of Al:**

$$0.00597\text{ mol }H_2 - X\text{ g Al} \times \frac{1\text{ mol Al}}{26.98\text{ g Al}} \times \frac{3\text{ mol }H_2}{2\text{ mol Al}} + (0.2500 - X)\text{ g Zn} \times \frac{1\text{ mol Zn}}{65.39\text{ g Zn}} \times \frac{1\text{ mol }H_2}{1\text{ mol Zn}}$$

**$0.00597 = 0.0556 \times (X) + (0.2500 - X) \times 0.0153$;** **$X = 0.00215/ 0.0403 = 0.0533$ g Al**

**g Zn = $0.2500 - 0.0533 = 0.1967$ g;** **% Zn = $(0.1967 \div 0.2500) \times 100 = 78.68\%$**

Problem 5.91 The complete combustion of octane can be used as a model for the burning of gasoline:

$$2\,C_8H_{18} + 25\,O_2 \rightarrow 16\,CO_2 + 18\,H_2O$$

Assuming that this equation provides a reasonable model of the actual combustion process, what volume of air at 1 atm and 25°C must be taken into an engine to burn 1 gallon of gasoline? (The partial pressure of oxygen in air is 0.21 atm, and the density of liquid octane is 0.70 g/mL.)

**The first step is to determine the moles present in a gallon of gasoline.**

$$1\text{ gallon} \times \frac{4\text{ quarts}}{1\text{ gallon}} \times \frac{1\text{ L}}{1.056\text{ quarts}} \times \frac{1000\text{ mL}}{1\text{ L}} \times \frac{0.70\text{ g}}{\text{mL}} \times \frac{1\text{ mol }C_8H_{18}}{114.2\text{ g }C_8H_{18}} = 23.2\text{ mol }C_8H_{18}$$

**Next, we do a stoichiometry calculation to determine the moles of $O_2$ required to react with the moles of octane.**

$$23.2\text{ mol }C_8H_{18} \times \frac{25\text{ mol }O_2}{2\text{ mol }C_8H_{18}} = 290.\text{ mol }O_2\text{ required.}$$

**Air is not pure oxygen: $P_{O2}$ in air = 0.21 atm; or there are $\frac{0.21\text{ mol }O_2}{1\text{ mol air}}$.**

**Therefore we need 290 mol $O_2$ $\times \frac{1\text{ mol air}}{0.21\text{ mol }O_2}$ = 1380 mol of "air" to supply the required amount of $O_2$.**

**Finally, we use the ideal gas law to calculate the volume of air.**

**$T = 25°C + 273 = 298\ K$** **$R = 0.08206\ L{\cdot}atm/mol{\cdot}K$**

$$PV = nRT \text{ or } V = \frac{nRT}{P} = \frac{(1380\ \text{mol})(0.08206\ \text{L}\cdot\text{atm/mol}\cdot\text{K})(298\ \text{K})}{(1\ \text{atm})} = 34{,}000\ \text{L air}$$

Problem 5.93 Some engineering designs call for the use of compressed air for underground work. If water containing iron(II) ions is present, oxygen in the compressed air may react according to the following net ionic equation.

$$Fe^{2+} + H^+ + O_2 \rightarrow Fe^{3+} + H_2O$$

(a) Write the balanced net ionic equation. Remember that the amounts of each substance and the charges must balance. (b) Assume all of the oxygen from 650 L of compressed air at 15°C and 6.5 atm is lost by this reaction. What mass of water would be produced? (The mole fraction of oxygen in air is about 0.21.) (c) What will be the final pressure after the loss of the oxygen?

**(a) Balanced equation:** $\underline{4}\ Fe^{2+} + \underline{4}\ H^+ + O_2 \rightarrow \underline{4}\ Fe^{3+} + \underline{2}\ H_2O$

**(b) The volume of $O_2 = 0.21 \times 650\ L = 137\ L$**

$$\text{Moles of } O_2 = n = \frac{PV}{RT} = \frac{(6.5\ \text{atm})(137\ \text{L})}{(0.08206\ \text{L}\cdot\text{atm/mol}\cdot\text{K})(288\ \text{K})} = 37.7\ \text{mol}$$

$$37.7\ \text{mol}\ O_2 \times \frac{2\ \text{mol}\ H_2O}{1\ \text{mol}\ O_2} \times \frac{18.02\ \text{g}\ H_2O}{1\ \text{mol}\ H_2O} = 1400\ \text{g}\ H_2O$$

**(c) Using Dalton's Law: $(1 - 0.21) \times 6.5\ atm = 5.1\ atm$**

Problem 5.95 ■ Homes in rural areas where natural gas service is not available often rely on propane to fuel kitchen ranges. The propane is stored as a liquid, and the gas to be burned is produced as the liquid evaporates. Suppose an architect has hired you to consult on the choice of a propane tank for such a new home. The propane *gas* consumed in 1 hour by a typical range burner at high power would occupy roughly 165 L at 25°C and 1 atm, and the range chosen by the client will have 6 burners. If the tank under consideration holds 500 gallons of *liquid* propane, what is the minimum number of hours it would take for the range to consume an entire tankful of propane? The density of *liquid* propane is 0.5077 kg/L.

**Propane is $C_3H_8$.**
**The strategy here is to calculate the moles of propane burned per hour, then the total moles propane in the tank. Finally, dividing moles by moles per hour will give the total hours of burn time on the range.**

$$\text{Mol/hr (one burner)} = n = \frac{PV}{RT} = \frac{(1\ \text{atm})(165\ \text{L})}{(0.08206\ \text{L}\cdot\text{atm/mol}\cdot\text{K})(298\ \text{K})} = 6.75\ \text{mol/hr}$$

**6 burners × 6.75 mol/hr = 40.5 mol/hr**

$$\text{Total mol in tank} = 500\text{ gal} \times \frac{3.785\text{ L}}{1\text{ gal}} \times \frac{0.5077\times10^3\text{ g}}{\text{L}} \times \frac{1\text{ mol }C_3H_8}{44.10\text{ g }C_3H_8} = 21790\text{ mol}$$

$$\text{Hours} = \frac{21790\text{ mol}}{\frac{40.5\text{ mol}}{\text{hr}}} = 538\text{ hrs burn time for the range on high power.}$$

Problem 5.97 ■Pure gaseous nitrogen dioxide ($NO_2$) cannot be obtained, because $NO_2$ *dimerizes*, or combines with itself, to produce a mixture of $NO_2$ and $N_2O_4$. A particular mixture of $NO_2$ and $N_2O_4$ has a density of 2.39 g/L at 50°C and 745 torr. What is the partial pressure of $NO_2$ in this mixture?

**Using the density, the average molar mass of the mixture may be calculated:**

$PV=nRT$; $n = \frac{m}{MM}$ **where** $m$ **= grams and** $MM$ **= the molar mass, so:**

$PV = \frac{m}{MM}RT.$ **Solving for** $MM$, $MM = (\frac{m}{V})\frac{RT}{P}$ **where** $\frac{m}{V}$ **is the density.**

$$MM = (\frac{2.39\text{ g}}{1\text{ L}})\frac{(62.4\text{ L}\cdot\text{torr/mol}\cdot\text{K})(323\text{ K})}{(745\text{ torr})} = 64.659\text{ g/mol}$$

**Let X = the mole fraction $NO_2$ and (1 – X) = the mole fraction $N_2O_4$**
**The average molar mass can be expressed using the individual molar masses:**
**(X)46.01 g/mol + (1 – X)92.02 g/mol = $MM$ = 64.658 g/mol**
**X = 0.5947**

**The partial pressure of $NO_2$ is equal to the mole fraction times the total pressure:**
$P_{NO2} = X_{NO2}P_T = 0.5946(745\text{ torr}) = 443\text{ torr}$

Problem 5.99 ■A mixture of $CH_4(g)$ and $C_3H_8(g)$ with a total pressure of 425 torr is prepared in a reaction vessel. Just enough $O_2(g)$ is added to allow complete combustion of the two hydrocarbons. The gas mixture is ignited, and the resulting mixture of $CO_2(g)$ and $H_2O(g)$ is cooled back to the original temperature. If the total pressure of the products is 2.38 atm, what was the mole fraction of $C_3H_8$ in the original mixture?

**The two combustion reactions are:**

$CH_4 + 2\,O_2 \rightarrow CO_2 + 2\,H_2O$ $C_3H_8 + 5\,O_2 \rightarrow 3\,CO_2 + 4\,H_2O$

**Let X = the mole fraction $C_3H_8$, and let (1 – X) represent the mole fraction $CH_4$.**
**Let $n_i$ = the total moles of the $C_3H_8$ and $CH_4$ reacted and let $n_f$ = the total moles of $CO_2$ and $H_2O$ formed.**
**So the total moles $C_3H_8$ = X$n_i$ and total moles $CH_4$ = (1 – X)$n_i$.**

**From the equations, we know there are 1 mole of $CO_2$ and 2 moles of $H_2O$ formed for every mole of $CH_4$ reacted and 3 moles of $CO_2$ and 4 mole s of $H_2O$ formed for every mole of $C_3H_8$ reacted.**

**Therefore,** $n_f = 3(\text{moles } CH_4) + 7(\text{moles } C_3H_8)$
**Or** $n_f = 3((1 - X)n_i) + 7(Xn_i)$

**The initial and final moles of gas in the reactor are proportional to the pressures before ($P_i$) and after the reaction ($P_f$):**

$$n_f = \frac{P_f}{P_i} n_i = \left(\frac{2.38\ \text{atm} \times \frac{760\ \text{torr}}{1\ \text{atm}}}{425\ \text{atm}}\right) n_i = 4.25 n_i$$

**Substituting this into the previous equation:**
$4.25\, n_i = 3((1 - X)n_i) + 7(Xn_i)$
**The moles of reactants, $n_i$, cancels out leaving:** $4.25 = 3(1 - X) + 7X;\ X = 0.313$

**The mole fraction of propane is 0.313.**

Problem 5.101 Aerospace engineers sometimes write the gas law in terms of the mass of the gas rather than the number of moles.

$$PV = mR_{\text{specific}}T$$

In such a formulation, the molar mass of the gas must be incorporated into the value of the gas constant, which means that the gas constant would differ from one gas to another. (We have written the gas constant here as $R_{\text{specific}}$ to emphasize this point. Many engineering texts use different notations to distinguish between the universal and specific gas constants.) (a). Suggest a reason why this approach might be particularly attractive in aerospace engineering.

**In aerospace engineering, they are most often concerned with gases in the atmosphere and therefore can use $R_{\text{specific}}$ to obtain more accurate characterization of gas properties (air).**

(b). Assume that the mole fractions of $O_2$ and $N_2$ in air are 0.21 and 0.79, respectively. Calculate the average molar mass of air (*i.e.*, the mass of one mole of air).

**The average molar mass of air can be calculated as follows:**

**0.21(32.00 g/mol) + 0.79(28.02 g/mol) = 28.856 g/mol**

(c). Use your result from (b) to determine the value of $R_{\text{specific}}$ for air, and express it in SI units of $m^2\ s^{-2}\ K^{-1}$.

$$R_{specific} = \frac{R}{MM} \quad \text{where } MM = \text{molar mass of the gas}$$

**The universal gas constant, $R$, can be expressed as 8.314 $\frac{J}{mol \bullet K}$ or**

**$8.314 \frac{kg \bullet m^2}{s^2 \bullet mol \bullet K}$ (recall $J = \frac{kg \bullet m^2}{s^2}$)**

**So the specific gas constant is $R_{specific} = \frac{R}{MM} = \frac{8.314\, kg \bullet m^2}{(28.856 \times 10^{-3}\, kg/mol)\ \ s^2 \bullet mol \bullet K}$**

**$= 0.288 \frac{m^2}{s^2 \bullet K}$.**

Problem 5.103 The decomposition of mercury (II) thiocyanate produces an odd brown snake-like mass that is so unusual the process was once used in fireworks displays. There are actually several reactions that take place when the solid $Hg(SCN)_2$ is ignited:

$$2Hg(SCN)_2(s) \rightarrow 2HgS(s) + CS_2(s) + C_3N_4(s)$$
$$CS_2(s) + 3O_2(g) \rightarrow CO_2(g) + 2SO_2(g)$$
$$2C_3N_4(s) \rightarrow 3(CN)_2(g) + N_2(g)$$
$$HgS(s) + O_2(g) \rightarrow Hg(\ell) + SO_2(g)$$

A 42.4-g sample of $Hg(SCN)_2$ is placed into a 2.4-L vessel at 21°C. The vessel also contains air at a pressure of 758 torr. The container is sealed and the mixture is ignited, causing the reaction sequence above to occur. Once the reaction is complete, the container is cooled back to the original temperature of 21°C.
(a) Without doing numerical calculations, predict whether the final pressure in the vessel will be greater than, less than, or equal to the initial pressure. Explain your answer.

**The overall net equation,**

$$2Hg(SCN)_2(s) + 4O_2(g) + C_3N_4(s) \rightarrow HgS(s) + CO_2(g) + 3SO_2(g) + 3(CN)_2(g) + N_2(g) + Hg(\ell)$$

**shows 4 moles of gas on the reactant side versus 8 moles gas on the product side. A rough estimate is that the pressure will increase. Pressure is proportional to moles of gas.**

(b) Calculate the final pressure and compare your result with your prediction. (Assume that the mole fraction of $O_2$ in air is 0.21.)

**First, the moles of $O_2$ and $N_2$ in the air originally present can be calculated using the partial pressures.**

$$n_{O2} = \frac{PV}{RT} = \frac{(0.21)(758\ \text{torr})(2.4\ \text{L})}{(62.4\ \text{L} \cdot \text{torr/mol} \cdot \text{K})(294\ \text{K})} = 0.02082 \text{ moles } O_2$$

$$n_{N2} = \frac{PV}{RT} = \frac{(0.79)(758\ \text{torr})(2.4\ \text{L})}{(62.4\ \text{L} \cdot \text{torr/mol} \cdot \text{K})(294\ \text{K})} = 0.07834 \text{ moles } N_2$$

**and the initial moles of $Hg(SCN)_2$: 42.4 g × $\frac{1 \text{ mol } Hg(SCN)_2}{316.77 \text{ g } Hg(SCN)_2}$ = 0.1339 moles of $Hg(SCN)_2$**

**To react all the $Hg(SCN)_2$ would require**
**0.1339 moles of $Hg(SCN)_2$ × $\frac{4 \text{ mol } O_2}{2 \text{ mol } Hg(SCN)_2}$ = 0.268 moles of $O_2$,**
**therefore $O_2$ is the limiting reactant.**
**Next, the total moles of gas left after the reaction must be found.**
**0.02082 moles $O_2$ × $\frac{8 \text{ mol product gases}}{4 \text{ mol } O_2}$ = 0.04164 moles gas formed plus 0.07834 moles $N_2$ originally present = 0.1200 moles total in the reactor.**

$$P = \frac{nRT}{V} = \frac{(0.1200\ \text{mol})(62.4\ \text{L} \cdot \text{torr/mol} \cdot \text{K})(294\ \text{K})}{(2.4\ \text{L})} = 920 \text{ torr}$$

**The pressure increased but did not double because the nitrogen present still represents most of the gas in the reactor.**

Problem 5.105 A soft drink can label indicates that the volume of the soda in contains is 12 oz or 355 mL. There is probably some empty space at the top of the can. Describe what you can measure and how that measurement allows you to determine the actual density of the soda.

**For this problem, you need to determine whether air is present by measuring various densities and determine if there are differences. First, the unopened can must be weighed and its volume determined (by water displacement if possible). Next, the soda must have its density determined separately, perhaps by measuring the mass of a volumetric container such as a graduated cylinder and then measuring the volume. There's some chance that the volume of the aluminum can be ignored in this problem, but if necessary, it could be measured. If the density measured for the whole system varies from that measured for the soda alone, the most logical source for the difference would be the gas in the "head space".**

Problem 5.107 An ore sample with a mass of 670 kg contains 27.7% magnesium carbonate, $MgCO_3$. If all the magnesium carbonate in this ore sample is decomposed to form carbon dioxide, describe how to determine what volume of $CO_2$ is evolved during the process. What would have to be measured to predict the needed volume in advance?

**Using the percentage of magnesium carbonate in the ore, the mass of actual $MgCO_3$ present can be determined. Next we need to write an equation showing the decomposition of magnesium carbonate to form $CO_2$ and MgO. Finally, a stoichoimetry calculation can be performed to determine the moles of $CO_2$ formed from the mass of $MgCO_3$ reacted. We would need to measure the temperature and pressure of the gas to calculate the volume of $CO_2$ formed.**

Problem 5.109 Table 5.1 provides the mole percentage composition of dry air. What is the mass percentage composition of dry air?

**The moles of each gas must be converted to mass, then summed and expressed as percentages**

$N_2$ $$31.929 \text{ mol} \times \frac{28.01 \text{ g } N_2}{1 \text{ mol } N_2} = 894 \text{ g } N_2$$

$O_2$ $$8.567 \text{ mol} \times \frac{32.00 \text{ g } O_2}{1 \text{ mol } O_2} = 274 \text{ g } O_2$$

Ar $$0.382 \text{ mol} \times \frac{39.95 \text{ g Ar}}{1 \text{ mol Ar}} = 15.3 \text{ g Ar}$$

$CO_2$ $$0.013 \text{ mol} \times \frac{44.01 \text{ g } CO_2}{1 \text{ mol } CO_2} = 0.572 \text{ g } CO_2$$

**Other** **0.002 (this will be neglected)**

**Total:** **1183 g**

**Percentages:**

$N_2$ **894 g $N_2$ ÷ 1183 g total × 100 = 75.5%**

$O_2$ **274 g $O_2$ ÷ 1183 g total × 100 = 23.2%**

Ar **15.3 g Ar ÷ 1183 g total × 100 = 1.3%**

$CO_2$ **0.572 g $CO_2$ ÷ 1183 g total × 100 = 0.04%**

Problem 5.111 ■ Consider a room that is 14 ft × 20 ft with an 8-ft ceiling. **(a)** How many molecules of air are present in this room at 20°C and 750 torr? **(b)** If a pollutant is present at 2.3 ppm, how many pollutant molecules are in this room?

**First, find the volume of the room in liters:**

$$\mathbf{14\ ft \times 20\ ft \times 8\ ft = 2240\ ft^3 \times \frac{28.316\ L}{1\ ft^3} = 63{,}400\ L}$$

**(a) Use the ideal gas law to find moles of air, then molecules.**

$$\mathbf{T = 20^{o}C + 273 = 293\ K \qquad P = 750\ torr \times \frac{1\ atm}{760\ torr} = 0.987\ atm}$$

$\mathbf{R = 0.08206\ L{\cdot}atm/mol{\cdot}K}$

$$\mathbf{PV = nRT \text{ or } n = \frac{PV}{RT} = \frac{(0.987\ atm)(63{,}400\ L)}{(0.08206\ L \cdot atm/mol \cdot K)(293\ K)} = 2.6 \times 10^3\ moles\ of\ air}$$

$$\mathbf{2600\ mol \times \frac{6.022 \times 10^{23}\ molecules}{1\ mol} = 1.6 \times 10^{27}\ molecules\ of\ air\ in\ the\ room.}$$

**(b) 2.3 ppm means there are 2.3 parts pollutant per 1 million parts of air. The parts we refer to would be the molecules present in air. Therefore, there are 2.3 molecules of pollutant for every one million molecules in air. We'll use the result from (a) to find our answer.**

$$\mathbf{\frac{2.3\ molecules\ pollutant}{1 \times 10^6\ molecules\ of\ air} \times \frac{1.6 \times 10^{27}\ molecules\ air}{room} = 3.7 \times 10^{21}\ molecules\ of\ pollutant\ are}$$
**in the room.**

Problem 5.113 ■ A 0.0125-g sample of a gas with an empirical formula of $CHF_2$ is placed in a 165-mL flask. It has a pressure of 13.7 mm Hg at 22.5 °C. What is the molecular formula of the compound?

$$\mathbf{Density,\ d = \frac{0.0125\ g}{0.165\ L} = 0.0758\ g/L \qquad 13.7\ mm\ Hg \times \frac{1.00\ atm}{760\ mm\ Hg} = 0.0180\ atm}$$

$$\mathbf{MM = \frac{dRT}{P} = \frac{\left(0.0758\ \frac{g}{L}\right) \times \left(0.08206 \frac{L \cdot atm}{mol \cdot K}\right) \times (295.7\ K)}{(0.0180\ atm)} = 102\ \frac{g}{mol}}$$

**51 g/mol is the empirical molar mass; the true formula is a whole-number ratio of the actual and empirical molar masses.** $\mathbf{\frac{102\ g/mol}{51\ g/mol} = 2}$

**The molecular formula is 2 × ($CHF_2$) or $C_2H_2F_4$**

# CHAPTER OBJECTIVE QUIZ

This quiz will test your understanding of the basic text chapter objectives and give you additional practice problems. You should work this quiz after completing the end-of-chapter questions. The solutions to these questions are found at the end of this chapter.

1. Which of the following properties does **not** describe gases?
   a. Gases expand to completely fill their containers.
   b. Gases mix completely and uniformly with each other.
   c. Gases can be easily compressed.
   d. Gases have relatively low densities.
   e. Gases volumes are not usually affected by temperature.

2. Which of the following substances are not considered part of urban air pollution?
   a. $N_2$ b. NO c. $O_3$ d. CO e. $SO_2$

3. Calculate the new volume of 10.2 mL of oxygen if the temperature is increased from 35.0°C to 70.0°C. Pressure remains constant.

4. Lethal carbon dioxide gas pockets sometimes escape from underground. If a $5.0 \times 10^5$ liter pocket of $CO_2$ at 2.5 atm and 12.8°C escapes to the surface, what would its new volume be? Assume the gas at the surface is 1 atm and 25.0°C.

5. Calculate the volume of 0.361 moles of an ideal gas with pressure equal to 2.51 atm and temperature of 303 K.

6. What is the pressure (in torr) of 0.989 mol of argon in a 6.65 L container at 33.9°C.

7. What is the temperature of a sample of $SO_2$ gas if 0.0657 moles are occupying 8.4 quarts at a pressure of 14.99 psi?

8. What is the molar mass of a gas if a 0.987-g sample occupies 432 mL at a temperature of 31.2°C and a pressure of 0.951 atm?

9. A mixture of gas contains 4 moles $O_2$, 5 moles $N_2$, and 2 moles $CO_2$. The total pressure is 12.1 atmospheres. What is the partial pressure of each gas?

10. If the mole percent of argon in air is 0.93%, estimate the partial pressure of argon at sea level.

11. Calcium reacts with water to produce hydrogen:

$$Ca(s) + 2\,H_2O(\ell) \rightarrow H_2(g) + Ca(OH)_2(aq)$$

If 8.43 L of $H_2$ at 35.1°C and 774 torr are produced, how many grams of calcium reacted?

12. Calcium carbide reacts vigorously with water to produce acetylene gas.

$$CaC_2(s) + 2\,H_2O(\ell) \rightarrow C_2H_2(g) + Ca(OH)_2(aq)$$

If a 25.6-g sample of $CaC_2$ is placed in a 1.00-L closed container with 50.0 mL of water, what is the final pressure in the container? Assume water has a density of 1.00 g/mL.

13. If 50.5 grams of propane ($C_3H_8$) burns, what volume of $CO_2$ at STP forms? Assume complete combustion.

14. Which of the following is not a postulate of the kinetic molecular theory of gases?
    a. Gas molecules are in constant random motion.
    b. Gas molecules are attracted to each other.
    c. Gas molecules are infinitely small.
    d. All gas molecules are the same.
    e. Gas molecules have elastic collisions.

15. On a molecular level, explain why increasing temperature increases the pressure of a gas (assuming constant moles and volume).

16. TRUE or FALSE: At a given temperature, all gas molecules in a sample have the same velocity.

17. TRUE or FALSE: The mass of a real gas does not affect its velocity.

18. Which of the following conditions might we expect a gas to not behave like an ideal gas?
    a. High temperature
    b. Low temperature
    c. High pressure
    d. Both (b) and (c)
    e. All three

19. A storage tank containing nitrogen has a volume of 500.0 L and contains 98.6 kg of gas at a temperature of 29.4°C. Calculate the pressure using both the ideal gas law and the van der Waal's equation and compare the results. (Use Table 5.2 for van der Waal's constants.)

20. Match the pressure sensor with the correct description.

| | |
|---|---|
| ___ Ionization gauge | A. Gas pressure bends a diaphragm, changing the capacitance of a fixed plate. |
| ___ Thermocouple gauge | B. As gas molecules collide with a hot filament, heat is removed, lowering the temperature. |
| ___Capacitance manometer | C. Gas molecules are converted into ions, collected and counted. |

## ANSWERS TO THE CHAPTER OBJECTIVE QUIZ

1. e

2. a

3. This problem is most easily solved using Charles's law: $\frac{V_1}{T_1} = \frac{V_2}{T_2}$ (constant $n$, $P$)

   Temperature must be expressed in absolute units: Kelvin.
   $V_1 = 10.2$ mL $T_1 = 35.0°C + 273.15 = 308.2$ K $T_2 = 70.0°C + 273.15 = 343.2$ K

   $$\frac{V_1}{T_1} = \frac{V_2}{T_2}, \quad V_2 = T_2 \times \frac{V_1}{T_1} = 343.2\text{ K} \times \frac{10.2\text{ mL}}{308.2\text{ K}} = 11.4\text{ mL}$$

4. Pressure, volume and temperature are all changing in this problem. We'll start with the ideal gas law.
   $PV = nRT$ or $\frac{PV}{nT} = R$, a constant. Since this ratio will stay the same even after the variables are changed, we can write $\frac{P_1V_1}{n_1T_1} = \frac{P_2V_2}{n_2T_2}$. The number of moles stays constant so our gas law becomes $\frac{P_1V_1}{T_1} = \frac{P_2V_2}{T_2}$. This expression is sometimes known as the *Combined Gas Law.* Now we just need to insert our data (temperature units must be Kelvin) and solve for the new volume:

   $T_1 = 12.8°C + 273 = 285.8$ K; $T_2 = 25°C + 273 = 298$ K
   $P_1 = 2.5$ atm; $P_2 = 1.0$ atm; $V_1 = 5.0 \times 10^5$ mL

   $$V_2 = \frac{T_2P_1V_1}{P_2T_1} = \frac{(298\text{ K})(2.5\text{ atm})(5.0 \times 10^5\text{ mL})}{(1.0\text{ atm})(285.5\text{ K})} = 1.3 \times 10^6\text{ L}$$

   We could have also solved the problem by using the ideal gas law twice: first to calculate the moles of gas with the original conditions, then to calculate the volume after pressure and temperature is changed.

5. The ideal gas law relates $P$, $V$, $T$, and moles: $PV = nRT$.

   The value of the gas constant to use is $R = 0.08206$ L·atm/mol·K.

$$PV = nRT \text{ or } V = \frac{nRT}{P} = \frac{(0.361\text{ mol})(0.08206\text{ L}\cdot\text{atm/mol}\cdot\text{K})(303\text{ K})}{(2.51\text{ atm})} = 3.57\text{ L}$$

6. **The ideal gas law relates $P$, $V$, $T$, and moles: $PV = nRT$.**

**The value of the gas constant to use is $R$ = 62,400 mL·torr/mol·K.**
**$T = 33.9^{o}C + 273.15 = 307.1$ K**
**$V$ = 6.65 L = 6650 mL**

$$PV = nRT \text{ or } P = \frac{nRT}{V} = \frac{(0.989\text{ mol})(62{,}400\text{ mL}\cdot\text{torr/mol}\cdot\text{K})(307.1\text{ K})}{(6650\text{ mL})} = 2850\text{ torr}$$

7. **This is another ideal gas law problem the units for pressure, and volume must be converted into appropriate units.**

$R = 0.08206$ L·atm/mol·K, $\quad P = 14.99\text{ psi} \times \frac{1.00\text{ atm}}{14.7\text{ psi}} = 1.02\text{ atm}$

$$V = 8.4\text{ quarts} \times \frac{0.9463\text{ L}}{1\text{ quarts}} = 7.95\text{ L}$$

$$T = \frac{PV}{nR} = \frac{(1.02\text{ atm})(7.95\text{ L})}{(0.0657\text{ mol})(0.08206\text{ L}\cdot\text{atm/mol}\cdot\text{K})} = 1.50 \times 10^3\text{ K}$$

8. **The molar mass is the ratio of mass to moles: $\frac{\text{grams}}{\text{moles}}$. With the $P$, $V$, and $T$ of the gas, the moles can be calculated. Then the molar mass can be found.**

$R = 0.08206$ L·atm/mol·K, $\quad V = 432\text{ mL} = 0.432\text{ L}$,

$T = 31.2^{o}C + 273.15\text{ K} = 304.4\text{ K}$, $\quad P = 0.951\text{ atm}$

$$n = \frac{PV}{RT} = \frac{(0.951\text{ atm})(0.432\text{ L})}{(0.08206\text{ L}\cdot\text{atm/mol}\cdot\text{K})(304.4\text{ K})} = 0.0164\text{ moles}$$

$$\text{Molar mass} = \frac{\text{grams}}{\text{moles}} = \frac{0.987\text{ g}}{0.0164\text{ moles}} = 60.0\text{ g/mol}$$

9. **The partial pressure of each gas is related to its mole fraction: $P_A = X_A (P_T)$. We can calculate each mole fraction and use the total pressure to find each partial pressure.**

**Total moles = 4 moles $O_2$ + 5 moles $N_2$ + 2 mol $CO_2$ = 11 moles**

$X_{O2} = \dfrac{4\text{ moles }O_2}{11\text{ moles}} = 0.364$, $X_{N2} = \dfrac{5\text{ moles }N_2}{11\text{ moles}} = 0.454$, $X_{CO2} = \dfrac{2\text{ moles }CO_2}{11\text{ moles}} =$ 0.182

$P_{O2} = X_{O2}(P_T) = 0.364(12.1\text{ atm}) = 4.40\text{ atm}$, $P_{N2} = X_{N2}(P_T) = 0.454(12.1\text{ atm}) = 5.49\text{ atm}$

$P_{CO2} = X_{CO2}(P_T) = 0.182(12.1\text{ atm}) = 2.20\text{ atm}$

10. If the mole percent of argon is 0.93%, this can be expressed as a mole fraction by dividing by 100: $\dfrac{0.93\%}{100} = 0.0093$. We can assume the atmospheric pressure is 1.00 atm.
The partial pressure of argon = $X_{Ar}(P_T) = 0.0093\ (1.00\text{ atm}) = 0.0093\text{ atm}$.

11. From the volume, pressure and temperature of the hydrogen produced, we can find the moles $H_2$ using the ideal gas law. Moles of $H_2$ formed can be related to the moles of calcium formed in a stoichoimetry calculation.

$R = 0.08206\ \text{L·atm/mol·K}$, $V = 8.43\text{ L}$,

$T = 35.1^{o}\text{C} + 273.15\text{ K} = 308.3\text{ K}$, $P = 774\text{ torr} \times \dfrac{1\text{ atm}}{760\text{ torr}} = 1.02\text{ atm}$

$$n = \frac{PV}{RT} = \frac{(1.02\text{ atm})(8.43\text{ L})}{(0.08206\text{ L}\cdot\text{atm/mol}\cdot\text{K})(308.3)} = 0.339\text{ moles }H_2\text{ formed.}$$

$$0.339\text{ moles }H_2 \times \frac{1\text{ mol Ca}}{1\text{ mol }H_2} \times \frac{40.08\text{ g Ca}}{1\text{ mol Ca}} = 13.6\text{ moles Ca reacted.}$$

12. The strategy here is to determine how many moles of $C_2H_2$ will form, and how much water will be leftover. The volume of water leftover must be subtracted from the volume of the container to find the pressure of gas.

$$25.6\text{ g }CaC_2 \times \frac{1\text{ mol }CaC_2}{64.10\text{ g }CaC_2} \times \frac{1\text{ mol }C_2H_2}{1\text{ mol }CaC_2} = 0.399\text{ mol }C_2H_2$$

$$25.6\text{ g }CaC_2 \times \frac{1\text{ mol }CaC_2}{64.10\text{ g }CaC_2} \times \frac{2\text{ mol }H_2O}{1\text{ mol }CaC_2} \times \frac{18.02\text{ g }H_2O}{1\text{ mol }H_2O} = 14.4\text{ g }H_2O\text{ reacted}$$

Volume of water left = $50.0\text{ mL} - 14.4\text{ g} \times \dfrac{1\text{ mL}}{1.00\text{ g}} = 35.6\text{ mL }H_2O$

Volume of space in container occupied by $C_2H_2$: $1000\text{ mL} - 35.6\text{ mL} = 964.4\text{ mL}$

$$PV = nRT \text{ or } P = \frac{nRT}{V} = \frac{(0.399\text{ mol})(0.08206\text{ L}\cdot\text{atm/mol}\cdot\text{K})(298\text{ K})}{0.9644\text{ L}} = 10.1\text{ atm}$$

**13. If we assume complete combustion, the products are carbon dioxide and water:**

$$C_3H_8(g) + 5\,O_2(g) \rightarrow 3\,CO_2(g) + 4\,H_2O(\ell)$$

**From the moles of propane reacting, we can find the moles of $CO_2$ formed. Since we are assuming STP conditions, standard molar volume allows us to find the volume of $CO_2$.**

$$50.5\text{ g } C_3H_8 \times \frac{1\text{ mol } C_3H_8}{44.096\text{ g } C_3H_8} \times \frac{3\text{ mol } CO_2}{1\text{ mol } C_3H_8} \times \frac{22.4\text{ L @ STP}}{1\text{ mol } CO_2} = 77.0\text{ L } CO_2\text{ @ STP}$$

**14. b**

**15. The average kinetic energy of a gas is directly related to the absolute temperature: $T \propto KE_{ave} = ½\, mv^2$. As the temperature increases, the velocity, v, of the molecules increases. If the volume of the gas is constant, this means molecules traveling faster will collide with the walls of the container more frequently. These collisions result in the pressure in the container, as a small force is imparted with each collision. Thus, a higher temperature means a higher rate of collisions, producing a higher pressure.**

**16. FALSE**

**17. FALSE**

**18. d**

**19.** $T = 29.4°C + 273.15 = 302.6\text{ K}$ $\quad R = 0.08206\text{ L·atm/mol·K}$

$$\text{moles } N_2 = 98.6\text{ kg} \times \frac{1000\text{ g}}{1\text{ kg}} \times \frac{1\text{ mol } N_2}{28.013\text{ g } N_2} = 3520\text{ moles } N_2 \qquad V = 500.0\text{ L}$$

$$PV = nRT \text{ or } P = \frac{nRT}{V} = \frac{(3520\text{ mol})(0.08206\text{ L}\cdot\text{atm/mol}\cdot\text{K})(302.6\text{ K})}{(500.0\text{ L})} = 175\text{ atm}$$

**The van der Waals equation is:**

$$(P + \frac{an^2}{V^2})(V - nb) = nRT. \qquad \text{Rearranging to solve for } P,\ P = \frac{nRT}{(V - nb)} - \frac{an^2}{V^2}$$

**From Table 5.2, the van der Waals constants for nitrogen are:**

$$a = 1.390\ \frac{L^2\text{ atm}}{\text{mol}^2} \qquad b = 0.03913\ \frac{L}{\text{mol}}$$

$$P = \frac{(3520\ \text{mol})(0.08206\ \text{L}\cdot\text{atm/mol}\cdot\text{K})(302.6\ \text{K})}{(500.0\ \text{L} - (3520\ \text{mol})(0.03913\ \text{L/mol}))} - \frac{(1.390\,\frac{\text{L}^2\ \text{atm}}{\text{mol}^2})(3520\ \text{mol})^2}{(500.0\ \text{L})^2}$$

$P = 241.4\ \text{atm} - 68.9\ \text{atm} = 172.5\ \text{atm}$

$P_{ideal} = 175\ \text{atm}$, $P_{\text{van der Waals}} = 172.5\ \text{atm}$

**These are very close; sometimes there isn't much difference between the result we get from the ideal gas law and the van der Waals equation.**

**20. <u>C</u> Ionization gauge**

**<u>B</u> Thermocouple gauge**

**<u>A</u> Capacitance manometer**

# CHAPTER 6

## *Study Goals*

The study goals outline specific concepts to be mastered in this section of the text chapter. Related problems at the end of the text chapter will also be noted. Working the questions noted should aid in mastery of each study goal and also highlight any areas that you may need additional help in.

**Section 6.1 *INSIGHT INTO* Trace Analysis**

1. Describe trace analysis and explain its role in material testing. *Work Problems 1–6.*

**Section 6.2 The Electromagnetic Spectrum**

2. Be familiar with the four central characteristics of waves: wavelength, frequency, amplitude, and velocity. *Work Problems 7–8.*
3. Use the relationship $\lambda \times \nu = c$ to relate wavelength and frequency for various forms of electromagnetic radiation. *Work Problems 9–11.*
4. Discuss how Einstein's photoelectric effect experiments led to the idea of quantized energy. *Work Problems 12–14, 22, 80, 82, 83.*
5. Calculate the energy of a photon for various forms of electromagnetic radiation. *Work Problems 15–21, 23, 24, 84, 87, 89, 95.*

**Section 6.3 Atomic Spectra**

6. Define atomic, emission, and absorption spectra and discuss how they support the idea of quantized energy of atoms. *Work Problems 25, 91–93.*
7. Use simple energy diagrams to calculate energy changes when photons are absorbed or emitted by atoms. *Work Problems 30 –33, 81.*
8. Identify how the Bohr model of an atom played a role in the development of the understanding of atomic structure. *Work Problems 26–29, 94.*

**Section 6.4 The Quantum Mechanical model of the Atom**

9. Understand why electrons can be viewed as waves and how their properties are described by Shrödinger wave equations. *Work Problems 34–36.*
10. Identify each of the four quantum numbers and what they describe about an electron. *Work Problems 37–42, 46, 47.*
11. Describe the shapes of the simpler orbitals. *Work Problems 43–45.*

**Section 6.5 The Pauli Exclusion Principle and Electron Configurations**

12. State the Aufbau principle, Pauli exclusion principle, and Hund's rule. *Work Problems 48 –53.*
13. Use quantum numbers to give the electron configuration of atoms and ions. *Work Problems 54–56.*

**Section 6.6 The Periodic Table and Electron Configurations**

14. Relate the structure of the periodic table to the periodic nature of filling orbitals with electrons. *Work Problems 57–62, 64.*

**Section 6.7 Periodic Trends in Atomic Properties**

15. Relate periodic trends in atomic radius, ionization energy, and electron affinity to periodicity in electron arrangement. *Work Problems 65–76, 85, 86, 88, 90.*

**Section 6.8 *INSIGHT INTO* Modern Light Sources: LEDs and Lasers**

16. Describe the basic principles of how lasers and LEDs operate. *Work Problems 77–79.*

## *Solutions to the Odd-Numbered Problems*

Problem 6.1 Trace analysis may be carried out via non-destructive testing. How does this differ from other types of trace analysis? Give an example of a situation in which non-destructive analysis might be important.

**Non-destructive testing does not destroy or alter the sample or material tested. If a sample must be tested later or with a different analysis or saved for evidence, nondestructive testing would be very useful.**

Problem 6.3 In analysis by atomic absorption spectroscopy, the wavelengths of light being absorbed reveal the identity of the elements present in the sample. What information is used to determine the relative amounts of the different elements?

**The *amount* of light absorbed provides information on how much of each element is present.**

Problem 6.5 The fluorescence emitted in XRF is also in the X-ray range of the spectrum, but is always at lower energies than the X-rays used to initiate the process. Explain why this must be true.

**The atoms in the sample must first be excited to higher energy states using X-rays. The excited atoms must release that excess energy but the resulting fluorescence is lower in energy than the initial X-rays because other forms of energy are released, such as heat.**

Problem 6.7 Explain why light is referred to as electromagnetic radiation.

**Electromagnetic radiation is a form of energy propagated by perpendicular electric and magnetic fields. It is also sometimes called wave energy. Visible light is just one portion of the electromagnetic radiation spectrum, with wavelengths ranging from about 400 – 700 nm.**

Problem 6.9 Arrange the following regions of the electromagnetic spectrum in order of increasing frequency: IR, UV, radio wave, and visible.

**IR is infrared radiation and UV is ultraviolet. Figure 6.6 in the text allows us to place the radiation in the correct order of increasing frequency: radio wave < IR < visible < UV.**

Problem 6.11 Decorative lights, such as those found on a Christmas tree, often achieve their colors by painting the glass of the bulb. If a string of lights includes bulbs with wavelengths of 480, 530, 580, and 700 nm, what are the frequencies of the lights? Use Figure 6.6 to determine which colors are in the set.

**Frequency and wavelength are inversely related, mathematically, $\lambda \times \nu = c$, the speed of light ($c = 3.0 \times 10^8$ m/s).**
**Consulting Figure 6.6 will allow us to estimate the color of the light.**

**480 nm:** $\nu = c \div \lambda$, $\nu = 3.0 \times 10^8 \text{ m/s} \div 480 \text{ nm} \times \dfrac{1\times10^{-9}\text{ m}}{1\text{nm}} = 6.3 \times 10^{14}\text{ s}^{-1}$ **(Hz), green**

**530 nm:** $\nu = 3.0 \times 10^8 \text{ m/s} \div 530 \text{ nm} \times \dfrac{1\times10^{-9}\text{ m}}{1\text{nm}} = 5.7 \times 10^{14}\text{ s}^{-1}$ **(Hz), yellow**

**580 nm:** $\nu = 3.0 \times 10^8 \text{ m/s} \div 580 \text{ nm} \times \dfrac{1\times10^{-9}\text{ m}}{1\text{nm}} = 5.2 \times 10^{14}\text{ s}^{-1}$ **(Hz), orange**

**700 nm:** $\nu = 3.0 \times 10^8 \text{ m/s} \div 700 \text{ nm} \times \dfrac{1\times10^{-9}\text{ m}}{1\text{nm}} = 4.3 \times 10^{14}\text{ s}^{-1}$ **(Hz), red**

Problem 6.13 Define the term *photon*.

**A *photon* refers to a small, discrete amount of light energy. It is sometimes referred to as a small packet of electromagnetic radiation that carries energy. Although light is not a particle, it delivers energy as though it were.**

Problem 6.15 Find the energy of a photon with each of the following frequencies: (a) 15.3 THz, (b) 1.7 EHz (see Table 1.2 if needed), (c) $6.22 \times 10^{10}$ Hz.

**Table 1.2 will help us with the metric prefixes for some of these frequencies.**
**Energy of a photon is related to frequency by the equation, $E_{photon} = h\nu$, where $h$ = Planck's constant ($h = 6.626 \times 10^{-34}$ J•s)**

**(a)** $E_{photon} = h\nu \quad = (6.626 \times 10^{-34}\text{ J·s}) \times (15.3\text{ THz} \times \dfrac{1\times10^{12}\text{ Hz}}{1\text{ THz}} \times \dfrac{1\text{ s}^{-1}}{1\text{Hz}}) = 1.01 \times 10^{-20}\text{ J}$

**(b)** $E_{photon} = h\nu \quad = (6.626 \times 10^{-34}\ \text{J·s}) \times (1.7\ \text{EHz} \times \dfrac{1 \times 10^{18}\ \text{Hz}}{1\ \text{EHz}} \times \dfrac{1\ \text{s}^{-1}}{1\text{Hz}}) = 1.1 \times 10^{-15}\ \text{J}$

**(c)** $E_{photon} = h\nu \quad = (6.626 \times 10^{-34}\ \text{J·s}) \times (6.22 \times 10^{10}\ \text{Hz}) \times (\dfrac{1\ \text{s}^{-1}}{1\text{Hz}}) = 4.12 \times 10^{-23}\ \text{J}$

Problem 6.17 For photons with the following energies, calculate the wavelength and identify the region of the spectrum they are from. (a) $3.5 \times 10^{-20}$ J, (b) $8.7 \times 10^{-26}$ J, (c) $7.1 \times 10^{-17}$ J, (d) 5.5 $\times 10^{-27}$ J

**Energy of a photon is related to the wavelength by the following equation:** $E_{photon} = \dfrac{hc}{\lambda}$, **where h and c are Planck's constant and the speed of light respectively.**

**(a)** $\lambda = \dfrac{hc}{E_{photon}} \quad = (6.626 \times 10^{-34}\ \text{J·s})(3.0 \times 10^{8}\ \text{m/s}) \div 3.5 \times 10^{-20}\ \text{J} = 5.7 \times 10^{-6}\ \text{m}$, **IR**

**(b)** $\lambda = \dfrac{hc}{E_{photon}} \quad = (6.626 \times 10^{-34}\ \text{J·s})(3.0 \times 10^{8}\ \text{m/s}) \div 8.7 \times 10^{-26}\ \text{J} = 2.3\ \text{m}$, **radio wave**

**(c)** $\lambda = \dfrac{hc}{E_{photon}} \quad = (6.626 \times 10^{-34}\ \text{J·s})(3.0 \times 10^{8}\ \text{m/s}) \div 7.1 \times 10^{-17}\ \text{J} = 2.8 \times 10^{-9}\ \text{m}$, **X-ray**

**(d)** $\lambda = \dfrac{hc}{E_{photon}} \quad = (6.626 \times 10^{-34}\ \text{J·s})(3.0 \times 10^{8}\ \text{m/s}) \div 5.5 \times 10^{-27}\ \text{J} = 36\ \text{m}$, **radio wave**

Problem 6.19 ■ Various optical disk drives rely on lasers operating at different wavelengths, with shorter wavelengths allowing a higher density of data storage. For each of the following drive types, find the energy of a single photon at the specified wavelength. (a) CD, $\lambda = 780$ nm, (b) DVD, $\lambda = 650$ nm, (c) Blu-ray® disc, $\lambda = 405$ nm

**Energy of a photon is related to the wavelength by the following equation:** $E_{photon} = \dfrac{hc}{\lambda}$, **where *h* and *c* are Planck's constant and the speed of light respectively. Wavelength must be expressed in meters.**

**(a) CD,** $\lambda = 780$ **nm** $\quad E_{photon} = \dfrac{(6.626 \times 10^{-34}\ \text{J}\cdot\text{s})(3.0 \times 10^{8})}{780\ \text{nm} \times \dfrac{1 \times 10^{-9}\ \text{m}}{1\ \text{nm}}} = 2.6 \times 10^{-19}$ **J per photon**

**(b) DVD,** $\lambda = 650$ **nm** $\quad E_{photon} = \dfrac{(6.626 \times 10^{-34}\ \text{J}\cdot\text{s})(3.0 \times 10^{8})}{650\ \text{nm} \times \dfrac{1 \times 10^{-9}\ \text{m}}{1\ \text{nm}}} = 3.1 \times 10^{-19}\ \dfrac{\text{J}}{\text{photon}}$

**(c) Blu-ray® disc, $\lambda$ = 405 nm** $E_{photon} = \dfrac{(6.626\times10^{-34}\ \text{J}\cdot\text{s})(3.0\times10^{8})}{405\ \text{nm}\times\dfrac{1\times10^{-9}\ \text{m}}{1\ \text{nm}}} = 4.9\times10^{-19}\ \dfrac{\text{J}}{\text{photon}}$

Problem 6.21 Assume that a microwave oven operates at a frequency of $1.00 \times 10^{11}\ \text{s}^{-1}$. (a) What is the wavelength of this radiation in meters? (b) What is the energy in joules per photon? (c) What is the energy per mole of photons?

**Given the frequency of electromagnetic radiation, determine the wavelength in meters, the energy of one photon, and the energy of one mole of photons. Use equations described in Sections 6.1 and 6.2.** $\nu = 1.00 \times 10^{11}\ \text{s}^{-1}$

**(a)** $\lambda = \dfrac{c}{\nu} = \dfrac{2.998\times10^{8}\ \text{m/s}}{1.00\times10^{11}\ \text{s}^{-1}} = 3.00\times10^{-3}\ \text{m}$

**(b)** $E = h\nu = (6.626\times10^{-34}\ \text{J·s})\times(1.00\times10^{11}\ \text{s}^{-1}) = 6.63\times10^{-23}$ **J for one photon**

**(c)** $\dfrac{6.63\times10^{-23}\ \text{J}}{1\ \text{photon}}\times\dfrac{6.022\times10^{23}\ \text{photons}}{1\text{mol photons}} = 39.9\ \text{J/mol}$

Problem 6.23 ■ When ultraviolet light with a wavelength of 58.5 nm strikes the surface of potassium metal, electrons are ejected with a maximum kinetic energy of $2.69 \times 10^{-18}$ J. What is the binding energy of the electrons to the metal?

**The binding energy is related to the energy of a photon:**

$E_{photon}$ = **Binding** $E$ + **Kinetic** $E$, **or** **Binding** $E$ = $E_{photon}$ − **Kinetic** $E$

**We'll calculate the energy of the photon with the wavelength, and then find the binding energy using the kinetic energy given.**

$$E_{photon} = \frac{hc}{\lambda}$$

$$E_{photon} = \frac{(6.626\times10^{-34}\ \text{J}\cdot\text{s})(3.0\times10^{8}\ \text{m/s})}{58.5\ \text{nm}\times\dfrac{1\times10^{-9}\ \text{m}}{1\ \text{nm}}} = 3.40\times10^{-18}\ \text{J per photon}$$

**Binding** $E = (3.40\times10^{-18}\ \text{J}) - (2.69\times10^{-18}\ \text{J}) = 7.1\times10^{-19}\ \text{J}$

Problem 6.25 What is the difference between continuous and discrete spectra?

**A continuous spectrum has no missing frequencies of light, similar to a rainbow produced by sunlight shining through water vapor. A discrete spectrum has only separate, specific frequencies that appear as thin colored lines, like the emission spectra of elements.**

Problem 6.27 Describe how the Bohr model of the atom accounts for the spectrum of the hydrogen atom.

**The Bohr model depicted electrons circling the nucleus in orbits that could only be at certain discrete distances from the nucleus. Because energy is related to the distance between charged particles, these orbits corresponded to energy levels. For the excited electron in the Bohr atom to return to the original orbit, the atom releases a photon of energy equal to the energy difference between the two orbits. The spectrum of the hydrogen atoms therefore corresponds to the specific energy released when electrons drop from a higher energy orbit to a lower energy one.**

Problem 6.29 Define the term *ground state*.

***Ground state*** **refers to the lowest possible energy state for an electron. An atom is said to be in its ground state if all its electrons occupy their lowest energy levels.**

Problem 6.31 Refer to the energy level diagram shown above in Problem 6.30, and find the wavelength of light that would be emitted when a hydrogen atom undergoes the transition from the state labeled as $n = 4$ to the state labeled as $n = 2$. Express your answer in nm.

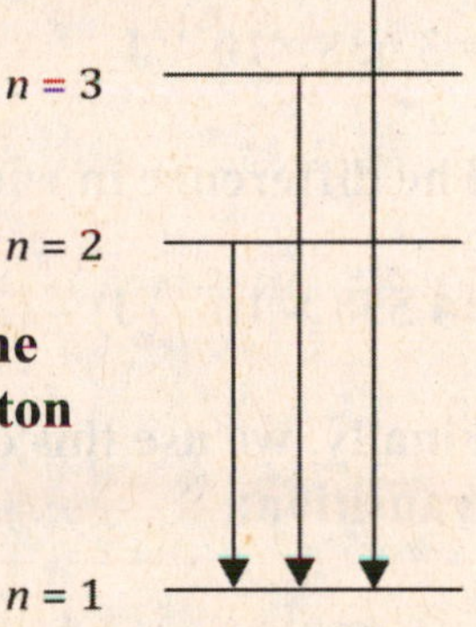


**The wavelengths given for the three transitions shown are 121.566 nm, 102.583 nm and 97.524 nm, respectively. With the given wavelengths, the energy change $(\Delta E = \frac{hc}{\lambda})$ for transitions $n = 4$ to $n = 1$ and $n = 2$ to $n = 1$ can be calculated. The difference between these is the energy change (energy of the photon released) for the transition $n = 4$ to $n = 2$. With this energy, the wavelength of the corresponding photon can be calculated.**

$$\Delta E_{4\rightarrow 1} = \frac{(6.626\times 10^{-34}\ \mathrm{J\cdot s})(2.998\times 10^{8}\ \mathrm{m/s})}{97.524\ \mathrm{nm}\times \frac{1\times 10^{-9}\ \mathrm{m}}{1\ \mathrm{nm}}} = 2.036\times 10^{-18}\ \mathrm{J}$$

$$\Delta E_{2\rightarrow 1} = \frac{(6.626\times 10^{-34}\ \mathrm{J\cdot s})(2.998\times 10^{8}\ \mathrm{m/s})}{121.566\ \mathrm{nm}\times \frac{1\times 10^{-9}\ \mathrm{m}}{1\ \mathrm{nm}}} = 1.634\times 10^{-18}\ \mathrm{J}$$

$$\Delta E_{4\rightarrow 2} = 2.036\times 10^{-18}\ \mathrm{J} - 1.634\times 10^{-18}\ \mathrm{J} = 4.020\times 10^{-19}\ \mathrm{J}$$

$$\lambda = \frac{hc}{\Delta E_{4\rightarrow 2}} = (6.626\times 10^{-34}\ \mathrm{J\cdot s})(2.998\times 10^{8}\ \mathrm{m/s}) \div 4.020\times 10^{-19}\ \mathrm{J} = 4.940\times 10^{-7}\ \mathrm{m}$$

$$4.940\times 10^{-7}\ \mathrm{m}\times \frac{1\mathrm{nm}}{1\times 10^{-9}\ \mathrm{m}} = 494.0\ \mathrm{nm}$$

Problem 6.33 ■ A mercury atom emits light at many wavelengths, two of which are at 435.8 nm and 546.1 nm. Both of these transitions are to the same final state. (a) What is the energy difference between the two states for each transition? (b) If a transition between the two higher energy states could be observed, what would be the frequency of the light?

**The same equation that we use to calculate the energy of a photon allows us to relate energy transitions of an electron to wavelength of light emitted. This is because the energy of the photon emitted is equal to the energy of the transition.**

$$\Delta E = E_{\text{photon}} = \frac{hc}{\lambda}$$

**For the first transition (435.8 nm),** $\Delta E = \frac{hc}{\lambda} = \frac{(6.626\times10^{-34}\ \text{J}\cdot\text{s})(2.998\times10^{8})}{435.8\ \text{nm}\times\frac{1\times10^{-9}\ \text{m}}{1\ \text{nm}}} =$

$= 4.558 \times 10^{-19}$ **J**

**For the second transition (546.1 nm),** $\Delta E = \frac{hc}{\lambda} = \frac{(6.626\times10^{-34}\ \text{J}\cdot\text{s})(2.998\times10^{8})}{546.1\ \text{nm}\times\frac{1\times10^{-9}\ \text{m}}{1\ \text{nm}}} =$

$= 3.638 \times 10^{-19}$**J**

**The difference in energy between these transitions is:**

$(4.558 \times 10^{-19}\ \text{J}) - (3.638 \times 10^{-19}\ \text{J}) = 9.20 \times 10^{-20}\ \text{J}$

**Finally, we use this energy to find the frequency of the photon corresponding to this energy transition:**

$$\nu = \frac{E_{\text{photon}}}{h} = (9.20 \times 10^{-20}\ \text{J}) \div (6.626 \times 10^{-34}\ \text{J}\bullet\text{s}) = 1.39 \times 10^{14}\ \text{s}^{-1}\ \text{(Hz)}$$

Problem 6.35 Why do we use a wave function to describe electrons?

**The wave function is used to describe electrons around the nucleus of the atom because electrons can be diffracted. Diffraction is usually associated with electromagnetic radiation (light), but electrons also produce diffraction patterns. This means that electrons also exhibit particle-wave duality. Very small, fast-moving particles can be described as waves distributed in space, instead of localized particles. In some instances, electrons are treated as particles; in other instances, they are treated as waves.**

Problem 6.37 What are the allowed values for the principle quantum number? For the secondary quantum number?

**For the principle quantum number the allowed values are integers beginning with 1 and increasing to infinity (theoretically), $n$ = 1, 2, 3, 4, 5, 6, 7.....**

**For the secondary quantum number, the allowed values are integers beginning with 0, with a maximum value restricted by the principle quantum number, $n$.**
**$\ell$ = 0, 1, 2, 3, 4,... maximum value ($n$ – 1)**

Problem 6.39 ■ A particular orbital has $n = 4$ and $\ell = 2$. What must this orbital be? (a) 3*p*, (b) 4*p*, (c) 5*d*, or (d) 4*d*

**The secondary quantum number, $\ell$, represents the shape of the subshell and $\ell = 2$ is a *d* subshell. The energy level is 4 ($n = 4$) so it must be a (d) 4*d* orbital.**

Problem 6.41 ■ What is the maximum number of electrons in an atom that can have the following quantum numbers? (a) $n = 2$; (b) $n = 3$ and $\ell = 1$; (c) $n = 3$, $\ell = 1$, and $m_\ell = 0$; (d) $n = 3$, $\ell = 1$, $m_\ell = -1$, and $m_s = -\frac{1}{2}$.

**Electron capacity in energy levels = $2n^2$.**

**(a) $2(2)^2 = 8$ electrons can have $n = 2$, 8 electrons**
**(b) $2(3)^2 = 18$ electrons can have $n = 3$ but only 6 of these can have $\ell = 1$, (a *p* sub-shell), 6 electrons**
**(c) Only 2 of the electrons in the 3*p* sub-shell can be in the $m_\ell = 0$ orbital, 2 electrons**
**(d) Only one electron can have the spin, $m_s = -\frac{1}{2}$, in that particular orbital of the 3*p* sub-shell, 1 electron**

Problem 6.43 The *p* and *d* orbitals are sometimes referred to as directional. Contrast them with *s* orbitals to offer an explanation for this terminology.

**Directional means that those shapes have an axis of orientation; they can be pointed in some particular direction. The *s* orbitals are spherical; they have no directionality because a sphere points in all directions at once. The *p* orbitals have an axis of orientation; they are aligned with the *x*, *y*, or *z*-axis. The *d* orbitals are also arranged in specific directions, although not all of them are aligned along the axes.**

Problem 6.45 Referring to Figure 6.15, draw a 4*p* orbital, showing all of its nodes.

The 4*p* orbitals are complicated. Each orbital has six lobes. There are two spherical nodes, one planar node, and two radial lobes.

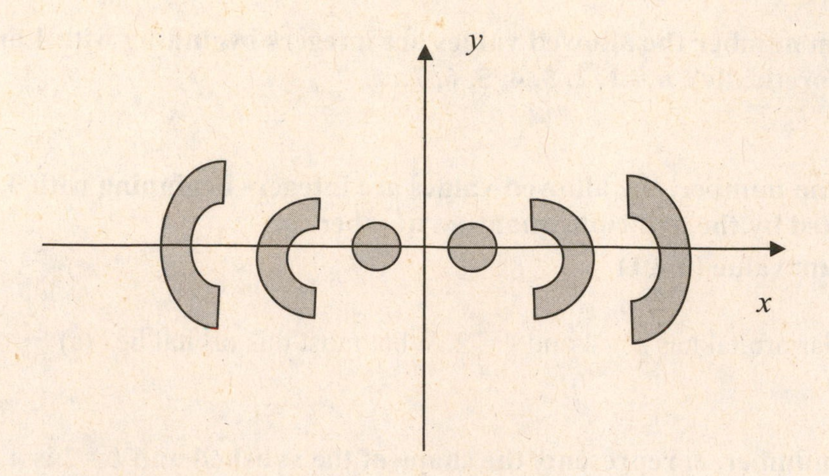


Problem 6.47 Define the term *spin-paired.*

**Spin-paired refers to two electrons occupying the same orbital with opposite spins. The spin of an electron can be represented by $m_s$ = + ½ or – ½ or with arrows: ⇅.**

Problem 6.49 Define the term *shielding*.

***Shielding* or *screening* describes the phenomenon where the inner or core electrons block the outer electrons from feeling the full effect of the nuclear charge. An analogy might be a basketball player setting a pick and blocking a defender so a teammate can have an open shot at the basket. The electrons in energy level one shield those in level two, those in level one and two shield electrons in level three, etc. The result is that electrons in outer orbitals are higher in energy and are easier to remove.**

Problem 6.51 What is effective nuclear charge and how does it relate to the size of an orbital?

**The *effective nuclear charge* is the apparent charge that the valence electrons experience as a result of shielding by the inner core electrons. The number of protons in the nucleus determines the *actual* nuclear charge. The effective nuclear charge is decreased because of the shielding effect. As effective nuclear charge increases, the outer electrons are drawn closer to the nucleus and the size of the atom decreases. The trend in this behavior is from increasing effective nuclear charge from left to right on the periodic table.**

Problem 6.53 How does the charge of electrons provide some rationale for Hund's rule?

**Electrons have negative charge, meaning they will repel each other when they are close to each other. Hund's rule says that electrons will singly occupy degenerate or equal energy orbitals before they begin pairing. Their electrostatic repulsions cause electrons to occupy separate orbitals, if possible. The only way two electrons will occupy the same space (same orbitals) is if they have opposite spins.**

Problem 6.55 ■ Which of these electron configurations are for atoms in the ground state? In excited states? Which are impossible?

(a) $1s^22s^1$, (b) $1s^22s^22p^3$, (c) [Ne] $3s^23p^34s^1$, (d) [Ne] $3s^23p^64s^33d^2$, (e) [Ne] $3s^23p^64f^4$, (f) $1s^22s^22p^43s^2$

**(a) $1s^22s^1$ is a ground state electron configuration. The 1*s* sublevel has the lowest energy and it is full. The 2*s* sublevel has the second lowest energy and the remaining electron is there. This is the lowest energy electron configuration, so it would be called the ground state.**

**(b) $1s^22s^22p^3$ could be ground state or excited state. If each of the *p* electrons is in a separate orbital with the same spin, that is the ground state. However, if any of the spins are reversed or if any of the electrons are paired, this is a different and higher energy state, hence an excited state. We would need to see an orbital energy diagram of this state to determine an unambiguous answer:**

**Ground state 2*p* orbital energy diagram:** 2p [↑][↑][↑]

**Some excited state 2*p* orbital energy diagrams:** 2p [↓][↑][↑]   2p [↑↓][↑][ ]

**(c) [Ne] $3s^23p^34s^1$ is an excited state. The lower-energy 3*p* sublevel needs to be full (i.e., $3p^6$) before the 4*s* sublevel gets any electrons; therefore, this possible electron configuration is not the lowest energy state, and would be called an excited state. The ground state electron configuration would be [Ne] $3s^23p^4$.**

**(d) [Ne] $3s^23p^64s^33d^2$ is impossible. The 4*s* sublevel has only one orbital, so the maximum number of electrons is 2. This electron configuration has three electrons in that orbital.**

**(e) [Ne]$3s^23p^64f^4$ is an excited state. Several lower-energy sublevels need to be full before the 4*f* sublevel gets any electrons; therefore, this possible electron configuration is not the lowest energy state, and would be called an excited state. The ground state electron configuration would be [Ne]$3s^23p^63d^24s^2$.**

**(f) $1s^22s^22p^43s^2$ is an excited state. The lower-energy 2*p* sublevel needs to be full ($2p^6$) before the 3*s* sublevel gets any electrons; therefore, this possible electron configuration is not the lowest energy state, and would be called an excited state. The ground state electron configuration would be $1s^22s^22p^6$.**

Problem 6.57 Halogen lamps rely on the predictable chemistry of the halogen elements. Compare the electron configuration of three halogens and use this comparison to justify the similarity of the chemistry of these elements.

**Let's look at F, Cl, and Br:**

**F: $1s^2 2s^2 2p^5$ or [He] $2s^2 2p^5$**

**Cl: $1s^2 2s^2$, $2p^6 3s^2 3p^5$ or [Ne] $3s^2 3p^5$**

**Br: $1s^2 2s^2 2p^6 3s^2 3p^6 4s^2 3d^{10} 4p^5$ or [Ar] $4s^2 3d^{10} 4p^5$**

**All three have seven valence electrons (electrons in filled *d* sets of orbitals are not considered valence electrons), with the last five occupying a *p* set of orbitals. We could therefore predict that they all would form an anion with a –1 charge to fill the *p* set: $F^-$, $Cl^-$, $Br^-$. We could also predict the formulas of compounds they would form: NaF, NaCl, NaBr. We can also explain why they form a single bond to each other and exist as diatomic elements: $F_2$, $Cl_2$, $Br_2$.**

Problem 6.59 Describe how valence electron configurations account for some of the similarities in chemical properties among elements in a group.

**See Problem 6.57. Within a group, elements have different total numbers of electrons, but have the same number of *valence* electrons (with a few exceptions). The valence electrons are responsible for most of the chemical properties of an atom, so they will behave very similarly. For example in Group 2:**

**Be [He] $2s^2$**
**Mg [Ne] $3s^2$**
**Ca [Ar] $4s^2$**

**All Group 2 elements have two valence electrons in an *s* orbital set, meaning they will all form a +2 cation, and form compounds with similar formulas: BeO, MgO, CaO.**

Problem 6.61 If another 50 elements were discovered, how would you expect the appearance of the periodic table to change?

**Starting with element number 113, the 7*p* orbital set would be filled (6 new elements), completing period 7. Period 8 would begin, filling the 8*s*, then 5*g*, then 6*f*, finally the 7*d* set of orbitals. So there would be an extra period, and an extra series in the *f*-block and a new *g*-block.**

Problem 6.63 Explain why the *s* block of the periodic table contains two columns, the *p* block contains six columns, and the *d* block contains ten columns.

**The *s* orbital can hold up to two electrons, the *p* orbitals up to six electrons, and the *d* orbitals up to ten electrons.**

Problem 6.65 How does the effective nuclear charge help explain the trends in atomic size across a period of the periodic table?

**The effective nuclear charge increases from left to right across a period. This is because as we move from the left-hand side of a period, more protons are being added to the nucleus, and an equal number of electrons are added. But the electrons are placed in the same energy level (across a period), meaning they are the same distance from the nucleus. These electrons therefore experience a stronger nuclear attraction as more protons are added. The result is that the valence electrons are drawn closer to the nucleus as we move from left to right, giving a smaller atom.**

Problem 6.67 ■ Using only a periodic table as a guide, arrange each of the following series of atoms in order of increasing size. (a) Na, Be, Li, (b) P, N, F, (c) I, O, Sn

**Size generally decreases as effective nuclear charge increases from left to right in a given period and generally increases as more shells are added going down a given group.**

**(a) Be < Li < Na**

**(b) F < N < P**

**(c) O < I < Sn**

Problem 6.69 At which ionization for chlorine would you expect the first large jump in ionization energy? Would this be the only large jump in energy if you continued to ionize the chlorine?

**Chlorine has seven valence electrons, so it wants to gain an eighth to complete its valence set of orbitals. Chlorine does not want to lose any electrons; so all the ionization energies would be large. The first large jump in ionization energy would be after the fifth ionization, when all the 3*p* electrons have been removed. It is more difficult to remove electrons from a completely filled orbital set (the 3*s*). The next large jump would be after the seventh ionization when we begin to remove electrons from the completely filled 2*p* subshell.**

Problem 6.71 How do we explain the fact that the ionization energy of oxygen is less than that of nitrogen?

**Oxygen has four electrons in its 2*p* set of orbitals, nitrogen has three (O: $1s^2 2s^2 2p^4$; N: $1s^2 2s^2 2p^3$. The spin-paired electrons for oxygen are slightly higher in energy and therefore it's easier to remove the <u>fourth</u> electron in a *p* orbital set than the <u>third</u>. Recall from Problem 6.53 that electrons would prefer not to be spin-paired. This means oxygen's first ionization energy is lower then nitrogen's, even though that contradicts our periodic trend.**

Problem 6.73 ■ Answer each of the following questions.
(a) Of the elements S, Se, and Cl, which has the largest atomic radius?

(b) Which has the larger radius, Br or $Br^-$?
(c) Which should have the largest difference between the first and second ionization energy: Si, Na, P, or Mg?
(d) Which has the largest ionization energy: N, P, or As?
(e) Which of the following has the largest radius: $O^{2-}$, $N^{3-}$, or $F^-$?

| | | |
|---|---|---|
| **(a)** | **Se** | **Selenium is in a higher period, with electrons in higher energy levels.** |
| **(b)** | **$Br^-$** | **Bromide is larger because there is an extra electron and greater electrostatic repulsions.** |
| **(c)** | **Na** | **Sodium will because it is very favorable to lose one electron but not at all to lose a second.** |
| **(d)** | **N** | **Ionization energies increase to the top of the periodic table as electrons occupy lower energy orbitals** |
| **(e)** | **$N^{3-}$** | **The isoelectronic ion with the fewest protons will be the largest as the effective nuclear attraction on the electrons is lower.** |

Problem 6.75 ■ Compare the elements Na, B, Al, and C with regard to the following properties. (a) Which has the largest atomic radius? (b) Which has the most negative electron affinity? (c) Place the elements in order of increasing ionization energy.

| | | |
|---|---|---|
| **(a)** | **Na** | **Sodium is farther left and lower than the other elements in the list.** |
| **(b)** | **C** | **Of the four elements, only carbon is a nonmetal. Nonmetals usually will gain electrons to form anions to achieve an octet, giving them a more negative electron affinity compared to metals or metalloids.** |
| **(c)** | **Na < Al < B < C** | **Ionization energies increase towards the top and right side of the periodic table.** |

Problem 6.77 ■ Several excited states of the neon atom are important in the operation of the helium-neon laser. In these excited states, one electron of the neon atom is promoted from the 2*p* level to a higher energy orbital.

An excited neon atom with a $1s^22s^22p^55s^1$ electron configuration can emit a photon with a wavelength of 3391 nm as it makes a transition to a lower energy state with a $1s^22s^22p^54p^1$ electron configuration. Other transitions are also possible. If an excited neon atom with a $1s^22s^22p^55s^1$ electron configuration makes a transition to a lower energy state with a $1s^22s^22p^53p^1$ electron configuration, it emits a photon with a wavelength of 632.8 nm. Find the wavelength of the photon that would be emitted in the transition from the $1s^22s^22p^54p^1$ electron configuration to the $1s^22s^22p^53p^1$ electron configuration. (It should help if you start by drawing an energy-level diagram.)

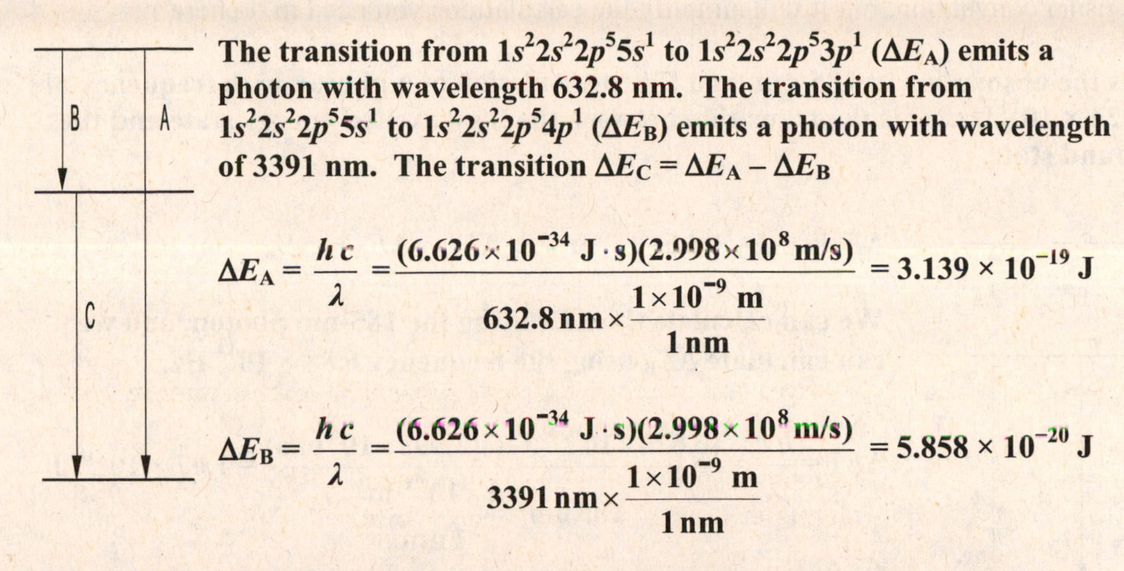


**The transition from $1s^22s^22p^55s^1$ to $1s^22s^22p^53p^1$ ($\Delta E_A$) emits a photon with wavelength 632.8 nm. The transition from $1s^22s^22p^55s^1$ to $1s^22s^22p^54p^1$ ($\Delta E_B$) emits a photon with wavelength of 3391 nm. The transition $\Delta E_C = \Delta E_A - \Delta E_B$**

$$\Delta E_A = \frac{hc}{\lambda} = \frac{(6.626\times10^{-34}\ \mathrm{J\cdot s})(2.998\times10^{8}\ \mathrm{m/s})}{632.8\ \mathrm{nm}\times\dfrac{1\times10^{-9}\ \mathrm{m}}{1\ \mathrm{nm}}} = 3.139\times10^{-19}\ \mathrm{J}$$

$$\Delta E_B = \frac{hc}{\lambda} = \frac{(6.626\times10^{-34}\ \mathrm{J\cdot s})(2.998\times10^{8}\ \mathrm{m/s})}{3391\ \mathrm{nm}\times\dfrac{1\times10^{-9}\ \mathrm{m}}{1\ \mathrm{nm}}} = 5.858\times10^{-20}\ \mathrm{J}$$

$$\Delta E_C = (3.139\times10^{-19}\ \mathrm{J}) - (5.858\times10^{-20}\ \mathrm{J}) = 2.555\times10^{-19}\ \mathrm{J}$$

$$\lambda = \frac{hc}{\Delta E_C} = \frac{(6.626\times10^{-34}\ \mathrm{J\cdot s})(2.998\times10^{8}\ \mathrm{m/s})}{2.555\times10^{-19}\ \mathrm{J}} = 7.780\times10^{-7}\mathrm{m}\ (778.0\ \mathrm{nm})$$

Problem 6.79 How much energy could be saved each year by replacing incandescent bulbs with LED bulbs? Assume that six 40-watt lamps are lit an average of three hours a night and they are each replaced with 8-watt LED bulbs. (See Problem 6.78.)

**The total hours lit for the year would be 3 hours/day × 365 day/year = 1095 hours.**

**A watt is defined as one J/second.**

$$6\ \text{bulbs} \times \frac{40\ \text{watts}}{\text{bulb}} \times \frac{\frac{1\ \mathrm{J}}{\mathrm{s}}}{1\ \text{watt}} \times \left(1095\ \text{hrs} \times \frac{3600\ \mathrm{s}}{1\ \text{hr}}\right) = 9.46\times10^{8}\ \mathrm{J}$$

$$6\ \text{bulbs} \times \frac{8\ \text{watts}}{\text{bulb}} \times \frac{\frac{1\ \mathrm{J}}{\mathrm{s}}}{1\ \text{watt}} \times \left(1095\ \text{hrs} \times \frac{3600\ \mathrm{s}}{1\ \text{hr}}\right) = 1.89\times10^{8}\ \mathrm{J}$$

**The savings would be:** $9.46\times10^{8}\ \mathrm{J} - 1.89\times10^{8}\ \mathrm{J} = 7.57\times10^{8}\ \mathrm{J}$

Problem 6.81 ■ A mercury atom is initially in its lowest possible (or ground state) energy level. The atom absorbs a photon with a wavelength of 185 nm and then emits a photon with a frequency of $4.924\times10^{14}$ Hz. At the end of this series of transitions, the atom will still be in an energy level above the ground state. Draw an energy-level diagram for this process and find the energy of this resulting excited state, assuming that we assign a value of $E = 0$ to the ground state. (This choice of $E = 0$ is *not* the usual convention, but it will simplify the calculations you need to do here.)

**A is the absorption at 185 nm, and B is the emission of a photon with frequency of 4.924 × 10[14] Hz. C is the transition between the final excited energy state and the ground state.**

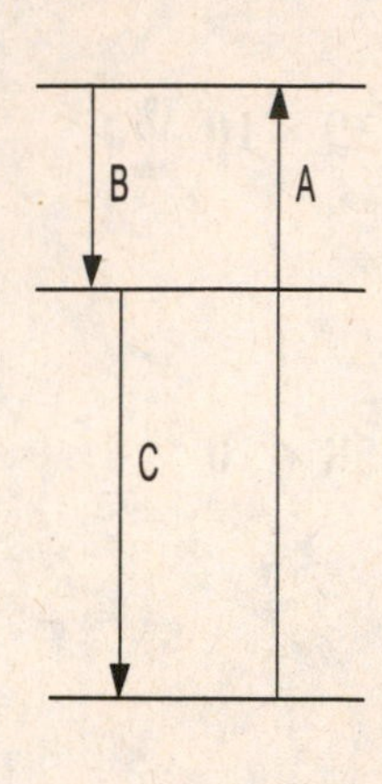


$\Delta E_A = \Delta E_B + \Delta E_C$ **or** $\Delta E_C = \Delta E_A - \Delta E_B$

**We can calculate the $\Delta E_A$ using the 185-nm photon, and we can calculate $\Delta E_B$ using the frequency 6.88 × 10[14] Hz.**

$$\Delta E_A = \frac{hc}{\lambda} = \frac{(6.626 \times 10^{-34}\ \text{J}\cdot\text{s})(2.998 \times 10^{8}\ \text{m/s})}{185\ \text{nm} \times \dfrac{1 \times 10^{-9}\ \text{m}}{1\ \text{nm}}} = 1.07 \times 10^{-18}\ \text{J}$$

$$\Delta E_B = h\nu = (6.626 \times 10^{-34}\ \text{J}\cdot\text{s})(4.924 \times 10^{14}\ \text{Hz} \times \frac{1\ \text{s}^{-1}}{1\text{Hz}}) =$$

$$= 3.26 \times 10^{-19}\ \text{J}$$

$$\Delta E_C = (1.07 \times 10^{-18}\ \text{J}) - (3.26 \times 10^{-19}\ \text{J}) = 7.48 \times 10^{-19}\ \text{J}$$

Problem 6.83 A metallic sample is known to be either barium, cesium, lithium, or silver. The electron binding energies for these metals are listed in the following table.

| Metal | Electron Binding Energy (J) |
|---|---|
| Barium | $4.30 \times 10^{-19}$ |
| Cesium | $3.11 \times 10^{-19}$ |
| Lithium | $3.94 \times 10^{-19}$ |
| Silver | $7.59 \times 10^{-19}$ |

One way to identify the element might be through a photoelectric effect experiment. The experiment was performed three times, each time using a different laser as the light source. The results are summarized below. (the kinetic energy of the ejected electrons was not measured.)

| Laser Wavelength | Photoelectrons Seen? |
|---|---|
| 532 nm | No |
| 488 nm | Yes |
| 308 nm | Yes |

Based on this information, what conclusions can be drawn as to the identity of the metal?

**The energy of the photon must be larger than the binding energy of the atom. Using the wavelengths given, we can calculate the energy of the three photons:**

$$E_{532\ \text{nm}} = \frac{hc}{\lambda} = \frac{(6.626\times10^{-34}\ \text{J}\cdot\text{s})(3.0\times10^{8}\ \text{m/s})}{532\ \text{nm}\times\dfrac{1\times10^{-9}\ \text{m}}{1\ \text{nm}}} = 3.74\times10^{-19}\ \text{J}$$

$$E_{488\ \text{nm}} = \frac{hc}{\lambda} = \frac{(6.626\times10^{-34}\ \text{J}\cdot\text{s})(3.0\times10^{8}\ \text{m/s})}{488\ \text{nm}\times\dfrac{1\times10^{-9}\ \text{m}}{1\ \text{nm}}} = 4.07\times10^{-19}\ \text{J}$$

$$E_{308\ \text{nm}} = \frac{hc}{\lambda} = \frac{(6.626\times10^{-34}\ \text{J}\cdot\text{s})(3.0\times10^{8}\ \text{m/s})}{308\ \text{nm}\times\dfrac{1\times10^{-9}\ \text{m}}{1\ \text{nm}}} = 6.45\times10^{-19}\ \text{J}$$

**The atom must be lithium because its binding energy is larger than the energy of the first photon (no photoelectrons) but its binding energy is smaller than the energy of the second and third photons. The atom cannot be silver, because none of the photons are larger than its binding energy. It cannot be barium because we should not see photoelectrons from the second laser; and it cannot be cesium because we would expect to see photoelectrons from all three lasers.**

Problem 6.85 ■ Arrange the members of each of the following sets of cations in order of increasing ionic radii. (a) $K^+$, $Ca^{2+}$, $Ga^{3+}$, (b) $Ca^{2+}$, $Be^{2+}$, $Ba^{2+}$, $Mg^{2+}$, (c) $Al^{3+}$, $Sr^{2+}$, $Rb^+$, $K^+$, (d) $K^+$, $Ca^{2+}$, $Rb^+$

**(a) Within an isoelectronic series, ionic radii increase with decreasing atomic number. Therefore, in order of increasing ionic radii, we have $Ga^{3+} < Ca^{2+} < K^+$.**

**(b) Ionic radii increase down a group. So, $Be^{2+} < Mg^{2+} < Ca^{2+} < Ba^{2+}$.**

**(c) $Al^{3+} < Sr^{2+} < K^+ < Rb^+$**

**(d) $Ca^{2+} < K^+ < Rb^+$**

Problem 6.87 ■ The photoelectric effect can be used in engineering designs for practical applications. For example, infrared goggles used in night-vision applications have materials that give an electrical signal with exposure to the relatively long wavelength IR light. If the energy needed for signal generation is $3.5\times10^{-20}$ J, what is the minimum wavelength and frequency of light that can be detected?

**Minimum Wavelength:**

$$\lambda = \frac{hc}{E_{IR}} = \frac{(6.626\times10^{-34}\ \text{J}\cdot\text{s})(3.0\times10^{8}\ \text{m/s})}{3.5\times10^{-20}\ \text{J}} = 5.7 \times 10^{-6}\text{m (5700 nm)}$$

**Minimum Frequency:**

$$\nu = \frac{E_{IR}}{h} = (3.5 \times 10^{-20}\ \text{J}) \div (6.626 \times 10^{-34}\ \text{J•s}) = 5.3 \times 10^{13}\ \text{s}^{-1}\ \text{(Hz)}$$

Problem 6.89 ■ Laser welding is a technique in which a tightly focused laser beam is used to deposit enough energy to weld metal parts together. Because the entire process can be automated, it is commonly used in many large-scale industries, including the manufacture of automobiles. In order to achieve the desired weld quality, the steel parts being joined must absorb energy at a rate of about $10^4$ W/mm$^2$. (Recall that 1 W = 1 J/s.) A particular laser welding system employs a Nd:YAG laser operating at a wavelength of 1.06 $\mu$m; at this wavelength steel will absorb about 80% of the incident photons. If the laser beam is focused to illuminate a circular spot with a diameter of 0.02 inch, what is the minimum power (in watts) that the laser must emit per second to reach the $10^4$ W/mm$^2$ threshold? How many photons per second does this correspond to? (For simplicity, assume that the energy from the laser does not penetrate into the metal to any significant depth.)

**Energy that must be absorbed = 10000 $\frac{\text{W}}{\text{mm}^2}$ but the steel will only absorb 80% of this energy. Therefore the total energy input is $10000\frac{\text{W}}{\text{mm}^2} \div 0.80 = 12500\frac{\text{W}}{\text{mm}^2}$.**

**Area of the circular spot = $\pi \times (d/2)^2$**

$$d = 0.02\ \text{in} \times \frac{25.4\ \text{mm}}{1\ \text{in}} = 0.508\ \text{mm} \qquad \text{area} = \pi \times (0.508/2)^2 = 0.203\ \text{mm}^2$$

$$\text{Total power for the circular spot} = 12500\ \frac{\text{W}}{\text{mm}^2} \times 0.203\ \text{mm}^2 = 2540\ \text{W} = 2540\ \text{J/s} \approx$$

**$\approx 3 \times 10^3$ J/s (one significant figure)**

**Energy of these photons:**

$$E_{1.06\mu\text{m}} = \frac{hc}{\lambda} = \frac{(6.626\times10^{-34}\ \text{J}\cdot\text{s})(3.0\times10^{8}\,\text{m/s})}{1.06\ \mu\text{m}\times\frac{1\times10^{-6}\ \text{m}}{1\ \mu\text{m}}} = 1.87 \times 10^{-19}\ \text{J/photon}$$

$$\frac{2540\ \text{J}}{\text{s}} \times \frac{\text{photon}}{1.87\times10^{-19}\ \text{J}} = 1.35 \times 10^{22}\ \frac{\text{photons}}{\text{s}} \approx 1 \times 10^{22}\ \frac{\text{photons}}{\text{s}}\ \text{(one significant figure)}$$

Problem 6.91 Atomic absorption spectroscopy is based on the spectra of the elements being studied. It can be used to determine the impurities in a metal sample. If an element is present, light at an appropriate wavelength is absorbed. You are working with a metal stamping company and the rolled steel you use to from panels for automobile doors is failing at an alarming rate. There is some chance that the problem is unacceptably high levels of manganese in the steel. Given that the atomic spectrum of manganese has three lines near 403 nm, how could you use a spectrometer to determine the amounts of manganese in the steel?

**The spectrometer would have to be calibrated by measuring the absorption of light near 403 nm in several samples of known concentration of Mn. The atoms in the steel would have to be excited somehow, most likely by heating. Then the steel in question could be analyzed, and based on how much light near 403 nm was being absorbed, the level of Mn determined.**

Problem 6.93 When we say that the existence of atomic spectra tells us that atoms have specific energy levels, are we using deductive or inductive reasoning?

**We are using inductive reasoning. If energy levels are not quantized, and any energy level can be occupied by electrons of an atom, the atomic spectra would be continuous not discrete (see Problem 6.25). We observe that atomic spectra are discrete and therefore we conclude that atoms have specific energy levels.**

Problem 6.95 The photochemical reaction that initiate the production of smog involves the decomposition of N–O molecules, and the energy needed to break the N–O bond is $1.04 \times 10^{-18}$ J. (a) Which wavelength of light is needed? (b) How many photons are needed to decompose 0.32 mg of NO?

**(a) From the bond energy we can calculate the wavelength of photon that is necessary.**

$$\lambda = \frac{hc}{E} = (6.626 \times 10^{-34}\ \text{J·s})(3.0 \times 10^{8}\ \text{m/s}) \div 1.04 \times 10^{-18}\ \text{J} = 1.91 \times 10^{-7}\ \text{m or 191 nm}$$

**(UV range)**

**(b) We need one photon per molecule, so we must calculate how many molecules are present.**

$$0.32\ \text{mg NO} \times \frac{1\ \text{g}}{1000\ \text{mg}} \times \frac{1\ \text{mol NO}}{30.01\ \text{g NO}} \times \frac{6.022 \times 10^{23}\ \text{molecules}}{1\ \text{mol}} \times \frac{1\ \text{photon}}{1\ \text{molecule}} =$$

$= 6.4 \times 10^{18}$ **photons**

# CHAPTER OBJECTIVE QUIZ

This quiz will test your understanding of the basic text chapter objectives and give you additional practice problems. You should work this quiz after completing the end-of-chapter questions. The solutions to these questions are found at the end of this chapter.

1. Which of the following characteristics of trace analysis is **incorrect**?
   a. Trace analysis means measuring substances in a sample at very low levels of concentration.
   b. Trace analysis can be either destructive or non-destructive.
   c. Atomic Absorption Spectroscopy (AAS) is a form or non-destructive trace analysis.
   d. In AAS, the frequencies of light absorbed depend on the chemical identity of the elements represent.
   e. In X-ray fluorescence (XRF), atoms gain energy from X-rays, causing them to fluoresce.

2. Which statement about wave characteristics is **false**?

   a. Amplitude is the height of the wave.
   b. Wavelength is the distance between successive crests.
   c. Frequency is the number of waves occurring per second.
   d. Frequency can be measured in m/s.
   e. Wavelength can be measured in m, nm, or Å.

3. TRUE or FALSE: Amplitude of a wave is related to the intensity of light.

4. What is the wavelength (in Å) of blue light with a frequency of $6.3 \times 10^{14}$ Hz?

5. What is the frequency of yellow light with wavelength of 595 nm?

6. Describe how Einstein's results in his photoelectric effect contradicted the conventional thought of light energy.

7. What is the energy of a photon of red light with a wavelength of 7000 Å?
8. If a photon of microwave radiation has energy of $1.99 \times 10^{-22}$ J, what is the frequency?

9. An atomic emission spectrum shows discrete colored lines where certain wavelengths of light are emitted by excited electrons. What would these spectra look like if the energy of atoms were not quantized?

10. Consider an atom with three energy levels:

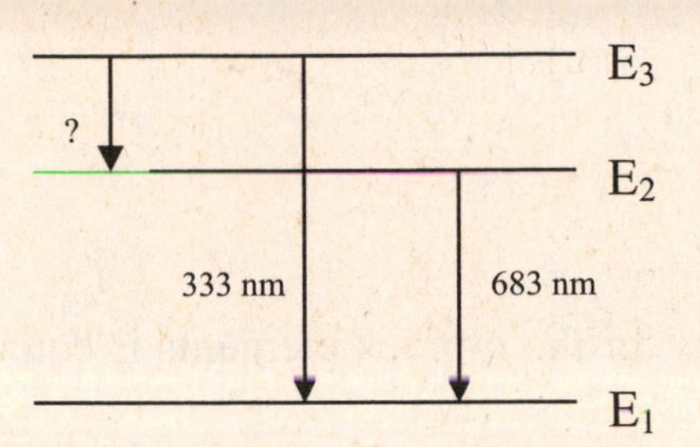


A photon with wavelength 333 nm is emitted when an electron goes from level 3 to level 1, and a photon of wavelength 683 nm is emitted when an electron moves from level 2 to level 1. What is the wavelength of a photon moving from level 3 to level 2?

11. TRUE or FALSE: In the Bohr model of the atom, electrons move between allowed orbits when an atom emits or absorbs light.

12. Which of the following statements about quantum mechanics theory is **incorrect**?
    a. Electrons are described as waves.
    b. The Shrödinger wave equations are fairly simple to solve.
    c. The four quantum numbers represent a way to express the information provided by the wave function solution.
    d. The *n* quantum number describes both potential energy and distance from the nucleus.
    e. The $\ell$ quantum number describes the region in space occupied by an electron.

13. Define an atomic orbital.

14. Identify the following subshells from their quantum numbers and list them in order of increasing energy:

(a) $n = 3, \ell = 1$ (b) $n = 3, \ell = 2$ (c) $n = 2, \ell = 0$ (d) $n = 4, \ell = 0$

15. How many orbitals are there in an *s* subshell? In a *p* subshell? In a *d* subshell? In an *f* subshell?

16. Describe the basic shape of an *s* and a *p* orbital.

17. Write the Aufbau order (increasing energy) of the first 12 orbital sets.

18. Write the full electron configurations for the following species:

(a) S (b) Zn (c) $Fe^{3+}$ (d) $Br^{-}$

19. Write the noble gas core electron configuration for the following species:

(a) Mg (b) Ti (c) $Zn^{2+}$ (d) $Se^{2-}$

20. Identify what is incorrect about the following subshells.

(a) $1p$
(b) $3s^3$
(c) $3f^7$
(d) $1s^2 2s^1 2p^6 3s^2$

21. TRUE or FALSE: The number of valence electrons for the *p*-block elements is equal to the Group number minus 10.

22. Match the following terms and descriptions:

____ atomic radius a. The energy required to remove the most loosely held electron.

____ ionization energy b. The change in energy when an electron is added.

____ electron affinity c. A measure of the size of an atom.

23. How does the atomic radius vary from left to right on the periodic table? From top to bottom?

24. Why does the ionization energy of a metal decrease from top to bottom in a Group on the periodic table?

## ANSWERS TO THE CHAPTER OBJECTIVE QUIZ

1. **c**

2. **d**

3. **TRUE**

4. **Frequency and wavelength are inversely related, mathematically, $\lambda \times \nu = c$, the speed of light ($c = 3.0 \times 10^8$ m/s).**

$$\lambda = c \div \nu,\ \lambda = 3.0 \times 10^8 \text{ m/s} \div 6.3 \times 10^{14}\ \frac{1}{\text{s}} = 4.8 \times 10^{-7} \text{ m} \times \frac{1 \text{ Angstrom}}{1 \times 10^{-10} \text{ m}} = 4800 \text{ Å}$$

5. **$\lambda \times \nu = c$, the speed of light ($c = 3.0 \times 10^8$ m/s)**

$$\nu = c \div \lambda,\ \nu = 3.0 \times 10^8 \text{ m/s} \div 595 \text{ nm} \times \frac{1 \times 10^{-9} \text{ m}}{1\text{nm}} = 5.04 \times 10^{14} \text{ s}^{-1} \text{ (Hz)}$$

6. **The conventional thought was that light is purely wave in nature, and will therefore deliver its energy continuously. If this assumption is correct, then <u>any</u> frequency of light will cause the photoelectric effect to occur. What Einstein found was that only light of a certain minimum frequency (minimum energy) would cause the photoelectric effect to occur. He concluded that light must deliver its energy as a particle would, in small discrete amounts called photons. The photon would either possess enough energy to eject an electron from the atom or it would not. Einstein's description of quantized light energy was quite revolutionary at the time.**

7. **Energy of a photon is related to the wavelength by the following equation: $E_{\text{photon}} = \frac{hc}{\lambda}$, where h and c are Planck's constant and the speed of light, respectively.**

$$E_{\text{photon}} = \frac{hc}{\lambda} = \frac{(6.63 \times 10^{-34} \text{ J} \cdot \text{s})(3.00 \times 10^8 \text{ m/s})}{7000 \text{ angstroms} \times \frac{1 \times 10^{-10} \text{ m}}{1 \text{ angstrom}}} = 2.84 \times 10^{-19} \text{ J}$$

8. **Energy of a photon is related to frequency by the equation, $E_{\text{photon}} = h\nu$, where $h$ = Planck's constant ($h = 6.626 \times 10^{-34}$ J·s).**

$$\nu = \frac{E_{\text{photon}}}{h} = \frac{1.99 \times 10^{-22} \text{ J}}{6.626 \times 10^{-34} \text{ J} \cdot \text{s}} = 3.00 \times 10^{11}\ \frac{1}{\text{s}} = 3.00 \times 10^{11} \text{ Hz}$$

**9.** If an atom's energy were not quantized, its electrons would be able to occupy any and all energy states, allowing the atom to emit any and all types of photons. The emission spectrum for an atom would appear as a continuous colored spectrum showing all wavelengths of light. It would be the same for all atoms, also.

**10.** The same equation that we use to calculate energy of a photon, allows us to relate energy transitions of an electron to wavelength of light emitted.

$$\Delta E = E_{\text{photon}} = \frac{hc}{\lambda}$$

We will calculate the energy of the transition from level 3 to level 1 ($\Delta E_{3\to1}$) and for the transition from level 2 to level 1 ($\Delta E_{2\to1}$). Subtracting will give us the energy of the transition from level 3 to level 2 ($\Delta E_{3\to2}$ ) and we'll use this to calculate the wavelength of the photon emitted.

$$\Delta E_{3\to1} = \frac{hc}{\lambda} = \frac{(6.626\times10^{-34}\ \text{J}\cdot\text{s})(3.0\times10^{8}\ \text{m/s})}{333\ \text{nm}\times\dfrac{1\times10^{-9}\ \text{m}}{1\ \text{nm}}} = 5.97\times10^{-19}\ \text{J}$$

$$\Delta E_{2\to1} = \frac{hc}{\lambda} = \frac{(6.626\times10^{-34}\ \text{J}\cdot\text{s})(3.0\times10^{8}\ \text{m/s})}{683\ \text{nm}\times\dfrac{1\times10^{-9}\ \text{m}}{1\ \text{nm}}} = 2.91\times10^{-19}\ \text{J}$$

The difference in energy between these transitions is:

$$\Delta E_{3\to1} - \Delta E_{2\to1} = \Delta E_{3\to2} = (5.97\times10^{-19}\ \text{J}) - (2.91\times10^{-19}\ \text{J}) = 3.06\times10^{-19}\ \text{J}$$

Using this energy to calculate the wavelength:

$$\lambda = \frac{hc}{E_{\text{photon}}} = \frac{(6.626\times10^{-34}\ \text{J}\cdot\text{s})(3.0\times10^{8}\ \text{m/s})}{3.06\times10^{-19}\ \text{J}} = 6.50\times10^{-7}\ \text{m}\times\frac{1\times10^{9}\ \text{nm}}{1\ \text{m}} = 650\ \text{nm}$$

**11.** TRUE

**12.** b

**13.** An atomic orbital is a region in space where there is a very high probability of finding an electron. Solving the wave functions for the electrons in an atom generates these regions.

**14.** The *n* or principal quantum number represents the energy level occupied by the electron. The secondary quantum number, $\ell$, represents the general shape of the region in space occupied by the electron, or the shape of the orbital. Orbital shape is also designated by letters *s, p, d,* and *f* representing $\ell = 0, 1, 2,$ and $3$, respectively.

(a) 3*p* (b) 3*d* (c) 2*s* (d) 4*s*

**Using the Aufbau order the arrangement in order of increasing energy is:**

**$2s < 3p < 4s < 3d$**

**15. An *s* subshell has only one orbital. A *p* subshell has three orbitals, a *d* has five, and an *f* has seven.**

**16. An *s*-orbital is generally spherical. At higher energy levels, the orbital expands outward farther from the nucleus. The 1*s* orbital has no nodes, or areas of zero probability of finding electrons. The 2*s* orbital has one node; the 3*s* has two nodes.**

**The *p*-orbitals are lobed-shaped or look somewhat like a dumbbell. These lobes are oriented along the *x*, *y*, and *z*-axes. These three spatial orientations of the *p*-orbital make up the *p*-subshell for a given energy level.**

**17. The Aufbau order lists the subshells in increasing order of energy. There are some subshells that appear to be out of order due to shielding. An *f* orbital is shielded more than a *d*, which is shielded more than a *p*. This means orbitals of a lower shell can be at a higher energy level.**

**The first 12 subshells are:**

$$1s < 2s < 2p < 3s < 3p < 4s < 3d < 4p < 5s < 4d < 5p < 6s$$

**18. The full electron configuration shows all the subshells and the number of electrons in them as a superscript.**

**(a) S: $1s^2 2s^2 2p^6 3s^2 3p^4$ (b) Zn: $1s^2 2s^2 2p^6 3s^2 3p^6 4s^2 3d^{10}$**

**(c) $Fe^{3+}$: $1s^2 2s^2 2p^6 3s^2 3p^6 3d^5$ (d) $Br^-$: $1s^2 2s^2 2p^6 3s^2 3p^6 4s^2 3d^{10} 4p^6$**

**19. The Noble gas core electron configuration represents the inner or core electrons with the symbol of a Noble gas.**

**(a) Mg [Ne] $3s^2$ (b) Ti [Ar] $4s^2 3d^2$ (c) $Zn^{2+}$ [Ar] $3d^{10}$**

**(d) $Se^{2-}$ [Ar] $4s^2 3d^{10} 4p^6$ = [Kr]**

**20. (a) There is no *p* subshell in the first energy level, only an s subshell.**
**(b) An *s* subshell cannot hold three electrons, only two.**
**(c) There is no *f* subshell in the third energy level, only an *s*, *p* and *d*.**
**(d) The 2*s* subshell must be filled before the 2*p*, 3*s*, etc. This is the Aufbau principle.**

**21. TRUE**

**22. c__ atomic radius a__ ionization energy b__ electron affinity**

**23. The trend in atomic radius is decreasing from left to right. This is because of increasing effective nuclear charge from left to right shrinking the orbitals, drawing the electrons closer to the nucleus. Atomic radii increase from top to bottom due to electrons occupying orbitals in higher energy levels, which are farther from the nucleus.**

**24. Ionization energy decreases from top to bottom on the periodic table because the valence electrons occupy orbitals in higher energy levels, farther from the nucleus, which experience an increasing amount of shielding. This means those electrons don't "feel" the full nuclear attraction and are easier to remove.**

# CHAPTER 7

## *Study Goals*

The study goals outline specific concepts to be mastered in this section of the text chapter. Related problems at the end of the text chapter will also be noted. Working the questions noted should aid in mastery of each study goal and also highlight any areas that you may need additional help in.

**Section 7.1 *INSIGHT INTO* Materials for Biomedical Engineering**

1. Recognize the importance of physical properties in determining biocompatibility and understand how knowledge of chemical bonding is essential in materials engineering. *Work Problems 1–6.*

**Section 7.2 The Ionic Bond**

2. Describe the formation of an ionic bond. Explain why metals tend to form cations and nonmetals tend to form anions using electron configurations. *Work Problems 7–16, 94, 97, 101.*

**Section 7.3 The Covalent Bond**

3. Define a covalent bond, bond length, and bond energy. *Work Problems 17–23, 28, 29, 96.*
4. Use Lewis dot symbols to represent the valence electrons of the main group elements. *Work Problems 24–27, 103, 104.*

**Section 7.4 Electronegativity and Bond Polarity**

5. Define electronegativity and use it to explain the existence of different bond types, such as nonpolar covalent, polar covalent, and ionic bonds. *Work Problems 30–42, 90, 91, 102.*

**Section 7.5 Keeping Track of Bonding: Lewis Structures**

6. Draw Lewis structures for various molecules and polyatomic ions. *Work Problems 43–46, 53–56, 86, 89, 99, 100.*
7. Define resonance and draw resonance structures for molecules when they exist. *Work Problems 47–52.*

**Section 7.6 Orbital Overlap and Chemical Bonding**

8. Describe the valence bond theory and how orbital overlap leads to sigma and pi bonds. *Work Problems 57–64.*

**Section 7.7 Hybrid Orbitals**

9. Predict the type of hybrid orbitals used by the central atom in a molecule by drawing the molecule's Lewis structure. *Work Problems 65–70, 92, 105.*

**Section 7.8 Shapes of Molecules**

10. State the valence shell electron pair repulsion (VSEPR) theory and be able to determine the number of electron pair groups associated with the central atom in a molecule. *Work Problems 71, 72, 78.*
11. Predict the shape and bond angles for molecules by applying the VSEPR theory. *Work Problems 73–76, 85, 95.*
12. Describe the geometry, hybridization, and bond angles of atoms involved in double or triple bonds. Distinguish between the sigma and pi bonds found in double and triple bonds. *Work Problems 77, 79, 84, 87, 88, 93, 98.*

**Section 7.9 *INSIGHT INTO* Molecular Scale Engineering for Drug Delivery**

13. Understand the importance of chemical bonding in design of drug delivery systems employing novel materials. *Work Problems 80–83.*

## *Solutions to the Odd-Numbered Problems*

Problem 7.1 Define the term *biocompatibility*.

***Biocompatibility* is the ability of materials to interact with the natural biological materials without triggering a response from the immune system.**

Problem 7.3 Describe how PMMA functions as bone cement. How does chemistry play a role in this process?

**Poly(methyl methacrylate) (PMMA) is a very strong cement; think "super glue". The glue can be applied to the area of bone that requires mending and then the polymerization reaction takes place, causing it to harden.**
**Chemistry affects both the hardness and durability of the cement as well as the biocompatibility.**

Problem 7.5 Why do biomedical engineers sometimes need to use coatings on materials they use? When coatings are generated, what are the materials used as reactants called?

**The materials biomedical engineers use may have the desired properties but not have acceptable biocompatibility. One method of handling this problem is to establish a biocompatible coating on top of the materials by chemically reacting small molecules on the surface. These molecules are referred to as precursors. An extensive knowledge of the properties of the precursors and the biomaterials is essential to produce a biocompatible situation in the body.**

Problem 7.7 Why is the $Na^{2+}$ ion not found in nature?

**Neutral sodium atoms have one valence electron. By losing it and becoming a +1 cation, sodium assumes the electron configuration of Ne, a noble gas. This gives a very stable, low energy arrangement of electrons for $Na^+$. Losing a second electron would give a much higher energy arrangement because the second electron is an inner or core electron. This would never happen in nature.**

Problem 7.9 ■ Select the smaller member of each of the following pairs. (a) N and $N^{3-}$, (b) Ba and $Ba^{2+}$, (c) Se and $Se^{2-}$, (d) $Co^{2+}$ and $Co^{3+}$

**Cations are smaller than the corresponding atoms, anions are larger.**

**(a) N**

**(b) $Ba^{+2}$**

**(c) Se**

**(d) $Co^{+3}$**

Problem 7.11 ■ Arrange the following sets of anions in order of increasing ionic radii: (a) $Cl^-$, $P^{3-}$, $S^{2-}$, (b) $S^{2-}$, $O^{2-}$, $Se^{2-}$, (c) $Br^-$, $N^{3-}$, $S^{2-}$, (d) $Br^-$, $Cl^-$, $I^-$

**(a) In an isoelectronic series, ionic radii increase with decreasing atomic number because of decreasing nuclear charge. Therefore, in order of increasing ionic radii, we have $Cl^- < S^{2-} < P^{3-}$.**

**(b) Ionic radii increase down a group. So, $O^{2-} < S^{2-} < Se^{2-}$.**

**(c) $S^{2-} < N^{3-} < Br^-$**

**(d) Ionic radii increase down a group. So, $Cl^- < Br^- < I^-$.**

Problem 7.13 In a lattice, a positive ion is often surrounded by eight negative ions. We might reason, therefore, that the lattice energy should be related to eight times the potential of interaction between these oppositely charged particles. Why is this reasoning too simple?

**Because in a lattice, there are also neighboring atoms which will contribute repulsive interactions to the overall lattice energy.**

Problem 7.15 Figure 7.2 depicts the interactions of an ion with its first nearest neighbors, second nearest neighbors and third nearest neighbors in a lattice. (a) Would the interactions with the fourth nearest neighbors be attractive or repulsive?

**The fourth nearest neighbors would be positively charged, like the ion in question, so the interaction would be repulsive.**

(b) Based on Coulomb's law, how would the relative sizes of the terms compare if the potential energy were expressed as $V = V_{1st} + V_{2nd} + V_{3rd} + V_{4th}$?

**Coulomb's Law is expressed as $V = k\frac{q_1q_2}{r}$, so the potential energy, $V$, is inversely proportional to the distance between the ions ($r$). The potential energy would decrease proportionally between the nearest neighbors, second nearest, etc. $V_{1st} > V_{2nd} > V_{3rd} > V_{4th}$**

Problem 7.17 Describe the differences between a covalent bond and an ionic bond.

**An ionic bond is the electrostatic force of attraction between oppositely charged ions.**

**A covalent bond is a force of attraction between neutral atoms that occurs when the two atoms share a pair of electrons in overlapping atomic orbitals.**

**The main difference is the extent to which electrons are shared versus transferred between the two bonding atoms.**

Problem 7.19 Sketch a graph of the potential energy of two atoms as a function of the distance between them. On your graph, indicate how bond energy and bond distance are defined.

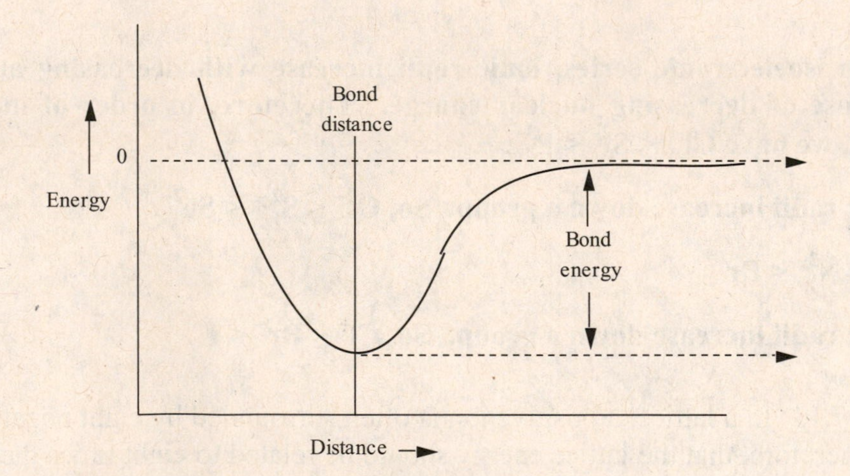


**Bond energy is the difference in energy between the neutral atoms and the bonded atoms. Bond distance is the distance between the bonded nuclei when their energy is at a minimum.**

Problem 7.21 Coulombic forces are often used to explain ionic bonding. Are coulombic forces involved in covalent bonding as well? Explain.

**In a different way, negatively charged electrons in the bond are attracted to the positively charged protons in both nuclei, creating coulombic forces.**

Problem 7.23 If the formation of chemical bonds always releases energy, why don't all elements form dozens of bonds to each atom?

**The number of bonds formed is usually limited by the number of electrons an atom needs to fill its valence shell of orbitals. Sometimes an atom can form more than this number of bonds, but a larger number of bonds creates significant repulsive forces because of the electron pairs. This means large numbers of bonds for a single atom are not possible.**

Problem 7.25 Theoretical models for the structure of atomic nuclei predict the existence of superheavy elements that have not yet been discovered and suggest that such elements might be fairly stable if they could be produced. So researchers are currently trying to synthesize those superheavy elements to test these theories. Suppose that element 117 has been synthesized and given the atomic symbol Lt. What would the Lewis dot symbol be for this new element?

: Lt ·

**The element would belong to Group 17, so it would have a total of seven valence electrons in the 7*s* and 7*p* orbitals.**

Problem 7.27 Define the term *lone pair*.

**A *lone pair* is sometimes referred to as an unshared pair. It is a pair of electrons occupying an orbital in the valence shell of an atom but not being shared in a bond.**

Problem 7.29 How does the bond energy of a double bond compare to that of two single bonds between the same elements? How does this relationship explain the types of reactions that compounds with double bonds undergo?

**A double bond is stronger than a single bond (higher bond energy) but not twice as strong. This means that two single bonds have more bond energy than one double bond. This leads to some reactions where the "second" bond of the double bond is broken and replaced by a single bond. Bond energy increases in this process as essentially a double bond is replaced by two single bonds.**

Problem 7.31 Distinguish between electron affinity and electronegativity.

**Electron affinity is defined as the energy change associated with the addition of an electron. It is related to formation of anions. Electronegativity is a property measuring the ability of an atom to attract electrons to it when the atom is involved in a bond.**

Problem 7.33 When two atoms with different electronegativities form a covalent bond, what does the electron distribution in the bond look like?

δ– δ+

**The shading represents a greater amount of electron density. The more electronegative atom on the left draws the shared electrons closer, creating a partial negative charge on the left and a partial positive charge on the right.**

Problem 7.35 Why is a bond between two atoms with different electronegativities called a polar bond?

**When atoms of dissimilar electronegativities bond, the shared electrons are pulled in the direction of the more electronegative atom. This causes a slightly negative charge to form on one side of the bond (the more electronegative atom) and a slightly positive charge on the other side. These opposite charges are referred to as poles, so the bond is called a polar bond. See Problem 7.31.**

Problem 7.37 In each group of three bonds, which bond is likely to be the most polar? Which will be the least polar?

**The least polar bond will be the one with the most similar electronegativties. We can consult a reference like Figure 7.8 in the text or use the general periodic trend: Electronegativity increases up and to the right on the periodic table.**

**(a)** C—H, O—H, S—H

**C and S both have the same electronegativity, which is less than that of O. The S – H and C – H bonds are similar and the least polar. The most polar is the O – H bond in which there is the largest difference in electronegativity.**

**(b)** C—Cl, Cl—Cl, H—Cl

**The Cl – Cl bond is the least polar (same electronegativity). The H – Cl is the most polar.**

**(c)** F—F, O—F, C—F

**The F – F bond is the least polar (same electronegativity). Carbon's electronegativity is less than that of oxygen; the C–F bond is the most polar.**

**(d)** N—H, N—O, N—Cl

**N and Cl have the same electonegativity, the N – Cl bond is the least polar. The N – H bond is the most polar.**

Problem 7.39 Can fluorine atoms ever carry a partial positive charge in a molecule? Why or why not?

**They cannot because fluorine has the highest electronegativity of all the elements: whenever F is involved in a bond, the electrons will be drawn to it.**

Problem 7.41 Using only halogen atoms, what would be the most polar diatomic molecule that could be formed? Explain your reasoning.

**The most polar molecule would be one with the biggest difference between the electronegativities of the atoms. Fluorine has the highest and iodine has the lowest electronegativity of the halogens (excluding astatine). F – I would be the most polar diatomic halogen molecule.**

Problem 7.43 ■ Draw the Lewis structures for each of the following molecules:

**(a)** CO

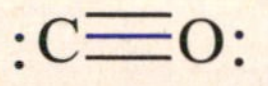

:C≡O:

**(b)** $H_2S$

H—S—H

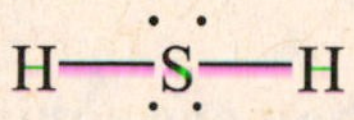

**(c)** $SF_6$

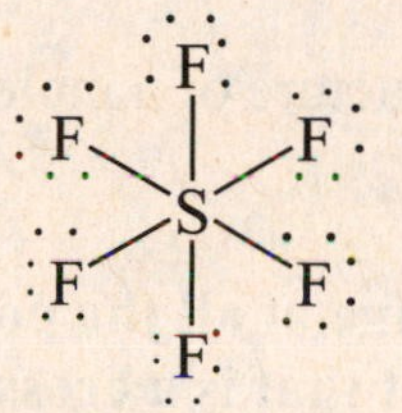

**(d)** $NCl_3$

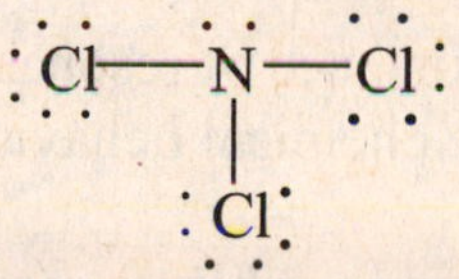

Problem 7.45 ■ Write Lewis structures for these molecules: (a) Tetrafluoroethylene, $C_2F_4$, the molecule from which Teflon is made, (b) Acrylonitrile, $CH_2CHCN$, the molecule from which Orlon® is made

**(a)** Tetrafluoroethylene, $C_2F_4$

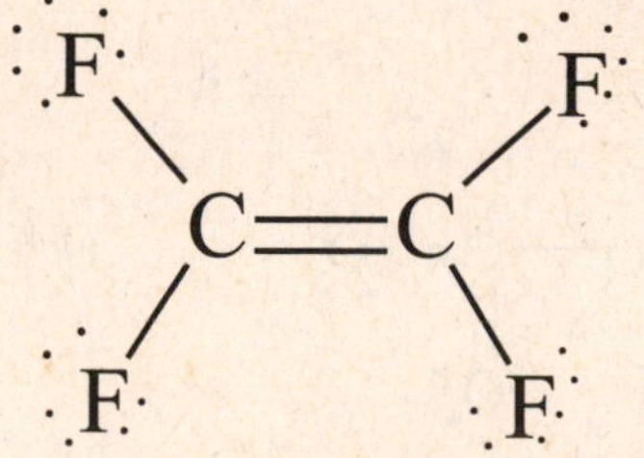

**(b)** Acrylonitrile, $CH_2CHCN$

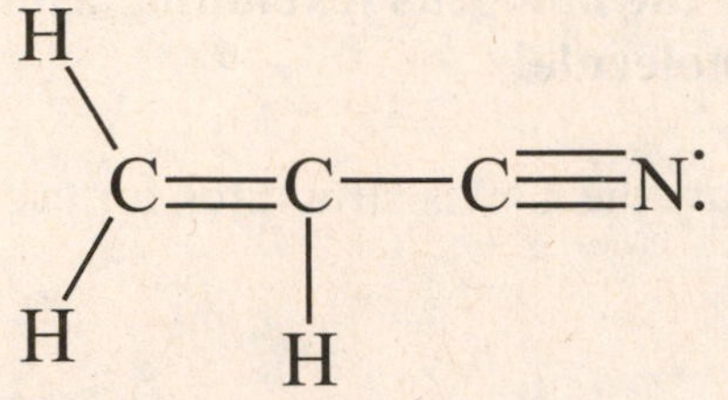

```
H
 \
  C==C—C≡N:
 /   |
H    H
```

<u>Problem 7.47</u> In the context of Lewis structures, what is resonance?

**Resonance occurs when more than one equivalent Lewis structure can be drawn, depending on where the double bond is placed. The double bond is said to "resonate" between the different positions. The actual structure is thought to be an average of the resonance structures.**

<u>Problem 7.49</u> How does the structure of a molecule that exhibits resonance justify the use of the term *resonance hybrid*?

**A *resonance hybrid* is an average of all the possible resonance structures. A *hybrid* is a term referring to an animal or plant that is a cross between species, so a resonance hybrid represents a "cross" between resonance structures.**

<u>Problem 7.51</u> How does the existence of resonance structures for a molecule such as benzene explain its unusually unreactive chemical behavior?

**When a molecule such as benzene exhibits resonance, the electrons in the resonating double bond are <u>delocalized</u>: that is they are not confined to the space of any single atoms but occupy space common to the entire molecule. Electrons occupying more space are lower in energy and less reactive.**

<u>Problem 7.53</u> Which of the species listed has a Lewis structure with only one lone pair of electrons? $F_2$, $CO_3^{2-}$, $CH_4$, $PH_3$

**We need to consider the Lewis structures for each molecule. $PH_3$ is the only molecule with one lone pair of electrons.**

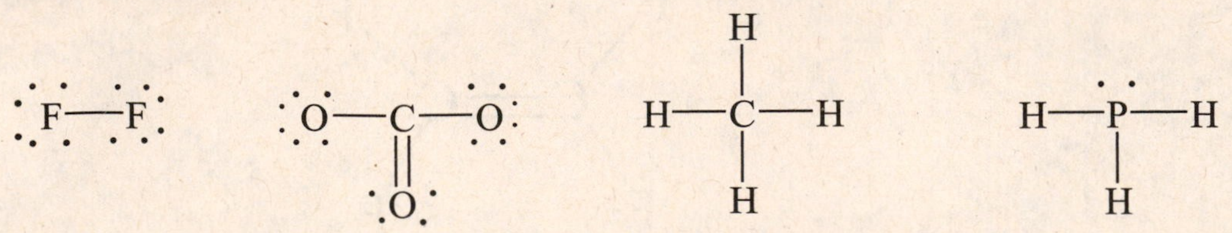

```
                                  H                    ..
:F̤—F̤:      :Ö—C—Ö:            |                 H—P—H
             ||              H—C—H                  |
             :O:                 |                   H
                                  H
```

Problem 7.55 Identify what is incorrect in the Lewis structures shown.

(a) $O_3$ **In the first structure there are too many valence (20 vs. 18) electrons.**

O—O—O O—O=O

(b) $XeF_4$ **In the first structure, there are too few valence (34 vs. 36) electrons.**

Problem 7.57 Explain the concept of wave interference in your own words.

**Wave interference refers to how waves are combined. When amplitudes of waves are in phase, constructive interference takes place. This means a bigger wave results. When the amplitudes of the combining waves are out of phase, the result is destructive interference and the wave is diminished.**

Problem 7.59 How is the concept of orbital overlap related to the wave nature of electrons?

**When orbitals overlap, they either add together or subtract. This is constructive interference and destructive interference; see Problem 7.57. This behavior is evidence of the wave nature of electrons.**

Problem 7.61 How do sigma and pi bonds differ? How are they similar?

**Sigma and pi bonds are both types of covalent bonds, where the electron density lies between the two bonded nuclei. They differ geometrically. Sigma bonds are formed by the direct or head-on overlap of orbitals. Pi bond result from the side-on overlap of orbitals.**

Problem 7.63 Draw the Lewis dot structure of the following compounds and identify the number of pi bonds in each.

**Pi bonds are found in double or triple bonds. Each double bond has one pi bond and each triple bond has two pi bonds.**

(a) $CS_2$ **Two pi bonds.**

S=C=S

**(b)** $CH_3Cl$ **Zero pi bonds.**

```
        H
        |       ..
    H—C—Cl:
        |       ..
        H
```

**(c)** $NO_2^-$ (N atom central) **One pi bond.**

$$\left[:\ddot{O}-\ddot{N}=\ddot{O}\right]^-$$

**(d)** $SO_2$ **One pi bond.**

$$\ddot{O}=\ddot{S}-\ddot{O}:$$

Problem 7.65 What observation about molecules compels us to consider the existence of hybridization of atomic orbitals?

**The geometry of molecules suggests hybridization is possible. For example, if the oxygen atom in the molecule used only *p* atomic orbitals to make bonds, the bond angle of the hydrogen atoms would be only 90°. We observe that the actual angle is close to 109.5°; supporting the existence of $sp^3$ hybrid orbitals that create 109.5° bond angles.**

Problem 7.67 What type of hybrid orbital is generated by combining the *s* orbital with all three *p* orbitals of an atom? How many hybrid orbitals result?

**In designating hybrid orbitals, a superscript show how many of a particular orbital were used. Combining one s and three *p* orbitals result in four hybrid $sp^3$ orbitals.**

Problem 7.69 ■ What hybrid orbitals would be expected for the central atom in each of the following molecules or ions? (a) $GeCl_4$, (b) $PBr_3$, (c) $BeF_2$, (d) $SO_2$

a. **$GeCl_4$ is a tetrahedral molecule. The Ge atom is $sp^3$ hybridized.**

b. **$PBr_3$ is a trigonal pyramidal molecule. The P atom is $sp^3$ hybridized.**

c. **$BeF_2$ is a linear molecule. The Be atom is *sp* hybridized.**

d. **$SO_2$ is a bent molecule. The S atom is $sp^2$ hybridized.**

Problem 7.71 What physical concept forms the premise of VSEPR theory?

**VSEPR (Valence Shell Electron Pair Repulsion) theory is based on the principle that like charges repel each other. Electron pair groups associated with the central atom will repel each other because they represent areas of negative charge.**

Problem 7.73 ■ Predict the shape of each of the following molecules or ions.

**The geometry or shape is dictated by the number of electron pair groups around the central atom, as well as which ones are bonded atoms versus lone pairs. First, the basic geometry of the electron pairs is determined (Table 7.3) and then the molecular shape can be found by considering only the positions of the bonded atoms in that basic geometry (Table 7.4).**

**(a)** $IF_3$

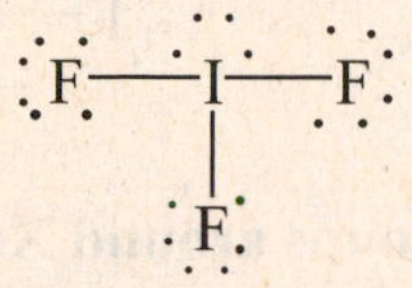

**There are five electron pair groups around the iodine atom, two lone pairs and three bonded atoms. The basic geometry would be trigonal bipyramidal. The molecular shape would be T-shape.**

**(b)** $ClO_3^-$

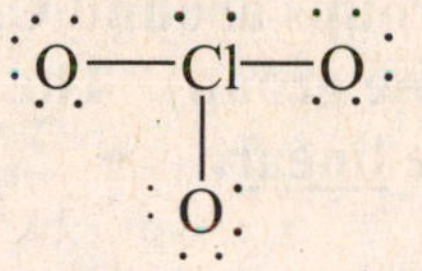

**There are four electron pair groups around Cl, one lone pair and three bonded atoms. The basic geometry would be tetrahedral. The molecular shape would be trigonal pyramidal.**

**(c)** $TeF_4$

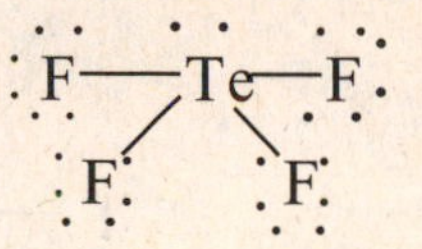

**There are five electron pair groups around Te, one lone pair and four bonded atoms. The basic geometry would be trigonal bipyramidal. The molecular shape would be seesaw.**

**(d)** $XeO_4$

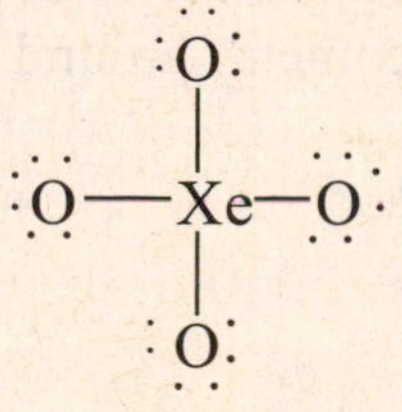

**There are four electron pair groups around Xe, zero lone pairs and four bonded atoms. The basic geometry would be tetrahedral. The molecular shape would also be tetrahedral.**

Problem 7.75 ■ Which of these molecules would be linear? Which have lone pairs around the central atom?

**We need to draw the Lewis structures and determine the molecular shape.**

**(a)** $XeF_2$

F—Xe—F

**There are five electron pair groups around Xe, three lone pairs and two bonded atoms. The basic geometry would be trigonal bipyramidal. The molecular shape would be linear.**

**(b)** $CO_2$

O═C═O

**There are two electron pair groups around C, zero lone pairs and two bonded atoms (double bonds only count as one group). The basic geometry would be linear. The molecular shape would also be linear.**

**(c)** $BeF_2$

F—Be—F

**There are two electron pair groups around Be, zero lone pairs and two bonded atoms. The basic geometry would be linear. The molecular shape would also be linear.**

**(d)** $OF_2$

F—O—F

**There are four electron pair groups around O, two lone pairs and four bonded atoms. The basic geometry would be tetrahedral. The molecular shape would be bent.**

Problem 7.77 Propene has the chemical formula $H_2C{=}CH{-}CH_3$. Describe the overall shape of the molecule by considering the geometry around each carbon atom.

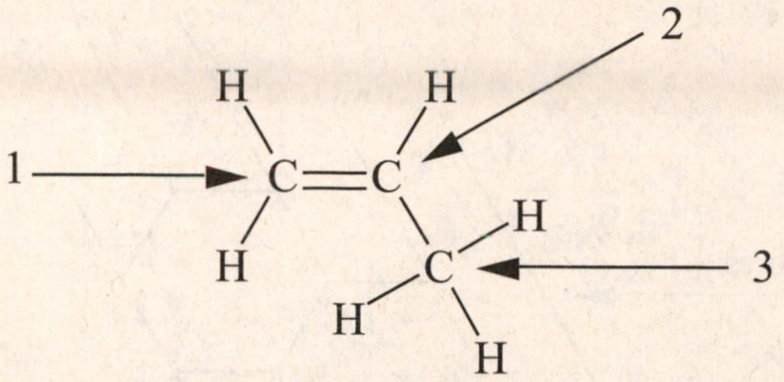

```
                          2
        H      H       /
         \    /      /
1 ——————> C==C <
         /    \   H
        H      \ /
                C <———————— 3
               / \
              H   H
```

**At carbon number 1, there are three bonded atoms (two single and one double) and zero lone pairs. The geometry would be trigonal planar.**
**At carbon number 2, there are three bonded atoms (two single and one double) and zero lone pairs. The geometry would be trigonal planar.**
**At carbon number 3, there are four bonded atoms and zero lone pairs. The geometry would be tetrahedral.**

Problem 7.79 Describe what happens to the shape about the carbon atoms when a C=C double bond undergoes an addition reaction in which it is converted into a C—C single bond.

**When a C=C exists, there are a total of three atoms bonded to each carbon. The geometry is trigonal planar. The addition reaction converts the double into a single bond and now there will be four bonded atoms for each carbon. The geometry is now tetrahedral.**

Problem 7.81 Use the fundamental structure of mesoporous silica nanoparticles (MSN) to explain why they have such a large surface area.

**As discussed earlier, silica is composed of networks of $SiO_4$ units. A mesoporous silica nanoparticle (MSN) builds these units in a honeycomb fashion, as shown in Figure 7.18. Because of this structure, these particles have enormous surface area. Although the overall surface area of the MSN is very large, the majority of that surface area represents the interior walls of the honeycomb pores. Relatively little of the surface area is actually on the outside of the particle itself.**

Problem 7.83 Draw a schematic picture of a MSN being used as a molecular engineering design selective drug delivery vehicle. Label all key components of the design.

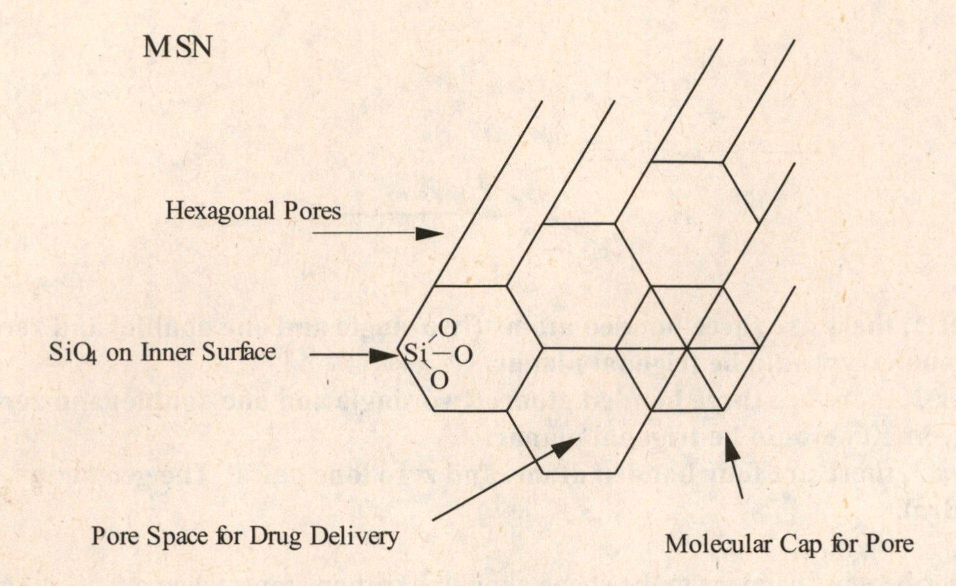


Problem 7.85 ■ The following molecules have similar formulas, but each has a different shape. Draw Lewis structures for these compounds, and then determine the expected hybridization of the central atom and the molecular shape.

**(a)** $SiF_4$

**There are four electron pair groups around Si, zero lone pairs and four bonded atoms. The molecular shape would be tetrahedral. Four electron pair groups require four hybrid *sp*$^3$ orbitals.**

**(b)** $KrF_4$

**There are six electron pair groups around Kr, two lone pairs and four bonded atoms. The basic geometry would be octahedral. The molecular shape would be square planar. Six electron pair groups require six hybrid *sp*$^3$*d*$^2$ orbitals.**

(c) $SeF_4$

There are five electron pair groups around Se, one lone pair and four bonded atoms. The basic geometry would be trigonal bipyramid. The molecular shape would be seesaw. Five electron pair groups require five hybrid $sp^3d$ orbitals.

Problem 7.87 ■ A Lewis structure for the oxalate ion is shown below. (One or more other resonance forms are also possible.)

What is the correct charge on the oxalate ion? What type of orbital hybridization is expected for each of the carbon atoms in this structure? How many sigma bonds and how many pi bonds does the structure contain?

**By adding the valence electrons of the two C atoms (four each) and four O atoms (six each) we get a total of 32 valence electrons. In the Lewis structure, there are a total of 34 valence electrons shown; meaning the charge of the oxalate ion is –2 due to the two extra electrons.**

**Each carbon atom has three bonded atoms and no lone pairs; they need to use three hybrid $sp^2$ orbitals.**

**Each single bond is a sigma bond; each double contains one sigma and one pi bond. There are a total of five sigma and two pi bonds in the oxalate ion.**

Problem 7.89 An unknown metal M forms a chloride with the formula $MCl_3$. This chloride compound was examined and found to have a trigonal pyramidal shape. Draw a Lewis structure for $MCl_3$ that is consistent with this molecular shape. Use your structure to propose an identity for the metal M, and explain how you made your choice.

**Trigonal pyramidal shape is formed when the central atom of the molecule has three bonded atoms and one lone pair (a total of four electron pair groups – see Table 7.4). An element that has one lone pair and needs to form three single bonds to complete its octet**

**would be found in Group 15 (e.g., nitrogen). The only true metal in that Group is bismuth, so the identity of M is most likely bismuth.**

Problem 7.91 Although we often classify bonds between a metal and a nonmetal as ionic, not all such bonds have the same degree of ionic character. Consider the series of tin(IV) halides: $SnF_4$, $SnCl_4$, $SnBr_4$, and $SnI_4$. Rank these compounds in terms of expected ionic character, from most ionic to least ionic. What property do you need to consider in making this ranking?

**The property needed to consider ionic character is electronegativity (EN). The greater the difference in electronegativity (ΔEN), the more ionic the bond character is. Usually, metals have low electronegativities and nonmetals high, but not always. This is why a bond between a metal and nonmetal is usually mostly ionic in character but not always. The electronegativity for Sn is 1.8 and for F, Cl, Br and I it is 4.0, 3.0, 2.8 and 2.5, respectively. The larger ΔEN corresponds to more ionic character, so $SnF_4 > SnCl_4 > SnBr_4 > SnI_4$.**

Problem 7.93 Consider the structure shown below for $N_2O_3$ as well as any other important resonance structures.

O=N—N(—O)=O

(a).What is the expected O–N–O bond angle in this structure?

**There are three electron pair groups (areas of high electron density) around the central atom, giving a trigonal planar geometry with approximately 120° bond angles.**

(b). The $N_2O_3$ molecule contains N–O bonds of two different lengths. How many *shorter* N–O bonds would be present?

**The shorter N–O bonds are the ones where the atoms are bonded with a double covalent bond, so there are two shorter N–O bonds. Double bonds are shorter than single because the two nuclei must be closer to allow the side-on overlap of *p* orbitals required to form the pi bond (the second bond).**

Problem 7.95 One of the "grand challenges" for engineering recently identified by the U.S. National Academy of Engineering is the capture of carbon dioxide produced by the burning of fossil fuels. Although carbon and oxygen have significantly different electronegativities, $CO_2$ cannot easily be separated from air by applying a voltage in the smokestack of power plant where it is generated. How does the geometry of carbon dioxide account for this observation?

O=C=O

**Carbon dioxide is a completely nonpolar molecule even though it is composed of very polar bonds. So in air, it behaves like the other nonpolar constituents: $O_2$, $N_2$, and argon. Therefore it is very difficult to separate from air.**

Problem 7.97 Lead selenide nanocrystals may provide a breakthrough in the engineering of solar panels to be efficient enough to be an economical source of electricity. Selenium is generally considered a nonmetal while lead is considered a metal. Is this distinction enough to suggest that this compound should be ionic? Explain your answer.

**The distinction of a compound being ionic is more complicated than just a metal and nonmetal forming a bond. Electronegativity largely determines the nature of the bond, regardless of whether the atoms fall into our arbitrary categories of metal or nonmetal. According to Figure 7.8, the electronegativity of Se is 2.4 and that of Pb is 1.7. The moderate difference in electronegativity indicates a polar covalent bond.**

Problem 7.99 Nitrogen is the primary component of our atmosphere. It is also used a in inert reagent to fill containers of chemicals that might react with the oxygen in air. Draw a Lewis structure of nitrogen and use this drawing to help explain why nitrogen does not react readily with other molecules.

$$:N\equiv N:$$

**To share the 10 valence electrons in the $N_2$ molecule and complete the octet for each N atom, a triple bond must exist between the two atoms. This is a very strong bond (large bond energy) and is difficult to break. This leads to the relative unreactivity of the nitrogen molecule. One example of where $N_2$ does react is at the higher temperature inside an automobile engine where unfortunately it reacts with oxygen to form nitrogen oxide (NO).**

Problem 7.101 If leads are attached to the opposite sides of a crystal of sodium chloride, it does not conduct electricity. Distilled water does not conduct electricity either. Yet, if the sodium chloride is dissolved in distilled water, the resulting solution is does conduct electricity. Use pictures at the molecular level of detail to describe why this occurs.

**Conduction of electricity requires movement of charge. Metals conduct electricity because the mobile "sea" of electrons is able to move through the empty conduction band of orbitals. In an ionic solid, the charged species are held in place within the crystal lattice and very little movement occurs. In distilled water, there are essentially no ions present and so again no movement of charge takes place. When the NaCl is dissolved in water, the $Na^+$ and $Cl^-$ become mobile in the solution and they are able to conduct electricity through the solution.**

**Solid – crystal lattice structure**

| | | | |
|---|---|---|---|
| $Na^+$ | $Cl^-$ | $Na^+$ | $Cl^-$ |
| $Cl^-$ | $Na^+$ | $Cl^-$ | $Na^+$ |
| $Na^+$ | $Cl^-$ | $Na^+$ | $Cl^-$ |
| $Cl^-$ | $Na^+$ | $Cl^-$ | $Na^+$ |

**Solution – random, mobile ions**

| | | | | | |
|---|---|---|---|---|---|
| $Na^+$ | $H_2O$ | $Cl^-$ | $H_2O$ | | $H_2O$ |
| $H_2O$ | $H_2O$ | $H_2O$ | $H_2O$ | $Na^+$ | $H_2O$ |
| | $Na^+$ | | $H_2O$ | | |
| $Cl^-$ | $H_2O$ | $H_2O$ | | $Cl^-$ | $H_2O$ |

Problem 7.103 How do the Lewis symbols for C, Si, and Ge reflect the similarity in their electron configurations?

**These three elements are in Group 14 and they all have four valence electrons, two in the n*s* orbital and two in the n*p* orbitals. Their Lewis symbols are all essentially the same; showing a total of four valence electrons.**

```
  •            •            •
• C •        • Si •       • Ge •
  •            •            •
```

Problem 7.105 When free radical polymerization occurs, how does the hybridization change the carbons involved?

```
                              1                                      2
     H   H            H          H                 H    H    H    H
     |   |             \        /                  |    |    |    |
R—O—C—C• -------->  C==C      ------->   R—O—C—C—C—C •
     |   |             /        \                  |    |    |    |
     H   H            H          H                 H    H    H    H
```

**When carbon is involved in two single and one double bond (C #1), it uses *sp*$^2$ hybrid orbitals which gives trigonal planar geometry. After the free radical polymerization, the double bond is opened to give an additional single bond. When carbon is involved in four single bonds (C #2) it uses *sp*$^3$ hybrid orbitals that give tetrahedral geometry.**

## CHAPTER OBJECTIVE QUIZ

This quiz will test your understanding of the basic text chapter objectives and give you additional practice problems. You should work this quiz after completing the end-of-chapter questions. The solutions to these questions are found at the end of this chapter.

1. Which of the following statements about Teflon® is incorrect?
   a. Teflon® is a polymer with –(– $CF_2$ – $CF_2$ –)– as the repeating unit.
   b. Teflon® is very resistant to corrosive chemicals.
   c. Most substances do not adhere to a Teflon® surface.
   d. Because $F_2$ is so reactive, Teflon® is chemically active.
   e. Teflon®'s properties owe to the very strong nature of the C – F bond.

2. Considering the electron configuration of the atom, explain why the following ions form.

   (a) $Ca^{2+}$ (b) $F^{-}$ (c) $S^{2-}$ (d) $Sc^{3+}$

3. In terms of energy, explain why an ionic solid exists in a regular, repeating lattice structure.

4. TRUE or FALSE: Covalent bonds result in lower energy of the bonding atoms.

5. How do the terms b*ond length* and *bond energy* describe a covalent bond?

6. TRUE or FALSE: The octet rule states that an atom will form covalent bonds to achieve a complement of eight valence electrons.

7. Draw Lewis dot symbols for:

   (a) Ar (b) Mg (c) P (d) $Ca^{2+}$

8. TRUE or FALSE: Excluding the noble gases, electronegativity increases up and to the right.

9. In terms of electronegativity, determine whether the following compounds contain nonpolar covalent, polar covalent, or ionic bonds.

   (a) I – I (b) NaI (c) Cl – I (d) H – I

10. Explain why the bond in $SnI_4$, *tin(IV) iodide*, is really more covalent than ionic, even though it is a bond between a metal and nonmetal.

11. TRUE or FALSE: In Lewis structures, if the central atom has fewer than eight electrons after all the valence electrons are placed in the molecule, lone pairs are used to create double or triple bonds to satisfy the octet rule.

12. Draw Lewis structures for the following.

    (a) $CS_2$ (b) $SCl_2$ (c) $SeI_4$ (d) $XeF_2$

13. Draw the resonance structures for $NO_2^-$.

14. Describe the formation of hybrid orbitals according to the valence bond theory.

15. For the molecules shown in Question 12, considering the central atom:

    (a) Identify the total number of electron pair groups
    (b) Identify the number of lone pairs

(c) Give the hybrid orbital type
(d) State the molecular shape
(e) Give the bond angle(s)

16. What are he expected bond angles in the following molecules? (a) $SiF_4$ (b) $OF_2$ (c) $KrF_4$

17. Sketch and name the molecular shape of the following theoretical molecules. A = central atom, B = bonded atom, E = lone pair of electrons

(a) $AB_2E$ (b) $AB_3E_2$ (c) $AB_4E_2$

18. How is a sigma bond different from a pi bond? How is each type of bond formed in a double bond?

19. State the hybrid orbitals used by each carbon atom in the following molecule.

$$CH_3-CH=CH-CH_2-\overset{\overset{\displaystyle O}{\|}}{C}-OH$$

a. b. c. d. e.

# ANSWERS TO THE CHAPTER OBJECTIVE QUIZ

1. **d**

2. **(a) Calcium's electron configuration shows two valence electrons: Ca [Ar] $4s^2$. Losing those two leaves a very stable noble gas electron configuration (full valence shell).**

   **(b) Fluorine's electron configuration shows seven valence electrons: F [He] $2s^2 2p^5$. Gaining one more electron makes a very stable full valence shell.**

   **(c) Sulfur has six valence electrons: S [Ne] $3s^2 3p^4$. Sulfur will gain two more electrons to fill the valence shell.**

   **(d) Scandium's electron configuration shows three valence electrons: Sc [Ar] $4s^2 3d^1$. Losing those three electrons leaves a very stable noble gas electron configuration.**

3. **The repeating array of ions within the lattice places oppositely charged ions closer to each other and like-charged ions farther from each other. This maximizes the attractive forces and minimizes the repulsive forces, creating a net force of attraction and lower energy.**

4. **TRUE**

5. ***Bond length* is the distance between the bonding nuclei when their energy is at a minimum. It represents the optimum distance between the atoms, a balance between a need for the atoms to move closer to overlap orbitals to form the bond and the nuclear repulsions when the atoms get too close.**
   ***Bond energy* is the difference in energy between the bonded atoms and the separate atoms. It represents the amount of lowering in energy that occurs when a covalent bond forms.**

6. **TRUE**

7. **Lewis dot symbols represent the *s* and *p* valence electron in the main group elements with dots.**

   **(a)** :Ar: (with dots above and below) **(b)** ·Mg· **(c)** ·P· (with two dots above and one below) **(d)** $Ca^{2+}$

8. **TRUE**

9. **(a) The bond in $I_2$ is a nonpolar covalent bond. The two iodine atoms have exactly the same electronegativity, so the electrons are shared perfectly equally.**

**(b) In NaI, there is a very large difference in electronegativity; so large that actual transfer of electrons occurs and ions form. This results in the formation of an ionic bond.**

**(c) The two bonded atoms in Cl – I are similar but not the same electronegativity. This produces a polar covalent bond, as the shared electrons are pulled closer to the more electronegative chlorine atom.**

**(d) The bond in H – I is also a polar bond. This bond is more polar than Cl – I because the electronegativity difference between the bonding atoms is greater.**

**10. The electronegativity difference between tin and iodine atoms is not that great. For tin it is 1.8 and for iodine it is 2.5. This difference of ΔEN = 0.7 would put the bond in the polar covalent range of bonding types.**

**11. TRUE**

**12. (a) (b) (c) (d)**

(a) :S=C=S:

(b) :Cl—S—Cl:

(c) $SeI_4$: Se bonded to four I atoms, with one lone pair on Se

(d) :F—Xe—F:

**13. Resonance structures are equivalent Lewis structures for a molecule that differ only in placement of the double bond. There are two resonance structures for the nitrite ion.**

[:O—N=O:]⁻ ⟷ [:O=N—O:]⁻

**14. The octet rule tells us that most main group elements will form a number of covalent bonds in order to share eight electrons and fill their valence shell. This does not explain how this occurs; many atoms don't have the necessary number of singly occupied orbitals to make these bonds. The valence bond theory says the central atom in a molecule will combine and mix a certain number of atomic orbitals, forming an equal number of identical hybrid orbitals. Since the hybrid orbitals are identical in shape and equal in energy, electrons can be promoted to the hybrid orbitals in such a way as to make the necessary number of bonds.**

**15. $CS_2$**

**(a) Identify the total number of electron pair groups: two**

(b) Identify the number of lone pairs: zero
(c) Give the hybrid orbital type: *sp*
(d) State the molecular shape: linear
(e) Give the bond angle(s): 180°

$SCl_2$

(a) Identify the total number of electron pair groups: four
(b) Identify the number of lone pairs: two
(c) Give the hybrid orbital type: $sp^3$
(d) State the molecular shape: bent
(e) Give the bond angle(s): ~109.5°

$SeI_4$

(a) Identify the total number of electron pair groups: five
(b) Identify the number of lone pairs: one
(c) Give the hybrid orbital type: $sp^3d$
(d) State the molecular shape: seesaw
(e) Give the bond angle(s): 90°, 120°, 180°

$XeF_2$

(a) Identify the total number of electron pair groups: five
(b) Identify the number of lone pairs: three
(c) Give the hybrid orbital type: $sp^3d$
(d) State the molecular shape: linear
(e) Give the bond angle(s): 180°

16. (a) The Lewis structure shows a tetrahedral shape which would give bond angle of 109.5 °.

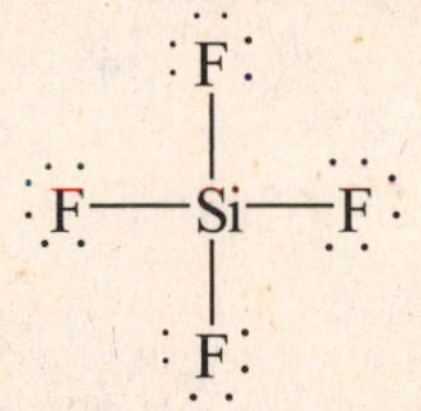

F
|
F—Si—F
|
F

(b) The Lewis structure shows that $OF_2$ would have a bent shape with approximately 109.5 ° bond angles.

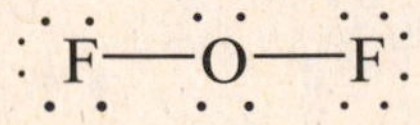

F—O—F

**(c)** **The Lewis structure For $KrF_4$ shows four bonded atoms and two lone pairs for the central atom. This would produce a seesaw shape with bond angle of 90, 120, and 180 °.**

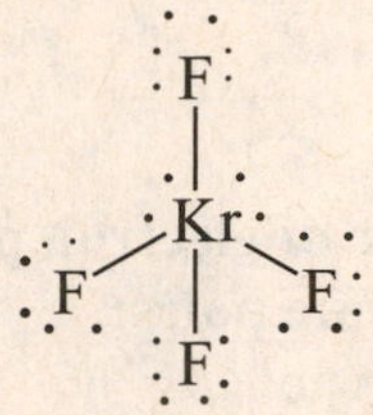

**17. (a) $AB_2E$** **(b) $AB_3E_2$** **(c) $AB_4E_2$**

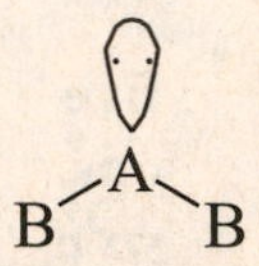

**bent, ~120°** 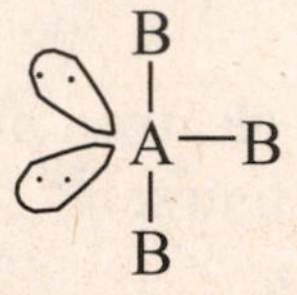 **T – shape** 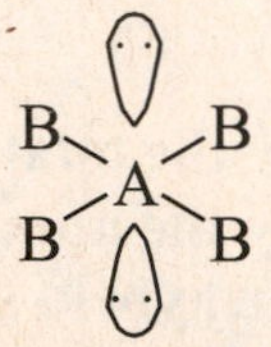 **square planar**

**18. A sigma (σ) bond is a covalent bond that results from the direct overlap of orbitals. In the formation of a double bond, these are *sp*$^2$ hybrid orbitals. A pi (π) bond is a covalent bond that results from overlap of *p*-atomic orbitals from the side. In a double bond, the second of the two covalent bonds is a pi bond that results from the side-on overlap of the unhybridized *p*-atomic orbitals left after the *sp*$^2$ hybridization. These *p* orbitals are perpendicular to the hybrid orbitals; therefore they must overlap from the side.**

**19.**

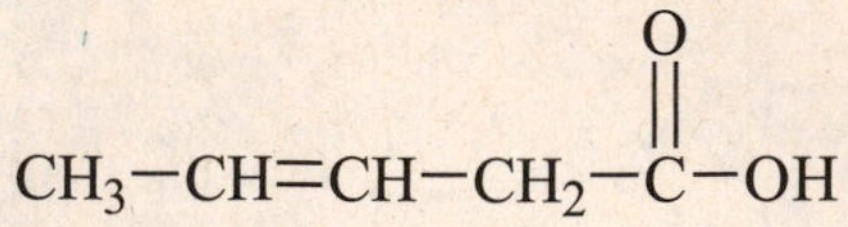

$CH_3-CH=CH-CH_2-C(=O)-OH$

**a. b. c. d. e.**

**a. *sp*$^3$** **b. *sp*$^2$** **c. *sp*$^2$** **d. *sp*$^3$** **e. *sp*$^2$**

# CHAPTER 8

## *Study Goals*

The study goals outline specific concepts to be mastered in this section of the text chapter. Related problems at the end of the text chapter will also be noted. Working the questions noted should aid in mastery of each study goal and also highlight any areas that you may need additional help in.

**Section 8.1 *INSIGHT INTO* Carbon**

1. Describe the differences between graphite and diamond on a molecular level and how that relates to their physical properties. *Work Problems 1–7, 16, 52.*

**Section 8.2 Condensed Phases – Solids**

2. Using a molecular perspective, describe the solid state. *Work Problems 8–12.*
3. Perform calculations using a basic unit cell of a crystalline solid, including packing efficiency, cell edge length, atomic radius, and density. *Work Problems 13–15, 17–21, 98.*

**Section 8.3 Bonding in Solids: Metals, Insulators, and Semiconductors**

4. Apply the band theory of bonding to metals, nonmetals, and metalloids. *Work Problems 22–30.*
5. Use the band theory to explain electrical conductivity in insulators, semiconductors, and conductors. Describe the process of producing n-type and p-type semiconductors. *Work Problems 31–38, 93, 96, 97, 104, 105.*

**Section 8.4 Intermolecular Forces**

6. Describe the most common types of intermolecular forces and use Lewis structures to identify which ones are present in molecular compounds. *Work Problems 39–51, 99, 102, 103.*

**Section 8.5 Condensed Phases – Liquids**

7. Use the molecular perspective to describe the liquid state.
8. Relate the strength of intermolecular forces to some common liquid properties and be able to make predictions of properties of molecules based on analysis of intermolecular forces. *Work Problems 53–64, 87–92, 94, 100.*

**Section 8.6 Polymers**

9. Compare and contrast the two most common methods of producing polymers: addition and condensation reactions. List several common types of polymers that are produced in this manner. *Work Problems 65–72.*
10. Discuss how the physical properties of polymers can be engineered through cross-linking, copolymers, grafts, and additives. *Work Problems 73–80, 95, 101.*

**Section 8.7** ***INSIGHT INTO*** **Micro–Electrical–Mechanical Systems (MEMS)**

11. Define Micro–Electrical–Mechanical Systems (MEMS) and describe how these systems are produced using chemical reactions (wet etching) or plasma etching (glow discharge). *Work Problems 81–86.*

## *Solutions to the Odd-Numbered Problems*

Problem 8.1 How many solid forms of carbon are known?

**There are three solid forms; these are also known as allotropes. Originally, diamond and graphite were the only solid forms of carbon known to exist. Now a third, Buckminsterfullerene (Buckyballs) and related fullerenes and nanotubes, is accepted.**

Problem 8.3 What property of diamond leads to the most engineering applications? Which types of applications would benefit most from this property?

**The extraordinary hardness of diamond is exploited in most engineering applications. Most of these applications would involve cutting, drilling, or machining of hard materials. Since diamond is harder than just about any other material, it is very useful for cutting or drilling these substances.**

Problem 8.5 What is the relationship between the structure of buckminsterfullerene and carbon nanotubes?

**The unique arrangement of carbon atoms in alternating pentagons and hexagons in buckminsterfullerene is also present in carbon nanotubes. Nanotubes are hollow cylinders as opposed to spheres. Carbon nanotubes are capped on the ends with hemispheres of buckminsterfullerene molecules.**

Problem 8.7 Use the web to look up the experimental conditions required to synthesize buckminsterfullerene in an electric arc. Would running an arc in air work?

**Answers will vary according to the websites found. Typically, a sharp graphite rod is touched to the surface of a graphite disk as a current of 100-200 amperes is passed between them. This occurs under conditions of a vacuum; and a soot of buckminsterfullerene is created as the vaporized carbon atoms form the $C_{60}$ molecules.**

**This method would not work in air because the carbon atoms would react with oxygen and nitrogen instead of forming the buckminsterfullerene molecules.**

Problem 8.9 Define *packing efficiency*.

**Packing efficiency measures the amount of space actually occupied by particles in a solid. The more space filled, the higher the packing efficiency. A packing efficiency of 100% means there is no open space at all. Packing efficiency is related directly to the density of a solid.**

Problem 8.11 Using pentagons, draw arrangements that demonstrate low packing efficiency and higher packing efficiency.

**Lower efficiency**

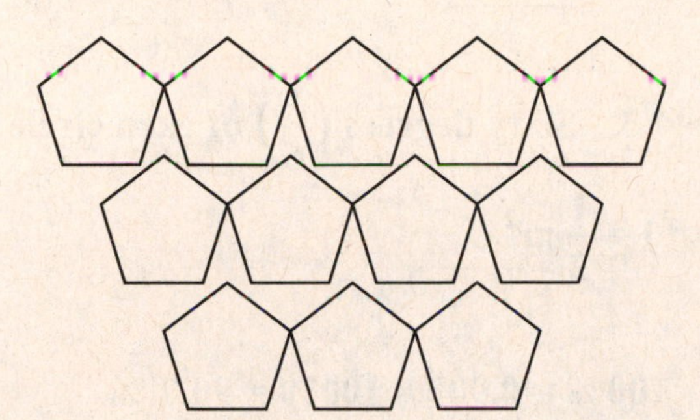

**Higher efficiency**

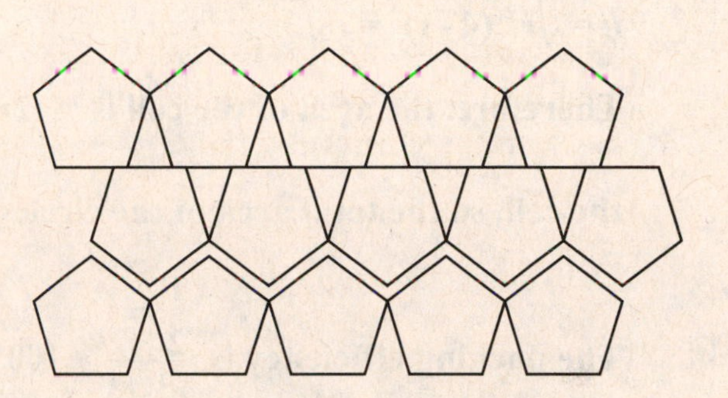

Problem 8.13 ■ Calculate the packing efficiencies for both two-dimensional arrangements shown in Figure 8.3.

**The packing efficiency for the two-dimensional figures can be defined as**

$$\frac{\textbf{Area of the marbles}}{\textbf{Area of the unit cell}} \times 100\ \%$$

**For (a) in Figure 8.3, the unit cell is a square with marbles at each of the four corners, and the unit cell edge = 2*r*. The unit cell in (b) is a hexagon but can be represented by an equilateral triangle with unit cell edge = 2*r***

**(a)**

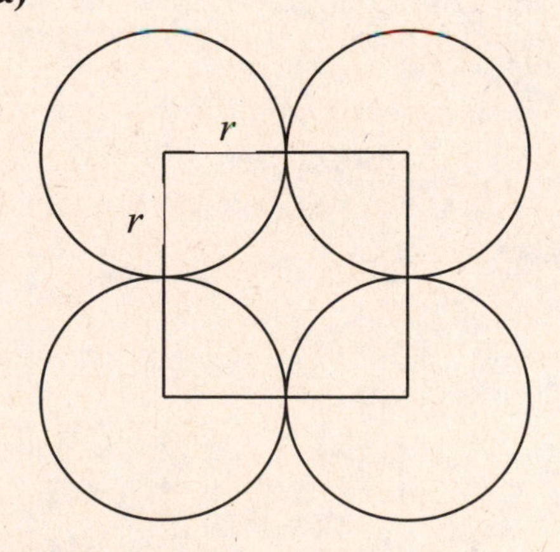


**(b)**

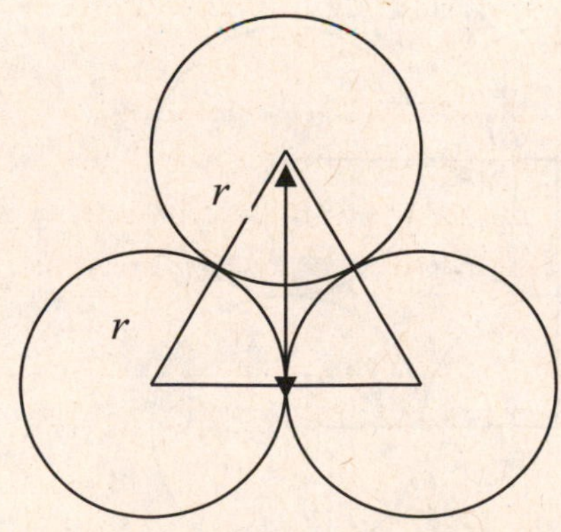

**In (a), the area of the cell = $2r \times 2r = 4r^2$. The area of the marbles is the area of one full circle (4 × ¼ circle) = $\pi r^2$. The packing efficiency is $\frac{\pi r^2}{4r^2} \times 100\% = \frac{\pi}{4} \times 100\% = 0.785 \times 100\% = 78.5\ \%$**

**In (b), the area of the cell = ½ (base) × (height) of the triangle. The base = $2r$. The height can be found using the Pythagorean theorem:**

$h^2 + r^2 = (2r)^2$, $\quad h = \sqrt{4r^2 - r^2}$

$h = \sqrt{r^2(4-1)} = r\sqrt{3}$

**Therefore, the area of the cell is $\frac{1}{2}(2r)(r\sqrt{3}) = r^2\sqrt{3}$. Sixty degrees $\left(\frac{1}{6}\right)$ of each circle is in the cell, so the total area of the circles is $3\left(\frac{1}{6}\pi r^2\right) = \frac{1}{2}\pi r^2$.**

**The packing efficiency is $\frac{\frac{1}{2}\pi r^2}{r^2\sqrt{3}} \times 100\% = \frac{\pi}{2\sqrt{3}} \times 100\% = 0.907 \times 100\% = 90.7\ \%$**

Problem 8.15 What is the coordination number of atoms in the diamond structure?

**Coordination number is the number of atoms bonded to the central atom in a molecule.**

**In the diamond structure, each carbon atom is bonded to four other carbons atoms. The coordination number is four.**

Problem 8.17 ■ Polonium is the only metal that forms a simple cubic crystal structure. Use the fact that the density of polonium is 9.32 g/cm$^3$ to calculate its atomic radius.

**In the simple cubic crystal structure there are four Po atoms at the corners of each unit cell.**

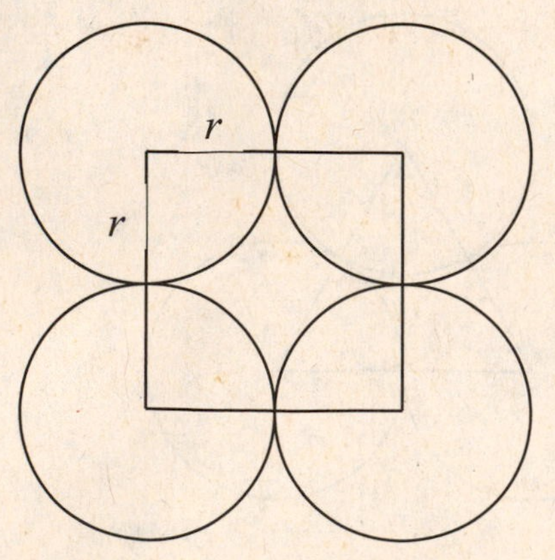

**The length of a unit cell, $a$, equals two times its radius, $a = 2r$. The number of Po atoms present in the simple cubic unit cell is $8 \times \frac{1}{8} = 1$ Po atom. The mass of one Po atom is:**

$$1 \text{ Po atom} \times \frac{1 \text{ mol Po}}{6.022 \times 10^{23} \text{ atoms Po}} \times \frac{209 \text{ g Po}}{1 \text{ mol Po}} = 3.47 \times 10^{-22} \text{ g}$$

**The volume of a unit cell is: $V_{cell} = (a)^3$**

**Using density, we can solve for the volume in terms of the cell edge length, $a$, and then finally find $a$.**

$$d = \frac{m}{V}; \qquad V = \frac{m}{d} = \frac{3.47 \times 10^{-22} \text{ g}}{9.32 \text{ g/cm}^3} = 3.72 \times 10^{-23} \text{ cm}^3$$

$$V_{cell} = (a)^3 = 3.72 \times 10^{-23} \text{ cm}^3; \qquad a = \sqrt[3]{3.72 \times 10^{-23} \text{ cm}^3} = 3.34 \times 10^{-8} \text{ cm}$$

**The cell edge $a = 2r$, so $r = a \div 2 = 3.34 \times 10^{-8} \text{ cm} \div 2 = 1.67 \times 10^{-8}$ cm.**

Problem 8.19 ■ Europium forms a body-centered cubic unit cell and has a density of 4.68 g/cm$^3$. From this information, determine the length of the edge of the cubic cell.

**The body-centered cubic unit cell contains $8 \times \frac{1}{8} + 1 = 2$ Eu atoms.**
**The mass of these two atoms is:**

$$2 \text{ Eu atoms} \times \frac{1 \text{ mol Eu}}{6.022 \times 10^{23} \text{ atoms Eu}} \times \frac{152.97 \text{ g Eu}}{1 \text{ mol Eu}} = 5.080 \times 10^{-22} \text{ g}$$

**The volume of this unit cell is, $V_{cell} = (a)^3$.**

**Using density, we can solve for the volume in terms of the cell edge length, a, and then finally find a.**

$$d = \frac{m}{V}; \qquad V = \frac{m}{d} = \frac{5.080 \times 10^{-22} \text{ g}}{4.68 \text{ g/cm}^3} = 1.09 \times 10^{-22} \text{ cm}^3$$

$$V_{cell} = (a)^3 = 1.09 \times 10^{-22} \text{ cm}^3; \qquad a = \sqrt[3]{1.09 \times 10^{-22} \text{ cm}^3} = 4.78 \times 10^{-8} \text{ cm}$$

Problem 8.21 The spacing in a crystal lattice can be measured very accurately by X-ray diffraction, and this provides one way to determine Avogadro's number. One form of iron has a body-centered cubic lattice, and each side of the unit cell is 286.65 pm long. The density of this crystal at 25°C is 7.874 g/cm$^3$. Use these data to determine Avogadro's number.

**We first need to determine the volume of a unit cell:**

$V_{cell}$ **= (286.65 pm)$^3$ = 2.3554 × 10$^7$ pm$^3$**

**Converting pm$^3$ to cm$^3$:**

$$2.3554 \times 10^7 \text{ pm}^3 \times \left(\frac{1\times10^{-10}\text{ cm}}{1\text{ pm}}\right)^3 = 2.3554 \times 10^{-23}\text{ cm}^3$$

**The mass of the unit cell is its volume times the density:** $m = V \times d$,

**Mass = 7.874 g/cm$^3$ × (2.3554 × 10$^{-23}$cm$^3$) = 1.855 × 10$^{-22}$ g**

**The body-centered cubic unit cell contains** $8 \times \frac{1}{8} + 1 = 2$ **Fe atoms, so the mass we just calculated is the mass of two iron atoms.**

**The mass of a single Fe atom is** $\frac{1.855\times10^{-22}\text{ g}}{2\text{ atoms}} = 9.273 \times 10^{-23}$ **g/atom.**
**Using the molar mass of iron from the periodic table (Fe, 55.847 g/mole), we know the mass of one mole of atoms and we know that Avogadro's number ($N_A$) represents the number of atoms in one mole.**
**We can now solve for Avogadro's number: 55.847 g/mole = $N_A$ × 9.273 × 10$^{-23}$ g/atom;**

$$N_A = \frac{55.847\text{ g/mol}}{9.273\times10^{-23}\text{ g/atom}} = 6.022 \times 10^{23}\text{ atoms/ molecule.}$$

Problem 8.23 One important physical property of metals is their malleability. How does the sea of electrons model account for this property?

**Our model of metallic bonding tells us that we have positive metal atom cores surrounded by a mobile sea of negative electrons. The atoms are held together by this attraction. When a deformation (change in shape) occurs in the solid, the forces of attraction holding it together don't change because the electrons are mobile: the atoms remain bonded. In contrast, when a deformation occurs in an ionic solid, the alignment of positive and negative ions shifts causing strong repulsive forces. The result is a fracture.**

Problem 8.25 How does the sea of electrons model of metallic bonding explain why metals are good conductors of electricity?

**Good conductors of electricity allow movement of charge, specifically electrons. In our model of metallic bonding, the "sea" of electrons is mobile, moving through an empty conduction band of molecular orbitals. When an electric potential is applied to a metal, electrons easily move: the metal conducts electricity.**

Problem 8.27 What is the difference between a bonding orbital and an antibonding orbital?

**In a bonding orbital the highest electron density lies between the two nuclei. This strengthens the bond between the two atoms when electrons occupy this type of orbital. In an antibonding orbital, the highest electron density is outside the two nuclei. There is actually a node – an area where electron cannot be found – between the two atoms. This weakens the bond when electrons occupy this type of orbital.**

Problem 8.29 Draw a depiction of the band structure of a metal. Label the valence band and conduction band.

**In a metal, there is essentially no gap in energy between the valence band of orbitals and the conduction band.**

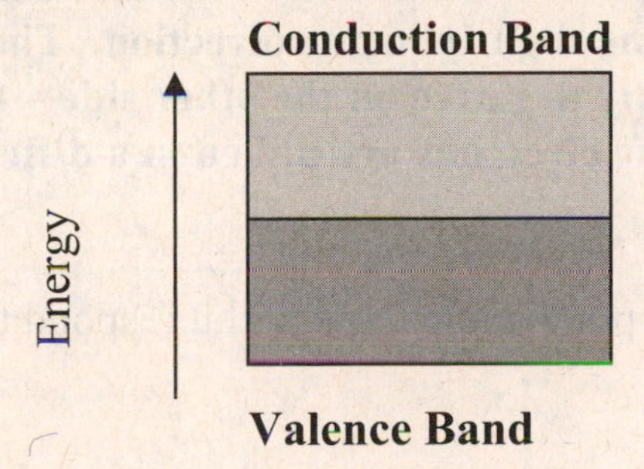


Problem 8.31 The conductivity of semiconductors increases as temperature is increased. Using band theory, explain this observation.

**Semiconductors have a small to moderate band gap between the conduction and valence bands. This means that some energy has to be added to the atoms of the semiconductor to allow electrons to reach the conduction band. The higher the temperature is, the greater the energy of the atoms, so a larger number of them are capable of reaching the conduction band. Therefore the semiconductor becomes more conductive at higher temperatures.**

Problem 8.33 Explain how doping affects the electronic properties of a semiconductor.

**Depending on what type of atom is used to produce the doping, either extra electrons are added (n-type) or electrons are missing (p-type). Both of these result in increased mobility of electrons in the semiconductor and increased conductivity.**

Problem 8.35 What type of atom is needed as a dopant in a p-type semiconductor? Why is it called p-type?

**The type of atom required for p-type doping is one that has fewer valence electrons than the semiconductor atoms being replaced. These dopant atoms produce missing electrons or "holes" in the lattice of semiconductor atoms, which increases the mobility of the electrons and therefore the conductivity of the material. The term *p-type* semiconductor refers to the electron "holes" which can be viewed as positive areas in the sea of electrons.**

Problem 8.37 How does a p-n junction serve as a voltage-gated switch?

**For a voltage-gated device, the flow of charge across some plane (in this case a junction between two types of semiconductors) is regulated by a change in voltage. Because an n-type semiconductor acts like a material with negative charges and a p-type semiconductor acts like a material with positive charges, the charge carriers of these materials will flow toward the opposite voltage polarity. With one polarity, the charge will flow across the junction, and when it is switched, the charge will flow away from the junction.**

Problem 8.39 What is an *instantaneous dipole*?

**An instantaneous dipole is created by the attraction of an atom (or molecule) towards a neighboring atom's electrons. This causes a distortion in the electron cloud of the neighboring atom, pulling the electrons slightly in one direction. The atom becomes slightly positive on one side and slightly negative on the other side – forming a dipole. This only occurs for an instant and then the electrons are pulled in a different direction by another neighboring atom.**

Problem 8.41 If a molecule is not very polarizable, how will it respond to an external electric field?

**Polarizability is the extent to which a molecule or atom responds to an external electric field. A material that is not very polarizable will not respond very strongly to an electric field.**

Problem 8.43 Most gaseous compounds consist of small molecules, while polymers are never gaseous at room temperature. Explain this observation based on intermolecular forces.

**Boiling point is related to strength of intermolecular forces; the stronger the intermolecular forces, the higher the boiling point. Most polymers are symmetrical molecules so even if they contain polar bonds, overall they are nonpolar molecules. The most significant intermolecular forces in polymers therefore are dispersion forces (induced-dipole forces), which get stronger as the size of the molecule increases. Since polymers are very large molecules, these are very strong forces. Small molecules usually have much weaker intermolecular forces. Polymers do not exist as gases at room temperature because of the very strong dispersion forces.**

Problem 8.45 Under what circumstances are ion-dipole forces important?

**When an ion is dissolved in a polar solvent, the permanent dipole of the solvent molecules can interact with the electrical charge of the ion.**

Problem 8.47 What is the unique feature of N, O, and F that causes them to play a role in hydrogen bonding?

**They are all three small, highly electronegative atoms. When hydrogen is bonded to one of these atoms an especially strong dipole is created.**

Problem 8.49 ■ Identify the kinds of intermolecular forces (London dispersion, dipole-dipole, or hydrogen bonding) that are the most important in each of the following substances. (a) methane ($CH_4$), (b) methanol ($CH_3OH$), (c) chloroform ($CHCl_3$), (d) benzene ($C_6H_6$), (e) ammonia ($NH_3$), (f) sulfur dioxide ($SO_2$)

**Hydrogen bonding forces are the strongest intermolecular forces, followed by dipole-dipole forces, and then dispersion forces.**

| | Molecule | Comment | Dominant Intermolecular Forces |
|---|---|---|---|
| (a) | $CH_4$ | nonpolar | London dispersion forces |
| (b) | $CH_3OH$ | polar with O-H bonds | hydrogen bonds |
| (c) | $CHCl_3$ | polar | dipole-dipole and London dispersion forces |
| (d) | $C_6H_6$ | nonpolar | London dispersion forces |
| (e) | $NH_3$ | polar with N-H bonds | hydrogen bonds |
| (f) | $SO_2$ | polar | dipole-dipole and London dispersion forces |

Problem 8.51 Carbon tetrachloride ($CCl_4$) is a liquid at room temperature and pressure, whereas ammonia ($NH_3$) is a gas. How can these observations be rationalized in terms of intermolecular forces?

**Carbon tetrachloride is a nonpolar molecule and therefore only has dispersion forces present. However, it is much more massive than ammonia and strength of dispersion forces increases with increasing size. Hydrogen-bonding forces are present in ammonia but it is a very small molecule and they are not strong enough to cause it to be liquid at room temperature.**

Problem 8.53 Describe how interactions between molecules affect the vapor pressure of a liquid.

**Vapor pressure is created when some liquid molecules (that have sufficient energy) leave the liquid state and enter the gaseous state. The stronger the intermolecular forces, the greater the strength of attractive forces holding molecules together. Molecules must overcome these forces to become gaseous. Therefore, the stronger the intermolecular forces, the lower the vapor pressure.**

Problem 8.55 ■ Answer each of the following questions with *increases, decreases,* or *does not change.*
(a) If the intermolecular forces in a liquid increase, the normal boiling point of the liquid ______.
(b) If the intermolecular forces in a liquid decrease, the vapor pressure of the liquid ______.
(c) If the surface area of a liquid decreases, the vapor pressure ______.
(d) If the temperature of a liquid increases, the equilibrium vapor pressure ______.

**(a) increases**
**(b) increases**
**(c) does not change**
**(d) increases**

Problem 8.57 ■ Which member of each of the following pairs of compounds has the higher boiling point? (a) $O_2$ or $N_2$, (b) $SO_2$ or $CO_2$, (c) HF or HI, (d) $SiH_4$ or $GeH_4$

**To vaporize a liquid, sufficient heat energy must be supplied to the molecules to overcome the intermolecular forces. The stronger the intermolecular forces, the higher the boiling point will be. To answer the question then, we need to determine which intermolecular forces are present and their relative strength.**

**(a) $O_2$: Only dispersion forces are present and oxygen is larger.**
**(b) $SO_2$: $CO_2$ is nonpolar and has only weak dispersion forces. $SO_2$ is polar and has stronger dipole-dipole forces.**
**(c) HF: Both are polar but relatively strong hydrogen bonding forces are present in HF.**
**(d) $GeH_4$: Both are nonpolar, only dispersion forces are present and $GeH_4$ is larger.**

Problem 8.59 A substance is observed to have a high surface tension. What predictions can you make about its vapor pressure and boiling point?

**Strong intermolecular forces cause higher surface tension; the molecules of the substance are more highly attracted to each other than some other surface. Stronger intermolecular forces also cause lower vapor pressure and higher boiling point as the more the molecules are attracted to each other, the more difficult it will be form molecules to enter the gaseous state.**

Problem 8.61 ■ Suppose that three unknown substances are liquids at room temperature. You determine that the boiling point of substance A is 53°C, that of substance B is 117°C, and that of substance C is 77°C. Based on this information, rank the three substances in order of their vapor pressures at room temperature.

**The stronger the intermolecular forces of a substance, the higher the boiling point will be. We can examine the boiling points of the three substances and determine that substance B has the strongest intermolecular forces, substance A the weakest.**
**Stronger intermolecular forces also causes lower vapor pressure, so substance B will have the lowest vapor pressure and substance A will have the highest.**

Problem 8.63 Draw the meniscus of a fluid in a container where the interactions among the liquid molecules are weaker than those between the liquid and the container molecules.

**The meniscus will be curved downward because the liquid molecules are more highly attracted to the container than each other.**

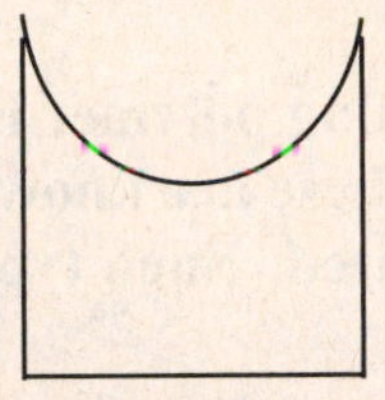

Problem 8.65 If shown structures of monomers that might polymerize, what would you look for to determine if the type of polymerization would be addition?

**There must be a carbon-carbon double bond present in the monomer for addition polymerization to take place.**

Problem 8.67 What is meant by the term *chain termination* in a polymerization reaction?

**A polymerization reaction can be thought of as a process where monomers are added to a growing "chain". This process is referred to as *propagation*. When a reaction occurs that stops this propagation process from continuing, it's called *chain termination*. This ends the growth of the polymer.**

Problem 8.69 Why is there no isotactic or syndiotactic form of polyethylene?

**These two terms refer to the geometry of the molecule. Isotactic polymers have the methyl groups or some other unique atom or group bonded to the carbon chain all pointed forward or all pointed backward. Syndiotactic polymers have these groups alternately pointed either forward or backward. In polyethylene, all the atoms bonded to the carbon chain are the same, so there is no difference geometrically.**

Problem 8.71 What are the products of a condensation polymerization reaction?

**There are always two products: the polymer formed and some other small molecule such as water.**

Problem 8.73 Distinguish between a block copolymer and a graft copolymer.

**In a block copolymer, the different monomers present form the polymer chain in alternating groups.**
**In a graft copolymer, the different groupings of monomers are attached to the polymer chain as side chains.**

Problem 8.75 What is the key physical property that characterizes a thermoplastic polymer?

**A thermoplastic polymer will soften and eventually melt as it is heated to high temperatures.**

Problem 8.77 What happens molecularly in a thermosetting polymer when it is heated?

**When a thermosetting polymer is heated, a reaction occurs that connects the polymer chains together. These are known as crosslinks. This process prevents the polymer from softening when heated. Such types of molecules will actually decompose before they melt.**

Problem 8.79 Use the web to find out what pencil erasers are made of. Would it be possible to form a new shape of an eraser by melting it and pouring it into a mold?

**Answers will vary according to the websites found. Erasers are crosslinked polymers, making them thermosetting substances. This means that they cannot be melted and remolded.**

Problem 8.81 Define the terms *substrate* and *mask* in the context of microfabrication of MEMS devices.

**In the etching of materials like silicon, the original material is called the *substrate*. It is covered with a layer of nonreactive material called the *mask*. During etching, the substrate beneath the mask is not exposed to the reactive species (or plasma discharge) and is therefore not removed.**

Problem 8.83 Silicon dioxide is also used as a masking material in microfabrication. If a silicon substrate with a $SiO_2$ mask is wet etched by $HNO_3$/HF, one of the reactions that takes place will be between $SiO_2$ and HF. Provide a reasonable explanation as to why $SiO_2$ can be used as a mask even though it will react with the etchant.

**The rate at which $SiO_2$ reacts (5 nm/hr) is so much slower than the rate Si reacts (1 μm/min), that for all practical purpose it doesn't react. (The silicon reacts 12,000 times faster). Significant depth of etching and substrate removal can quickly be achieved even though technically the mask is also being etched.**

Problem 8.85 Estimate the number of atoms that would have to be removed from a silicon substrate to produce a rectangular channel 5 $\mu$m wide, 10 $\mu$m deep, and 20 $\mu$m long. (The density of silicon is 2.330 g cm$^{-3}$.)

**The volume of silicon removed is equal to the width × depth × length of the channel. Multiplying the volume times the density gives the mass. Using the molar mass of silicon and Avogadro's number, the number of atoms is found. 1 $\mu$m = 1 × $10^{-6}$ m**

$$(5 \times 10^{-6}\ \text{m}) \times (10 \times 10^{-6}\ \text{m}) \times (20 \times 10^{-6}\ \text{m}) = 1 \times 10^{-15}\ \text{m}^3$$

$$1 \times 10^{-15}\ \text{m}^3 \times \frac{(100\ \text{cm})^3}{\text{m}^3} \times \frac{2.330\ \text{g}}{\text{cm}^3} \times \frac{1\ \text{mole Si}}{28.09\ \text{g}} \times \frac{6.022 \times 10^{23}\ \text{atoms}}{1\ \text{mole Si}} = 4.995 \times 10^{13}\ \text{atoms}$$

Problem 8.87 ■ Use the vapor pressure curves illustrated here to answer the questions that follow.

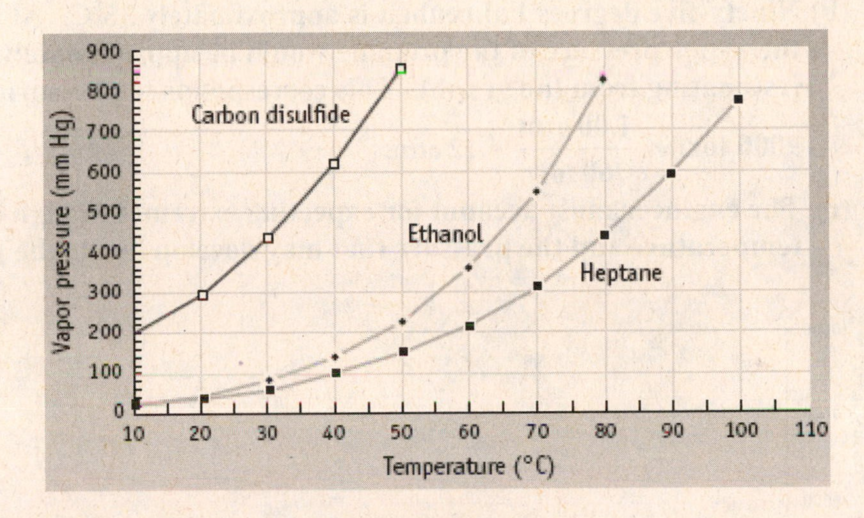


(a) What is the vapor pressure of ethanol ($C_2H_5OH$) at 60°C?
(b) Considering only carbon disulfide ($CS_2$) and ethanol, which has the stronger intermolecular forces in the liquid state?
(c) At what temperature does heptane ($C_7H_{16}$) have a vapor pressure of 500 mm Hg?
(d) What are the approximate normal boiling points of each of the three substances?
(e) At a pressure of 400 mm Hg and a temperature of 70°C, is each substance a liquid, a gas, or a mixture of liquid and gas?

**(a) 350 mm Hg**

**(b) ethanol (lower vapor pressure at every temperature)**

**(c) 84°C**

**(d) $CS_2$: 46°C $C_2H_2OH$: 78°C $C_7H_{16}$: 99°C**

**(e) $CS_2$: gas $C_2H_2OH$: gas $C_7H_{16}$: liquid**

Problem 8.89 ■ The following data below show the vapor pressure of liquid propane as a function of temperature. (a) Plot a vapor pressure curve for propane and use it to estimate the normal boiling point. (b) Use your curve to estimate the pressure (in atm) in the propane tank supplying fuel for a gas barbecue grill on a hot summer day when the temperature is 95°F. (c) What implications might your answer to (b) have for an engineer designing propane storage tanks?

| T (°C) | −100 | −80 | −60 | −40 | −20 | 0 | 20 |
|---|---|---|---|---|---|---|---|
| P (torr) | 22.4 | 100.6 | 328.8 | 856.9 | 1888.6 | 3665.6 | 6445.9 |

**(a) The normal boiling point would be the temperature where the vapor pressure of propane equals atmospheric pressure or 760 torr. This occurs at approximately −48°C.**

**(b) Ninety-five degrees Fahrenheit is approximately 35°C. At that temperature, the vapor pressure of the propane would be approximately 9000 torr (estimating from the graph). This corresponds to pressure in atmospheres =**

$$9000 \text{ torr} \times \frac{1.00 \text{ atm}}{760 \text{ torr}} \approx 12 \text{ atm.}$$

**(c) The engineer must account for expecting extremes in high daytime temperature and the pressure that may develop inside the propane tank.**

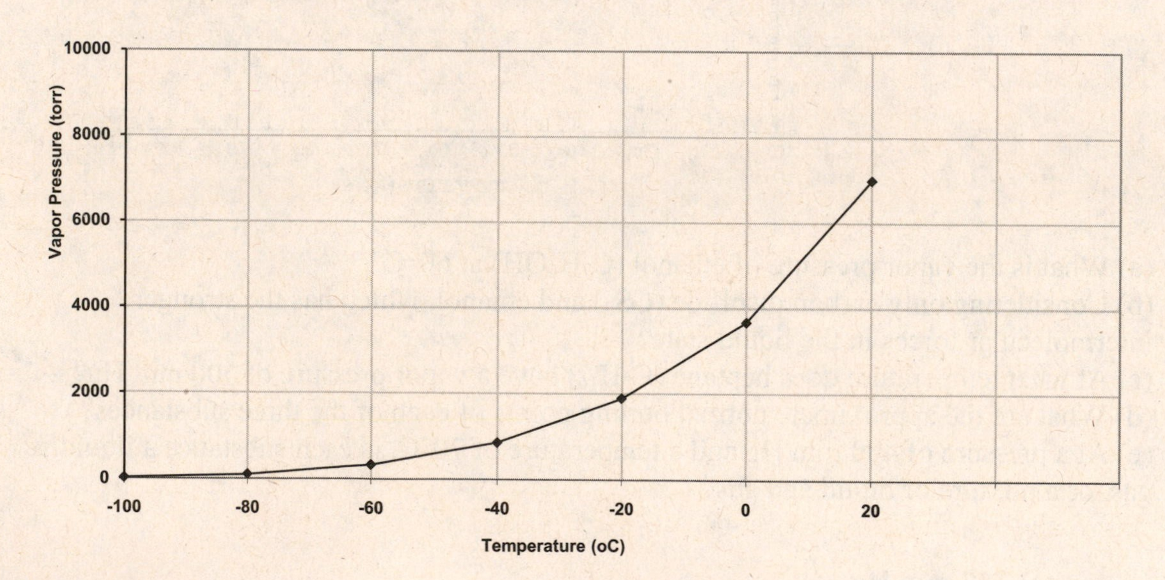


Problem 8.91 ■ If you place 1.0 L of ethanol ($C_2H_5OH$) in a room that is 3.0 m long, 2.5 m wide, and 2.5 m high, will all the alcohol evaporate? If some liquid remains, how much will there be? The vapor pressure of ethyl alcohol at 25°C is 59 mm Hg, and the density of the liquid at this temperature is 0.785 g/cm$^3$.

**We'll first calculate how much ethanol would evaporate in a room this size.**

$$1.0 \times 10^3 \text{ mL} \times \frac{1 \text{ cm}^3}{1 \text{ mL}} \times \frac{0.785 \text{ g}}{1 \text{ cm}^3} \times \frac{1 \text{ mol } C_2H_5OH}{46.07 \text{ g}} = 17 \text{ mol } C_2H_5OH$$

$$V_{room} = 3.0 \text{ m} \times 2.5 \text{ m} \times 2.5 \text{ m} \times \frac{(10^2 \text{ cm})^3}{(1 \text{ m})^3} \times \frac{1 \text{ mL}}{1 \text{ cm}^3} \times \frac{1 \text{ L}}{10^3 \text{ mL}} = 1.9 \times 10^4 \text{ L}$$

$$P = 59 \text{ mm Hg} \times \frac{1.00 \text{ atm}}{760 \text{ mm Hg}} = 0.0776 \text{ atm}$$

$$n = \frac{PV}{RT} = \frac{(0.0776 \text{ atm})(1.9 \times 10^4 \text{ L})}{(0.08206 \text{ L} \cdot \text{atm/mol} \cdot \text{K})(298 \text{ K})} = 60. \text{ moles ethanol}$$

**Less ethanol is available (17 mol) than would be required to completely fill the room with vapor (60. mol), so all of the ethanol will evaporate.**

Problem 8.93 Do a web search to find out how p- and n-type semiconductors can be used in designing solar panels. Write a short paragraph explaining what you learn. What new types of materials are being considered for use in solar panel systems?

**Answers will vary according to the website found.**

**Solar panels are made of a layer of n-type semiconductor beneath a layer of p-type semiconductor material. When sun energy (a light photon) strikes an atom of the n-type semiconductor, an electron can be ejected from the excited atom. The electron travels towards an atom in the n-type semiconductor across the p-n junction. The movement of electrons across the p-n junction in response to the photons creates the electrical current.**

**Some new innovations include adding titanium and vanadium to the semiconductors. This allows for absorption of infrared photons, boosting efficiency. Iron sulfide and copper sulfide are being investigated for replacing crystalline silicon in the solar cells.**

Problem 8.95 It is possible to make syndiotactic polystyrene, although most polystyrene is atactic. Syndiotactic polystyrene melts at 270°C while atactic polystyrene melts closer to 240°C. (a) Why might this difference in structure result in different boiling points? (b) If your supplier charges 20% more for syndiotactic polystyrene, what factors of a design might call for using this material despite the higher cost?

**Melting points (and many other properties) of polymers is related to how the molecules pack together in the solid state. Syndiotactic molecules can pack closer because of the regular arrangement of the side branches. This will increase the strength and melting point of the plastic.**
**If the polystyrene is being used in a high-temperature application then the higher cost of the syndiotactic polymer might be worthwhile.**

Problem 8.97 The doping of semiconductors can be done with enough precision to tune the size of the band gap in the material. Generally, in order to have a larger band gap, the dopant should be smaller than the main material. If you are a materials engineer and need a semiconductor that has lower conductivity than pure silicon, what element or elements could you use as your dopant? (You do not want either an n- or a p-type material.) Explain your reasoning.

**To avoid creating either an n- or p-type semiconductor, the dopant would have to have the same number of valence electrons as the silicon. The dopant must also be a smaller atom to produce a larger band gap, thus lowering conductivity. Carbon is really the only element that fits these requirements.**

Problem 8.99 If a pin is run across a magnet a number of times, it will become magnetized. How could this phenomenon be used as an analogy to describe polarization in atoms and molecules?

**The magnet is providing an external field that is inducing an alignment of $e^-$ in the pin. Polarization is similar in that an external field is used to alter the electron distribution of a molecule.**

Problem 8.101 A materials engineer, with an eye toward cost, wants to obtain a material whose degree of polymerization is high. What types of measurements must be made in the laboratory to determine whether the degree of polymerization is acceptable?

**To measure the degree of polymerization, the engineer would have to measure the molecular mass of the polymer and divide by the molecular mass of the monomer.**

Problem 8.103 In previous chapters, we have noted that cryolite is used in refining aluminum. Use the web to look up what this addition does for the process, and relate it to the concepts of intermolecular forces from this chapter.

**Answers will vary according to the website found. Cryolite is a very useful flux for separating the Al and O in aluminum oxide.**

Problem 8.105 Use the web to look up the percentage of dopant for a commercially available n-type semiconductor. Imagine that you were setting up a process for doping 1 metric ton of silicon with this dopant. **(a)** What mass would be required? **(b)** What would be the mole fraction of the dopant?

**Answers will vary according to the website found. A typical number might be 1 ppm As in Si.**

$$\frac{1\text{ g As}}{1\times10^{6}\text{ g Si}} \times (1000 \times 10^{3}\text{ g Si}) = 1\text{ g As}$$

$$1\text{ g As} \times \frac{1\text{ mol As}}{74.92\text{ g As}} = 0.0133\text{ g As} \qquad (1\times10^{6}\text{ g Si}) \times \frac{1\text{ mol Si}}{28.09\text{ g Si}} = 35600\text{ g Si}$$

**Total: 35600.0133g** $X_{As} = \dfrac{0.0133 \text{ mol As}}{35600.0133 \text{ mol}} \times 100 = 3.74 \times 10^{-5} \text{ mol \%}$

## CHAPTER OBJECTIVE QUIZ

This quiz will test your understanding of the basic text chapter objectives and give you additional practice problems. You should work this quiz after completing the end-of-chapter questions. The solutions to these questions are found at the end of this chapter.

1. Which of the following statements about the allotropes of carbon is false?
   a. In diamond, the carbons atoms are bonded together in a three-dimensional tetrahedral lattice.
   b. Graphite has similar properties to diamond because they are both made of only carbon atoms.
   c. Diamond is very hard and inert because of the strong covalent bonds holding all the atoms together.
   d. In graphite, carbon atoms are bonded together in two-dimensional planes.
   e. Graphite is soft because only weak forces hold the planes of C atoms together.

2. Consider one solid with a packing efficiency of 74% and another with packing efficiency of 68%. Which one would have a higher density? Why?

3. Which of the following statements about the solid state is incorrect?
   a. There is little atomic or molecular motion.
   b. Most solids exist in the amorphous (random) state.
   c. By arranging in a crystalline lattice structure, particles in the solid state exist at lower energy.
   d. Solids have relatively high densities because the particles are packed closely together.
   e. The volumes of solids are little affected by moderate changes in pressure and temperature.

4. TRUE or FALSE: When metal atoms bond together, they pool valence electrons and valence orbitals together, forming a filled "valence" band of orbitals and an empty "conduction" band.

5. TRUE or FALSE: Metals have a relatively large difference in energy between the valence band of orbitals and the conduction band.

6. Draw a band diagram for the following:
   (a) Ca
   (b) S
   (c) Si doped with As
   (d) Si doped with Ga

7. Rank the following elements in order of increasing conductivity by considering the band theory of bonding in solids.
   (a) Sr (b)Ti (c) C (d) Si doped with As (e) Si

8. Describe why a metal like magnesium is able to conduct electricity very well using the band theory of bonding in solids.

9. Give a brief description of the following intermolecular forces:
   (a) Dispersion forces
   (b) Dipole-dipole forces
   (c) Hydrogen bonding forces

10. For the following molecules, describe which intermolecular forces are present and rank them in order of increasing strength.
   (a) $C_4H_{10}$ (b) Br–I (c) $CH_4$ (d) F–Cl (e) $CH_3NH_2$

11. Which of the following statements about intermolecular forces and liquid properties is false?
   a. The hydrogen bonding force is the strongest intermolecular force.
   b. Stronger intermolecular forces cause a higher boiling point in the liquid state.
   c. Stronger intermolecular forces cause a higher vapor pressure in the liquid state.
   d. Water has a high surface tension because of its strong intermolecular forces.
   e. Gasoline is more volatile than water.

12. Why is a completely nonpolar substance like carbon tetrachloride, $CCl_4$, a liquid at room temperature while another nonpolar substance like silane, $SiH_4$, is a gas?

13. Generally, the larger a molecule is, the higher its boiling point because of increasing strength of dispersion forces in the larger molecules. Explain the following ranking of boiling points for the Group 16 hydrides: $H_2S < H_2Se < H_2Te < H_2O$

14. Describe the difference between a polymer produced by an addition reaction and one produced by a condensation reaction.

15. Which of the following statements about polymers is false?
   a. Addition polymers begin with the formation of a free radical.
   b. The degree of polymerization is defined as $\frac{\text{molar mass of the polymer}}{\text{molar mass of the monomer}}$.
   c. HDPE and polystyrene are examples of condensation polymers.
   d. A small molecule such as water is eliminated during a condensation reaction.
   e. Creating copolymers or graft polymers is a very useful way to engineer specific properties in the polymer.

16. TRUE or FALSE: In the context of microfabrication of MEMS devices, etching can be achieved by either using chemical reactions or a plasma discharge.

# ANSWERS TO THE CHAPTER OBJECTIVE QUIZ

1. b

2. The packing efficiency of a structure represents the percentage of space that is occupied by a given arrangement. Different crystal structures will produce different packing efficiencies and this is directly related to the density of the solid. Higher packing efficiency will produce a higher density, so the solid with the 74% efficiency will have a higher density.

3. b

4. TRUE

5. FALSE

6. (a) Calcium is a metal, so the band diagram would show a zero band gap (difference in energy) between the bands.

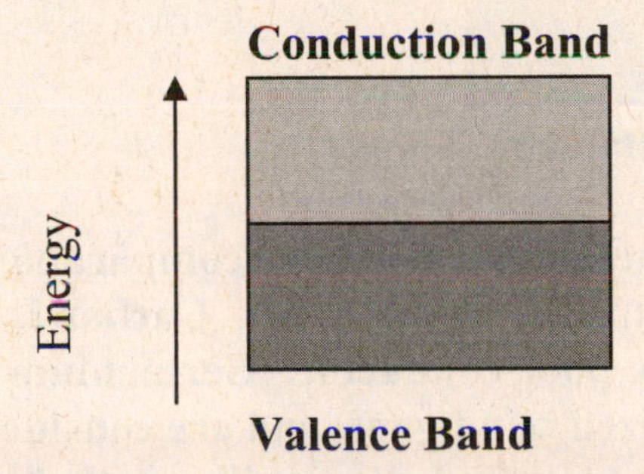


(b) Sulfur in a nonmetal, so there would be a large band gap.

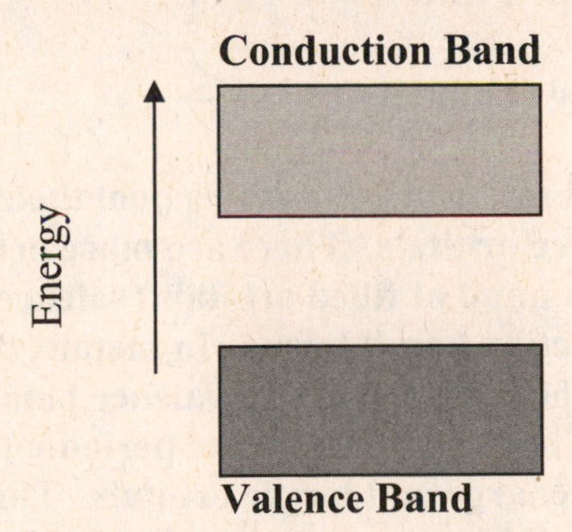


(c) Silicon doped with arsenic atoms would produce an n-type semiconductor. The band gap will be moderately sized and some electrons will occupy the conduction band.

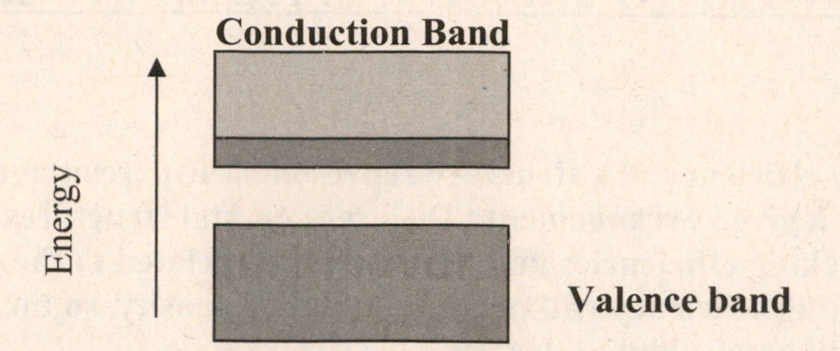


**(d) Silicon doped with gallium atoms will produce a p-type semiconductor, so the band gap will be <u>moderately sized</u> and there will be empty space in the valence band.**

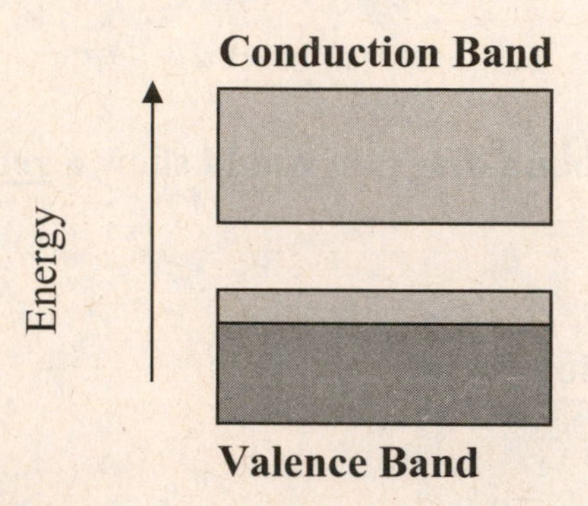


7. **Titanium and strontium are metals and will have comparable electrical conductivity because most metals have essentially no band gap. Carbon is a nonmetal and will have a large band gap and be a poor conductor. Germanium and silicon are both metalloids, have moderately-sized band gaps, and are considered semiconductors. Doping the silicon with arsenic (n-type doping) will greatly increase its conductivity.**

   **C < Si < Si doped with As < Ti, Sr**

   **→ increasing conductivity →**

8. **When a metal such as magnesium bonds, the atoms pool their valence electrons together, as well as their valance orbitals. There are more orbitals than are required for the electrons, so a band of filled orbitals (valence band) forms and a band of empty orbitals (conduction band) forms. In metals, there is essentially no difference in energy between the conduction and valence bands. This is because metals are located more to the left and down on the periodic table, meaning they have a lower effective nuclear charge and larger orbitals. This zero band gap allows metals to conduct electricity well because just a small electrical potential applied to the atoms causes electrons to travel through the conduction band.**

**9.** **(a) Dispersion forces. These are generally weak intermolecular forces caused by an atom's attraction to a neighboring atom's electrons. This causes a small temporary dipole to be induced by the distortion of the electron cloud. Although small and temporary, these induced dipoles form continually and create a net force of attraction. Dispersion forces are present in all form of matter and become increasingly stronger as the size of the atoms or molecules increases.**

**(b) Dipole-dipole forces. This is a net force of attraction between neighboring molecules with a permanent dipole moment. This means that this intermolecular force is present in only polar molecules and becomes stronger as the magnitude of the dipole moment increases.**

**(c) Hydrogen-bonding forces. This is an especially strong dipole-dipole force that forms between molecules that have hydrogen atoms bonded to N, O, or F atoms. These three atoms are all small, highly electronegative atoms that create strong dipoles when bonded to hydrogen atoms.**

**10.** **(a) $C_4H_{10}$ is a nonpolar molecule so only dispersion forces are present. (b) Br–I is a polar molecule so both dispersion and dipole-dipole forces are present. (c) $CH_4$ is also nonpolar. It is smaller than $C_4H_{10}$ so the dispersion forces are weaker. (d) F–Cl is also polar, the electronegativity difference is larger, meaning the dipole-dipole force is stronger. Dispersion forces are also present, but not that significant. (e) Hydrogen is bonded to nitrogen in $CH_3NH_2$, so hydrogen-bonding forces are present in addition to dispersion forces.**

$$CH_4 < C_4H_{10} < Br\text{–}I < F\text{–}Cl < CH_3NH_2$$

**→ increasing intermolecular strength →**

**11.** **c**

**12.** **Carbon tetrachloride is a much heavier molecule (molar mass = 153.81 g/mol compared to silane, molar mass = 32.12 g/mol). The only intermolecular forces affecting the boiling points are dispersion forces. Since $CCl_4$ is about five times larger, these are much stronger and are causing its boiling point to be much higher than silane's.**

**13.** **The ranking shows water apparently out of place in terms of boiling point. Excluding water, the hydrides follow the trend of increasing boiling point with increasing molecular size, due to increasing strength of intermolecular forces. Water has the highest boiling point because only $H_2O$ has hydrogen–bonding forces influencing its properties. Hydrogen bonding is much stronger than the other molecules' intermolecular forces, causing the boiling point of water to be much higher than expected.**

**14. In addition reactions, a radical is formed (a specie with an unpaired electron) which attacks the double bond of the monomer molecule. This forms another radical that reacts with another monomer molecule. The polymer chain continues to grow in this manner, adding one monomer at a time. This is why the reaction is called addition.**

**In condensation reactions, there are two different functional groups on the ends of two monomer molecules that react to form a small molecule. This small molecule splits off separately leaving the two monomer bonded together. This process continues as more monomers are bonded to the polymer chain.**

**15. C**

**16. TRUE**

# CHAPTER 9

## *Study Goals*

The study goals outline specific concepts to be mastered in this section of the text chapter. Related problems at the end of the text chapter will also be noted. Working the questions noted should aid in mastery of each study goal and also highlight any areas that you may need additional help in.

### Section 9.1 *INSIGHT INTO* Energy Use and the World Economy

1. Recognize the relationship between energy use and economic and social advancement. *Work Problem 1.*
2. List several sources of energy production in the U.S. along with consumptions. *Work Problems 2–5.*

### Section 9.2 Defining Energy

3. Define energy, heat, and work; be familiar with common energy units and be able to convert energy quantities in these units. *Work Problems 6–16, 102.*

### Section 9.3 Energy Transformation and Conservation of Energy

4. Define the thermodynamic terms system, surroundings, universe, and boundary; understand the first law of thermodynamics. *Work Problem 20.*
5. Express the energy change of a system, $\Delta E$, in terms of heat and work. *Work Problems 17–19, 21–28, 88.*

### Section 9.4 Heat Capacity and Calorimetry

6. Calculate heat flows using either molar heat capacity or specific heat. *Work Problems 30–35, 39, 40, 81, 87, 91, 94, 98–100, 104.*
7. Describe calorimetry and calculate heat flows or energy density in a calorimetry experiment. *Work Problems 29, 36–38, 77–80.*

### Section 9.5 Enthalpy

8. Define enthalpy and describe how it is related to heat flows. *Work Problems 41–45, 82.*
9. Calculate heat flows during phase changes using heats of fusion, vaporization, or solution. *Work Problems 46–50, 84, 89, 90, 96.*
10. Understand how the energy change of a reaction occurs because of the bonds breaking and forming. *Work Problems 60, 61, 86, 93.*
11. Describe the heat flows occurring in various chemical reactions using enthalpies of reaction. *Work Problems 51–53.*

### Section 9.6 Hess's Law and Heats of Reaction

12. Describe Hess's law and, given thermochemical equations, use it to find the heat of reaction for another reaction. *Work Problems 54–57.*

13. Express Hess's law mathematically in terms of formation reactions and calculate heats of reaction using this formula. *Work Problems 58, 59, 95.*

**Section 9.7 Energy and Stoichiometry**

14. Using a thermochemical equation, relate the heat of reaction to the stoichoimetry of the equation and calculate the heat flow when a specific amount of reaction occurs. *Work Problems 62–72, 83, 85, 92, 97, 101, 103.*

**Section 9.8 *INSIGHT INTO* Power Distribution and the Electrical Grid**

15. Be familiar with the role of the electrical grid system and how electricity moves from the generating station to the consumer. *Work Problems 73–76.*

## *Solutions to the Odd-Numbered Problems*

Problem 9.1 List reasons why there might be a connection between the amount of energy used by a country and its economic development.

**A large amount of energy used would indicate a large amount of industry, businesses, automobiles and heating and cooling. All of this would indicate an advanced economy in the country and a high standard of living.**

Problem 9.3 What differences (if any) would you expect to find between energy use patterns of the United States shown in Figure 9.2 and Figure 9.3 and use in **(a)** European countries, **(b)** developing countries?

**(a) In Europe, which is similar in development to the United States, we would expect similar energy production and consumption. One might expect the energy used in transportation to be less because the population is not spread as far apart in Europe as in the United States.**

**(b) Developing countries have much less industry, business, and transportation than in other countries. The energy consumption would probably not include nuclear or renewable sources of energy. Conversion losses would probably be higher in developing countries also.**

Problem 9.5 Using the value you calculate from **(a)** in the previous problem and the fact that on average barrel of petroleum has an energy equivalent of 5.85 million Btu, how many barrels of oil, on average, must be imported each day?

**From Figure 9.2, the amount of energy that was imported in the year 2011 was $28.59 \times 10^{15}$ Btu.**
**Solving for the amount of oil equivalent of that amount on a daily basis:**

$$\frac{28.59 \times 10^{15} \text{ Btu}}{\text{year}} \times \frac{1 \text{ year}}{365 \text{ days}} \times \frac{\text{barrel of oil}}{5.85 \times 10^6 \text{ Btu}} = 13.4 \text{ million barrels of oil / day}$$

Problem 9.7 Define the term *internal energy*.

***Internal energy*** **is all the various forms of energy contained within a specific amount of a substance. This includes kinetic, potential, heat, and all other forms of energy.**

Problem 9.9 What is the kinetic energy of a single molecule of oxygen if it is traveling at $1.5 \times 10^3$ m/s?

**Kinetic energy is defined as one-half the mass times velocity squared: $KE = ½ mv^2$. The mass of a single $O_2$ molecule is:**

$$\frac{32.00 \text{ g } O_2}{1 \text{ mol } O_2} \times \frac{1 \text{ mol } O_2}{6.022 \times 10^{23} \text{ molecules} O_2} = 5.31 \times 10^{-23} \text{ g/molecule}$$

**We want to express mass in kilograms when we calculate the kinetic energy (recall $J = \frac{\text{kg m}^2}{\text{s}^2}$).**

$$KE = ½ (5.31 \times 10^{-26} \text{ kg})(1.5 \times 10^3 \text{ m/s})^2 = 6.0 \times 10^{-20} \text{ kg m}^2/\text{s}^2 \text{ per molecule}$$

$$= 6.0 \times 10^{-20} \text{ J/molecule}$$

Problem 9.11 Analyze the units of the quantity (pressure × volume) and show that they are energy units, consistent with the idea of PV-work.

**Pressure is force divided by area: $P = \frac{F}{A}$, Force is mass times acceleration: $F = \text{kg} \times \frac{\text{m}}{\text{s}^2}$**

**Area is distance squared: $A = \text{m}^2$ Volume is distance cubed: $V = \text{m}^3$**

**Pressure can be expressed as** $P = \frac{F}{A} = \frac{\text{kg} \times \frac{\text{m}}{\text{s}^2}}{\text{m}^2}$

**Now multiplying pressure times volume:** $PV = \frac{\text{kg} \times \frac{\text{m}}{\text{s}^2}}{\text{m}^2} \times \text{m}^3 = \frac{\text{kg m}^2}{\text{s}^2}$

**Recall that a joule is defined by $J = \frac{\text{kg m}^2}{\text{s}^2}$, PV-work is energy.**

Problem 9.13 Define the term *hydrocarbon*.

**A *hydrocarbon* is a type of organic molecule that consists of only carbon and hydrogen atoms. Crude oil, coal, and natural gas are all rich in hydrocarbons and these are sources of fuels like propane, butane, and gasoline.**

Problem 9.15 Carry out the following conversions of energy units:

(a) 14.3 Btu into calories $\mathbf{14.3\ Btu \times \frac{1054.35\ J}{1\ Btu} \times \frac{1\ cal}{4.184\ J} = 3.60 \times 10^3\ cal}$

(b) $1.4 \times 10^5$ cal into joules $\mathbf{1.4 \times 10^5\ cal \times \frac{4.184\ J}{1\ cal} = 5.9 \times 10^5\ J}$

(c) 31.6 mJ into Btu $\mathbf{31.6\ mJ \times \frac{1\ J}{1000\ mJ} \times \frac{1\ Btu}{1054.35\ J} = 3.00 \times 10^{-5}\ Btu}$

Problem 9.17 If a machine does $4.8 \times 10^3$ kJ of work after an input of $7.31 \times 10^4$ kJ of heat, what is the change in internal energy for the machine?

**Internal energy change, $\Delta E$, is expressed in terms of heat flow, $q$, and work, $w$: $\Delta E = q + w$**

**When the system does work on the surroundings the sign is negative; work is negative in this problem. When heat is added or absorbed by a system, the sign is positive; heat is positive in this problem. Combining heat and work:**

$\mathbf{\Delta E = q + w = (+7.31 \times 10^4\ kJ) + (-\,4.8 \times 10^3\ kJ) = 6.83 \times 10^4\ kJ}$

Problem 9.19 If the algebraic sign of $\Delta E$ is negative, in which direction has energy flowed?

**The negative sign associated with any energy flow indicates that energy is being released from the system into the surroundings.**

Problem 9.21 Which type of energy, heat or work, is valued more by society? What evidence supports your judgment?

**Work, energy used to move an object against an opposing force, is more valued. This is evident in the fact that we often sacrifice heat energy to obtain work. For example, we burn gasoline in our automobiles to provide energy to do the work to move our car. We accept that much of that energy is going to be wasted in the form of heat and "thrown away" in the atmosphere.**

Problem 9.23 Which system does more work: **(a)** $\Delta E = -436$ J, $q = 400$ J; or **(b)** $\Delta E = 317$ J, $q = 347$ J?

**Internal energy change, $\Delta E$, is expressed in terms of heat flow, $q$, and work, $w$: $\Delta E = q + w$**

**We can subtract the heat flow to find work for the system.**

**(a) $w = \Delta E - q = -436\ J - 400\ J = -836\ J$**
**(b) $w = \Delta E - q = 317\ J - 347\ J = -30\ J$ System (a) does more work.**

Problem 9.25 Figure 9.5 shows projections for improved efficiency in many important technologies over the next two decades. Will these efficiency gains necessarily lead to reduced household energy demand? What factors might apply pressure for continued increases in household energy use?

**Household energy demand may still increase if there are simply more households or if existing households are using more electricity even if more efficient devices are used. Population growth and the introduction of more electronic devices contribute to increased household energy demand.**

Problem 9.27 When energy conservation programs are promoted, they sometimes include strategies such as turning out lights. Use the web to determine the average percentage of U.S. household energy usage that is attributable to lighting, and use the information to comment on how effective the "turn out the lights" strategy might be.

**Answers will vary. Geographic location and time of year can greatly affect the amount of lighting used in a home. One study provides the figure 9% for the percentage of energy used in lighting in a residence. This is not a very significant amount of the total energy used, so reducing the time lights are lit in a home will only have a marginal effect.**

Problem 9.29 Define the term *calorimetry*.

***Calorimetry* is the technique of measuring temperature changes associated with a physical or chemical system, for the purpose of determining heat flows, q.**

Problem 9.31 A metal radiator is made from 26.0 kg of iron. The specific heat of iron is 0.449 J $g^{-1}\ {}^{o}C^{-1}$. How much heat must be supplied to the radiator to raise its temperature from 25.0 to 55.0°C?

**Heat flow can be calculated by:**
**$q = mc\Delta T$ = (mass, g)(specific heat, J/g · °C)(temperature change, °C)**

**$m$ = 26.0 kg = 26,000 g**

**$q = (26000\ g)(0.449\ J/g \cdot {}^{o}C)(55.0^{o}C - 25.0^{o}C) = 3.50 \times 10^5\ J$**

Problem 9.33 Copper wires used to transport electrical current heat up because of the resistance in the wire. If a 140-g wire gains 280 J of heat, what is the change in temperature in the wire? Copper has a specific heat of 0.384 J $g^{-1}\ {}^{o}C^{-1}$.

**Heat flow can be calculated by:**
**$q = mc\Delta T$ = (mass, g)(specific heat, J/g · °C)(temperature change, °C)**

**Rearranging,** $\Delta T = \dfrac{q}{mc} = \dfrac{280\ \text{J}}{(140\ \text{g})(0.384\ \text{J/g}\cdot{}^{\circ}\text{C})} = 5.21^{\circ}\text{C}$

Problem 9.35 ■ A piece of titanium metal with a mass of 20.8 g is heated in boiling water to 99.5°C and then dropped into a coffee cup calorimeter containing 75.0 g of water at 21.7°C. When thermal equilibrium is reached, the final temperature is 24.3°C. Calculate the specific heat capacity of titanium.

**The final temperature ($T_f$) of the metal and the water is the same. The heat lost by the metal will be equal to the heat gained by the water. We'll set up an equation for these heat flows and solve for the specific heat of Ti. Remember that heat lost or released is negative by convention.**
**From Appendix D in the text, the specific heat of water is 4.18 J/g ·°C.**

**Heat flow can be calculated by:**
**$q = mc\Delta T$ = (mass, g) (specific heat, J/g · °C) (temperature change, °C)**

**Heat lost by titanium = heat gained by the water**
**– [(mass Ti) (specific heat, Ti) ($\Delta T$, Ti)] = (mass, water) (specific heat, water) ($\Delta T$, water)**

$-[(20.8\ \text{g})(c_{\text{Ti}})(24.3^{\circ}\text{C} - 99.5^{\circ}\text{C})] = (75.0\ \text{g})(4.18\ \text{J/g}\cdot{}^{\circ}\text{C})(24.3^{\circ}\text{C} - 21.7^{\circ}\text{C})$

$1.56 \times 10^3 (\text{g}\cdot{}^{\circ}\text{C})(c_{\text{Ti}}) = 815\ \text{J}$

$c_{\text{Ti}} = 0.52\ \text{J/g}\cdot{}^{\circ}\text{C}$

Problem 9.37 ■ A calorimeter contained 75.0 g of water at 16.95°C. A 93.3-g sample of iron at 65.58°C was placed in it, giving a final temperature of 19.68°C for the system. Calculate the heat capacity of the calorimeter. Specific heats are 4.184 J g$^{-1}$ °C$^{-1}$ for $H_2O$ and 0.444 J g$^{-1}$ °C$^{-1}$ for Fe.

**Plan: (1) Determine the heat gained by the calorimeter.**
**(2) Find the heat capacity of the calorimeter (calorimeter constant).**

**(1)$|\text{heat lost}|_{\text{iron}} = |\text{heat gained}|_{\text{water}} + |\text{heat gained}|_{\text{calorimeter}}$**

$|\text{specific heat} \times \text{mass} \times \Delta T|_{\text{iron}}$

$= |\text{specific heat} \times \text{mass} \times \Delta T|_{\text{water}} + |\text{heat gained}|_{\text{calorimeter}}$

$(0.444\ \text{J/g}\cdot{}^{\circ}\text{C})(93.3\ \text{g})(65.58^{\circ}\text{C} - 19.68^{\circ}\text{C})$

**= (4.184 J/g ·°C)(75.0 g)(19.68°C − 16.95°C) + |heat gained|$_{calorimeter}$**

**$1.90 \times 10^3$ J = $8.57 \times 10^2$ J + |heat gained|$_{calorimeter}$**

**Therefore, |heat gained|$_{calorimeter}$ = $1.90 \times 10^3$ J − 857 J = $1.04 \times 10^3$ J.**

**(2) heat capacity of calorimeter = |heat gained|$_{calorimeter}$ ÷ temp. change of water**

$$C_{calorimeter} = \frac{q}{\Delta T} = \frac{1040\text{ J}}{2.73°\text{C}} = 383\text{ J/°C}$$

Problem 9.39 ■ How much thermal energy is required to heat all of the water in a swimming pool by 1°C if the dimensions are 4 ft deep by 20 ft wide by 75 ft long? Report your result in megajoules.

**The specific heat is 4.184 J/g · °C for $H_2O$.**

**Find the volume of the pool in cubic centimeters and use the density of water (1.00 g/cm$^3$) to determine the mass of the water sample being heated. Then use the equation $q = c \times m \times \Delta T$ to calculate the heat energy required.**

**Mass of water:**

$$m = (4\text{ ft}) \times (20\text{ ft}) \times (75\text{ ft}) \times \left(\frac{12\text{ in}}{1\text{ ft}} \times \frac{2.54\text{ cm}}{1\text{ in}}\right)^3 \times \frac{1.00\text{ g H}_2\text{O}}{1\text{ cm}^3} = 1.70 \times 10^8\text{ g H}_2\text{O}$$

$$q_{water} = (4.184\text{ J/g}\cdot°\text{C}) \times (1.70 \times 10^8\text{ g}) \times (1\text{ °C}) = 7 \times 10^8\text{ J}$$

$$7 \times 10^8\text{ J} \times \frac{1\text{MJ}}{10^6\text{ J}} = 7 \times 10^2\text{ MJ}$$

Problem 9.41 Under what conditions does the enthalpy change equal the heat of a process?

**It is equal to the heat under conditions of constant pressure. Enthalpy change is defined as the heat flow that occurs for a process occurring under conditions of constant pressure.**

Problem 9.43 Define the terms *exothermic* and *endothermic*.

**An *exothermic* system is one that releases energy. An *endothermic* system will absorb or gain energy.**

Problem 9.45 What happens to the temperature of a material as it undergoes an endothermic phase change? If heat is added, how can the temperature behave in this manner?

**In an endothermic phase change (melting or boiling), energy is absorbed or gained by the system. The temperature, however, does not change during a phase change. The**

**temperature can remain constant because the energy being added is used to overcome the intermolecular forces holding the particles together as a solid or liquid.**

Problem 9.47 If 14.8 kJ of heat are given off when 1.6 g of HCl condenses from vapor liquid, what is $\Delta H_{cond}$ for this substance?

**$\Delta H_{cond}$ is the heat flow per mole of HCl condensing:**

$$\Delta H_{cond} = \frac{q}{\text{mol HCl}} = \frac{14.8\text{ kJ}}{1.6\text{ g HCl}} \times \frac{36.46\text{ g HCl}}{1\text{ mol HCl}} = -337\text{ kJ/ mol HCl}$$

**(The negative sign indicates that heat is released during condensation.)**

Problem 9.49 $\Delta H_{vap} = 31.3$ kJ/mol for acetone. If 1.40 kg of water were vaporized to steam in a boiler, how much acetone (in kg) would need to be vaporized to use the same amount of heat?

**First we'll calculate the heat flow when the water is vaporized.**

**From Appendix D, the heat of vaporization of water is 2260 J/g.**

$$q = 2260\text{ J/g} \times 1400\text{ g} = 3.16 \times 10^6\text{ J} = 3.16 \times 10^3\text{ kJ}$$

**Second we'll calculate the moles and then kg of acetone, $(CH_3)_2CO$, needed to be vaporized to produce the same amount of heat.**

$$q = 3.16 \times 10^3\text{ kJ} = (\text{mole acetone})(31.3\text{ kJ/mole}),\ \text{mole acetone} = \frac{3.16\times 10^3\text{ kJ}}{31.3\text{ kJ/mol}} = 101\text{ mole}$$

$$101\text{ mole acetone} \times \frac{58.08\text{ g acetone}}{1\text{ mol acetone}} = 5870\text{ g acetone} = 5.87\text{ kg acetone}$$

Problem 9.51 Define the term *formation reaction.*

**A *formation reaction* is defined as a reaction that shows one mole of a compound formed from its elements in their standard states. Standard state is the most stable form of the element under standard state conditions: 1.00 atm pressure, 298 K temperature, and 1.00 M concentration.**

Problem 9.53 Explain why each of the following chemical equations is *not* a correct formation reaction:

**(a)** $4\,Al(s) + 3\,O_2(g) \rightarrow 2\,Al_2O_3(s)$

**This reaction shows two moles being formed, not one.**

**(b)** $N_2(g) + \frac{3}{2} H_2(g) \rightarrow NH_3(g)$

**This reaction is not correctly balanced: the coefficient for $N_2$ should be $\frac{1}{2}$.**

**(c)** $2\ Na(s) + O(g) \rightarrow Na_2O(s)$

**This reaction does not use the correct standard state for oxygen. It should be $\frac{1}{2}\ O_2\ (g)$ instead of O (g).**

Problem 9.55 ■ Using these reactions, find the standard enthalpy change for the formation of 1 mol PbO(s) from lead metal and oxygen gas:

$PbO(s) + C(graphite) \rightarrow Pb(s) + CO(g)$ $\Delta H° = 106.8$ kJ
$2\ C(graphite) + O_2(g) \rightarrow 2\ CO(g)$ $\Delta H° = -221.0$ kJ

If 250 g of lead reacts with oxygen to form lead(II) oxide, what quantity of thermal energy (in kJ) is absorbed or evolved?

**Write the equation for the desired net reaction. After that, calculate the heat energy evolved or absorbed by finding the moles and using the thermochemical expression to provide the molar energy change.**

**The formation equation for PbO is written with the standard state elements as reactants and one mole of compound as the product:**

$$Pb(s) + \tfrac{1}{2}\ O_2(g) \rightarrow PbO(s)$$

**Look at the first reactant: Pb(s). The only given equation that has Pb(s) is the first one, but Pb(s) is a product, so the equation must be reversed. Therefore, we must write that equation reversed with the sign of its $\Delta H°$ changed. Look at the second reactant: $O_2$ (g). The only given equation that has $O_2$ (g) is the second one. It is a reactant, but the stoichiometric coefficient is not what we want it to be.**

$$\text{Multiplier} = \frac{\text{coefficient we want}}{\text{coefficient we have}} = \frac{\frac{1}{2}}{1} = \frac{1}{2}$$

**Therefore, we must include that reaction multiplied by $\frac{1}{2}$ going forward with $\frac{1}{2} \times \Delta H°$. We have now planned how we will use all the given reactions, so let's do it.**

**Add all the reactions as planned, eliminate reactants that also show up as products: the CO and the C(graphite). Add all the remaining reactants and products to form the net reaction, and add all the individual $\Delta H°$ values to get the $\Delta H°$ for the net reaction.**

$Pb(s) + CO(g) \rightarrow PbO(s) + C(graphite)$ $\Delta H° = -(106.8\ \text{kJ})$
$\frac{1}{2} \times [2\ C(graphite) + O_2\ (g) \rightarrow 2\ CO(g)]$ $\frac{1}{2} \times [\Delta H° = -221.0\ \text{kJ}]$

**Multiplying the second equation by $\frac{1}{2}$, then simplifying and adding the two equations gives:**

$$\begin{array}{llr} & Pb(s) + \cancel{CO(g)} \rightarrow PbO(s) + \cancel{C(graphite)} & \Delta H° = -106.8\ kJ \\ + & \cancel{C(graphite)} + \frac{1}{2}\ O_2(g) \rightarrow \cancel{CO(g)} & +\Delta H° = -110.5\ kJ \\ \hline & Pb(s) + \frac{1}{2}\ O_2(g) \rightarrow PbO(s) & \Delta H° = -217.3\ kJ \end{array}$$

**The enthalpy change is negative, so heat energy is evolved during the reaction:**

$$250\ g\ Pb \times \frac{1 mol\ Pb}{207.2\ g\ Pb} \times \frac{217.3\ kJ\ evolved}{1 mol\ Pb} = 2.62 \times 10^2\ kJ\ \ evolved$$

Problem 9.57 Hydrogen gas will react with either acetylene or ethylene gas. The thermochemical equations for these reactions are provided below. Write the thermochemical equation for the conversion of acetylene into ethylene by hydrogen gas.

$$C_2H_2(g) + 2\ H_2(g) \rightarrow C_2H_6(g) \qquad \Delta H° = -311\ kJ$$
$$C_2H_4(g) + H_2(g) \rightarrow C_2H_6(g) \qquad \Delta H° = -136\ kJ$$

**The equation we are interested in is:** $C_2H_2(g) + H_2(g) \rightarrow C_2H_4(g)$

**We will apply Hess's Law – we need to combine the first two equations in such a way that they add together to give the desired reaction. We will then combine the enthalpy changes in exactly the same way to find the enthalpy change of the desired reaction.**

$$\begin{array}{rlr} \text{Reaction 1:} & C_2H_2(g) + 2\ H_2(g) \rightarrow C_2H_6(g) & \Delta H° = -311\ kJ \\ \text{Reverse Reaction 2:} & \underline{C_2H_6(g) \rightarrow C_2H_4(g) + H_2(g)} & \underline{-(\Delta H° = -136\ kJ)} \\ \text{Summing:} & C_2H_2(g) + H_2(g) \rightarrow C_2H_4(g) & \Delta H°_{rxn} = -175\ kJ \end{array}$$

Problem 9.59 The heat of combustion of butane is −2877 kJ/mol. Use this value to find the heat of formation of butane. (You may also need to use additional thermochemical data found in Appendix E.)

**We start by writing the combustion reaction:**

$$2\ C_4H_{10}(g) + 13\ O_2(g) \rightarrow 8\ CO_2(g) + 10\ H_2O(\ell) \quad \Delta H = -2877\ kJ/mol\ butane$$

**Applying Hess's law, the net change in enthalpy for the reaction, $\Delta H°_{rxn}$, is equal to the sum of the heats of formation of the product (each times its coefficient) minus the sum of the heats of formation of the reactants (each times its coefficient). In equation form:**
$\Delta H°_{rxn} = \sum n\Delta H_f°{}_{products} - \sum n\Delta H_f°{}_{reactants}$ **where n = the stoichiometric coefficients in the equation.**

**In applying this formula to our equation, we already know $\Delta H^{o}_{rxn}$; by looking up the heats of formation of $CO_2$ and $H_2O$, and $O_2$, we can solve algebraically for $\Delta H^{o}_{f}$ of butane. Remember, the heat of combustion is expressed per <u>one mol</u> of butane.**

$$2\,(-2877\text{ kJ/mol}) = \Delta H^{o}_{rxn} = [8\,(\Delta H^{o}_{f,CO2}) + 10\,(\Delta H^{o}_{f,H2O})] - [2\,(\Delta H^{o}_{f,C4H10}) + 13\,(\Delta H^{o}_{f,O2})]$$

$$2\,(-2877\text{ kJ/mol}) = [8\,(-393.5\text{ kJ/mol}) + 10\,(-285.8\text{ kJ/mol})] - [2\,(\Delta H^{o}_{f,C4H10}) + 13\,(0)]$$

**Solving for $\Delta H^{o}_{f,C4H10}$,**

$$2\,(-2877\text{ kJ/mol}) + 3148\text{ kJ/mol} + 2858\text{ kJ/mol} = -2\,(\Delta H^{o}_{f,C4H10})$$

$$252\text{ kJ/mol} = -2\,(\Delta H^{o}_{f,C4H10})$$

$$-126\text{ kJ/mol} = \Delta H^{o}_{f,C4H10}$$

<u>Problem 9.61</u> When a reaction is exothermic, is the sum of bond energies of products or reactants greater?

**The sum of the bond energies of the products is greater than the products, meaning some of the excess energy is released when the products are formed. This sounds a little contradictory, but you must remember that bond energy is defined as the energy required to break one mole of bonds in the gaseous form. Mathematically,**

$$\Delta H^{o}_{rxn} = \sum B.E._{reactants} - \sum B.E._{products}$$

<u>Problem 9.63</u> For the reaction $N_2(g) + O_2(g) \rightarrow 2\,NO(g)$, $\Delta H^{o} = 180.5$ kJ, how much energy is needed to generate 35 moles of NO(g)?

**The basis for $\Delta H^{o}$ is "per mole of reaction", which means per the balanced equation in its smallest whole-number coefficients. There are 2 moles of NO formed per mole of reaction, so the formation of 2 moles of NO requires 180.5 kJ.**

$$35\text{ mol NO} \times \frac{1\text{ mol rxn}}{2\text{ mol NO}} \times \frac{180.5\text{ kJ}}{\text{mol rxn}} = 3200\text{ kJ required.}$$

<u>Problem 9.65</u> Silane, $SiH_4$, burns according to the reaction, $SiH_4 + 2\,O_2 \rightarrow SiO_2 + 2\,H_2O$, with $\Delta H^{o} = -1429$ kJ. How much energy is released if 15.7 g of silane is burned?

**Once again, the basis of $\Delta H^{o}$ is per mole of reaction. Our plan is to calculate the moles of silane burned and relate them to moles of reaction and enthalpy.**

$$15.7\text{ g SiH}_4 \times \frac{1\text{ mol SiH}_4}{32.12\text{ g SiH}_4} \times \frac{1\text{ mol rxn}}{1\text{ mol SiH}_4} \times \frac{-1429\text{ kJ}}{\text{mol rxn}} = -698\text{ kJ} \quad \textbf{(698 kJ released)}$$

Problem 9.67 ■ Reactions of hydrocarbons are often studied in the petroleum industry. One such reaction is $2\ C_3H_8(g) \rightarrow C_6H_6(\ell) + 5\ H_2(g)$, with $\Delta H^\circ = 698$ kJ. If 35 L of propane at 25°C and 0.97 atm is to be reacted, how much heat must be supplied?

**This is a two-part problem. First we use the ideal gas law to calculate the moles of propane reacted. Second, we relate moles propane to moles of reaction and enthalpy.**

$V = 35$ L, $P = 0.97$ atm, $T = 25°\text{C} + 273 = 298$ K, $R = 0.08206$ L•atm/mol•K

$$PV = nRT,\ n = \frac{PV}{RT} = \frac{(0.97\text{ atm})(35\text{ L})}{(0.08206\text{ L}\cdot\text{atm/mol}\cdot\text{K})(298\text{ K})} = 1.4\text{ moles C}_3\text{H}_8$$

$$1.4\text{ moles C}_3\text{H}_8 \times \frac{1\text{ mol rxn}}{2\text{ mol C}_3\text{H}_8} \times \frac{698\text{ kJ}}{\text{mol rxn}} = 4.9 \times 10^2\text{ kJ must be supplied.}$$

Problem 9.69 When 0.0157 kg of a compound with a heat of combustion of –37.6 kJ/mol is burned in a calorimeter, 18.5 kJ of heat are released. What is the molar mass of the compound?

**If we know the grams of a compound and the moles, then we just divide to find the molar mass: $MM = \frac{\text{grams}}{\text{mol}}$. Our plan is to use the heat released to find the moles of compound burned, and then calculate the molar mass.**

$$18.5\text{ kJ} \times \frac{\text{mol of compound burned}}{37.6\text{ kJ}} = 0.492\text{ mol compound}$$

$$\text{Molar mass } (MM) = \frac{15.7\text{ g}}{0.492\text{ mol}} = 31.9\text{ g/mol}$$

Problem 9.71 Why is the energy density so important in the transportation of fuels?

**Transportation costs are a major contribution to the overall cost of fuels. Fuels with a higher energy density contain more joules (more energy) per gram of mass. This makes transporting the "energy" less expensive.**

Problem 9.73 How are the roles of transmission substations and distribution substations in the electrical grid similar? How are they different?

**They are both responsible for changing the voltage of the electricity as it moves from the generating station to the consumer. The transmission substation steps the**

**voltage up to a very high value to reduce energy losses while moving electricity long distances. The distribution substation steps the voltage down to much lower values prior to delivering it to commercial and private customers.**

Problem 9.75 In recent years the notion of a "smart grid" has emerged. Do a web search and research the smart grid concept. How would the smart grid differ from the traditional grid?

**Answers will vary from the web. There are many ways to define a "smart grid" but it is essentially the modernization of the electrical system by integration of new information in real-time. This includes behaviors of both suppliers and consumers. Some of the differences and benefits of such a smart grid would include fault detection and repair without human intervention, ability to accommodate bidirectional energy flows, and direct communication with devices to prevent system overloads.**

Problem 9.77 ■ The figure below shows a "self-cooling" beverage can.

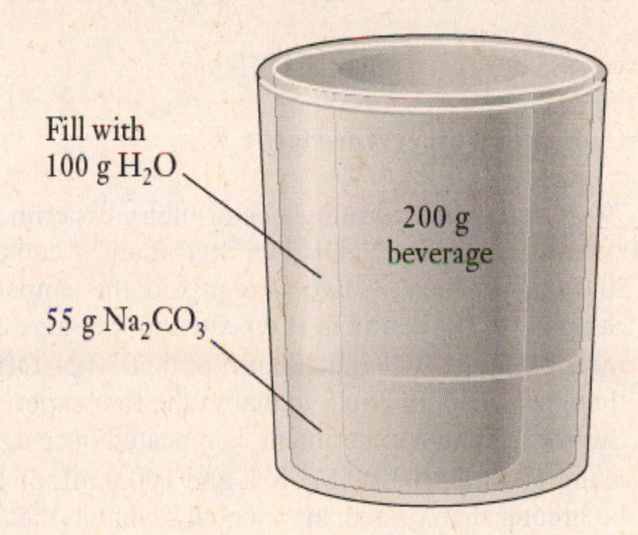


The can is equipped with an outer jacket containing sodium carbonate ($Na_2CO_3$), which dissolves in water rapidly and endothermically:

$$Na_2CO_3(s) \rightarrow 2\,Na^+(aq) + CO_3^{2-}(aq) \quad \Delta H^\circ = 67.7\text{ kJ}$$

The user adds water to the outer jacket, and the heat absorbed in the chemical reaction chills the drink. The can contains 200 g of drink, the jacket contains 55 g of $Na_2CO_3$, and 100 g of water are to be added. If the initial temperatures of the can and the water are both 32°C on a summer day, what is the coldest temperature that the drink can reach? The can itself has a heat capacity of 40 J/°C. Assume that the $Na_2CO_3$ solution and the drink both have the same heat capacity as pure water, 4.184 J $g^{-1}$ $°C^{-1}$. (HINT: Treat this like a calorimetry problem.)

**Treating this as a calorimetry problem, we need to set up an equation for the heat balance: Heat absorbed by the reaction = Heat lost by the drink and solution + heat lost by the can.**

**First we'll calculate the heat absorbed by the sodium carbonate dissolving.**

$$55 \text{ g } Na_2CO_3 \times \frac{1 \text{ mol } Na_2CO_3}{105.99 \text{ g } Na_2CO_3} \times \frac{67.7 \text{ kJ}}{1 \text{ mol } Na_2CO_3} = 35.1 \text{ kJ absorbed}$$

**Heat lost by the drink and solution = $mc\Delta T$**

**= (total mass)(specific heat)(temperature change)**

**= (200 g + 55 g + 100 g)( 4.184 J $g^{-1}$ $^{o}C^{-1}$)($T_f$ – 32°C)**

**Setting the heat flows equal (with the correct sign convention):**

$$35.1 \text{ kJ} = 35100 \text{ J} = -(355 \text{ g})(4.184 \text{ J g}^{-1}\ {}^{o}\text{C}^{-1})(T_f - 32^{o}\text{C})$$

$$35100 \text{ J} = -1485 \text{ J}\ {}^{o}\text{C}^{-1}(T_f\ {}^{o}\text{C} - 32^{o}\text{C}) = -1485 \text{ J } T_f + 47530 \text{ J}$$

$$-12430 \text{ J} = -1485 \text{ J } T_f \qquad T_f = 8.4\ {}^{o}\text{C}$$

**The coolest possible temperature is 8.4 °C.**

Problem 9.79 A student performing a calorimetry experiment combined 100.0 mL of 0.50 M HCl and 100.0 mL of 0.50 M NaOH in a Styrofoam™ cup calorimeter. Both solutions were initially at 20.0°C, but when the two were mixed, the temperature rose to 23.2°C.

(a) Suppose the experiment is repeated in the same calorimeter, but this time using 200 mL of 0.50 M HCl and 200 mL of 0.50 M NaOH. Will the $\Delta T$ observed be greater than, less than, or equal to that in the first experiment, and why?

(b) Suppose that the experiment is repeated once again in the same calorimeter, this time using 100 mL of 1.00 M HCL and 100.0 mL of 1.00 M NaOH. Will the $\Delta T$ observed be greater than, less than, or equal to that in the first experiment, and why?

**(a) The temperature change observed will be about the same in the second experiment. Even though twice as many moles of acid and base are reacting, there is twice the volume of solution absorbing the heat. More heat is released but it is dissipated in more solution. The result is the temperature increases by about the same amount.**

**(b) The temperature change in the third experiment will be larger than that in the first. A greater number of moles reacting releases a greater amount of heat and causes a larger temperature increase. The volume of solution is the same in the first and third experiment.**

Problem 9.81 ■ What will be the final temperature of a mixture made from equal masses of the following: water at 25.0°C, ethanol at 35.5°C, and iron at 95.0°C?

**The heat lost, $q$, by the substances with a temperature higher than the final temperature must be equal to the heat gained, $q$, by the substances with a temperature lower than the final temperature.**

$$q(\text{lost}) = q(\text{gained})$$

$$\Sigma\,(s \times m \times \Delta T) = \Sigma\,(s \times m \times \Delta T)$$

**Since the temperature of the iron, 95.0°C, is far larger than the temperatures of the other two substances, assume the final temperature will be greater than 35.5°C and greater than 25°C. Then, write the heat term for the iron on the left and the sum of the heat terms for water and ethanol on the right. Since the masses, $m$, are equal, divide both sides by $m$ to eliminate this term, and solve one equation in one unknown, letting T equal the final temperature. (To simplify the setup, only the temperature unit is retained.)**

$$\Sigma\,(s \times \Delta T) = \Sigma\,(s \times \Delta T)$$

$$0.444 \times (95.0°\text{C} - T) = 2.46 \times (T - 35.5°\text{C}) + 4.18 \times (T - 25.0°\text{C})$$

$$42.18°\text{C} - 0.444\,T = 2.46\,T - 87.33°\text{C} + 4.18\,T - 104.5°\text{C}$$

$$-7.084\,T = -234.01°\text{C}$$

$$T = 33.03 = 33.0°\text{C} \quad \text{(the final temperature)}$$

Problem 9.83 ■ A sample of natural gas is 80.0% $CH_4$ and 20.0% $C_2H_6$ by mass. What is the heat from the combustion of 1.00 g of this mixture? Assume the products are $CO_2(g)$ and $H_2O(\ell)$.

**The balanced equations are**

$$CH_4(g) + 2O_2(g) \rightarrow CO_2(g) + 2H_2O(\ell)$$

$$C_2H_6(g) + 3.5O_2(g) \rightarrow 2CO_2(g) + 3H_2O(\ell)$$

**First, calculate the moles of $CH_4$ and $C_2H_6$ (let 80.0% = 0.800 g and 20.0% = 0.200 g).**

$$\text{Moles } CH_4 = 0.800 \text{ g } CH_4 \times \frac{1 \text{ mol } CH_4}{16.04 \text{ g } CH_4} = 0.04987 \text{ mol } CH_4$$

$$\text{Moles } C_2H_6 = 0.200 \text{ g } C_2H_6 \times \frac{1 \text{ mol } C_2H_6}{30.07 \text{ g } C_2H_6} = 0.006651 \text{ mol } C_2H_6$$

**Now, use the enthalpies of formation (Appendix E) to calculate the total heat:**

| $CH_4(g)$ | + | $2O_2(g)$ | → | $CO_2(g)$ | + | $2H_2O(\ell)$ |
|---|---|---|---|---|---|---|
| 0.04987 | | excess | | 0.04987 | | 2(0.04987)(mol) |
| −74.81 | | 0 | | −393.5 | | −285.8 (kJ/mol) |
| −3.731 | | 0 | | −19.62 | | −28.51(kJ/mol·eqn) |

| $C_2H_6(g)$ | + | $3.5O_2(g)$ | → | $2CO_2(g)$ | + | $3H_2O(\ell)$ |
|---|---|---|---|---|---|---|
| **0.006651** | | **excess** | | **2(0.006651)** | | **3(0.006651)(mol)** |
| **−84.86** | | **0** | | **−393.5** | | **−285.8 (kJ/mol)** |
| **−0.5644** | | **0** | | **−5.234** | | **−5.703(kJ/mol·eqn)** |

**For the combustion of $CH_4$:**

$$\Delta H = [(-28.51) + (-19.62)] - [(-3.731) + 0] = -44.399 \text{ kJ} \approx -44.4 \text{ kJ}$$

**For the combustion of $C_2H_6$:**

$$\Delta H = [(-5.234) + (-5.703)] - [(-0.5644) + 0] = -10.373 \text{ kJ}$$

**For the combustion of the total (1.00 g) of $CH_4$ and $C_2H_6$:**

$$\Delta H = -44.399 + (-10.373) = -54.772 = -54.8 \text{ kJ}$$

**Heat evolved = 54.8 kJ**

Problem 9.85 ■ You want to heat the air in your house with natural gas ($CH_4$). Assume your house has 275 $m^2$ (about 2800 $ft^2$) of floor area and that the ceilings are 2.50 m from the floors. The air in the house has a molar heat capacity of 29.1 J/mol/K. (The number of moles of air in the house can be found by assuming that the average molar mass of air is 28.9 g/mol and that the density of air at these temperatures is 1.22 g/L.) What mass of methane do you have to burn to heat the air from 15.0°C to 22.0°C?

$$275 \text{ m}^2 \times 2.50 \text{ m} \times \frac{1 \text{ L}}{0.001 \text{ m}^3} \times \frac{1.22 \text{ g}}{1 \text{ L}} \times \frac{1 \text{ mol air}}{28.9 \text{ g}} = 2.90 \times 10^4 \text{ mol air}$$

$$q_{air} = (2.90 \times 10^4 \text{ mol})(29.1 \text{ J/mol·K})(295.2 \text{ K} - 288.2 \text{ K}) = 5.9 \times 10^6 \text{ J} = 5.9 \times 10^3 \text{ kJ}$$

$$CH_4(g) + 2\ O_2(g) \rightarrow CO_2(g) + 2\ H_2O(g)$$

$$\Delta H^\circ_r = \Delta H^\circ_f[CO_2(g)] + 2\ \Delta H^\circ_f[H_2O(g)] - \Delta H^\circ_f[CH_4(g)]$$

$$\Delta H^\circ_r = 1 \text{ mol } (-393.5 \text{ kJ/mol}) + 2 \text{ mol } (-241.8 \text{ kJ/mol}) - 1 \text{ mol } (-74.81 \text{ kJ/mol})$$

$\Delta H^\circ_r = -802.3$ **kJ/mol-rxn (recall we only need $-5.9 \times 10^3$ kJ of heat)**

$$-(5.9 \times 10^3 \text{ kJ}) \times \frac{1 \text{ mol}\cdot\text{rxn}}{-802.30 \text{ kJ}} \times \frac{1 \text{ mol } CH_4}{1 \text{ mol}\cdot\text{rxn}} \times \frac{16.04 \text{ g } CH_4}{1 \text{ mol } CH_4} = 1.2 \times 10^2 \text{ g } CH_4$$

Problem 9.87 ■ An engineer is designing a product in which a copper wire will carry large amounts of electricity. The resistive heating of a 65-g copper wire is expected to add 580 J of heat energy during a 10-minute operating cycle. The specific heat of copper is 0.385 J $g^{-1}$ $°C^{-1}$, the density is 8.94 g/$cm^3$, and the coefficient of thermal expansion is 16.6 $\mu$m $m^{-1}$ $K^{-1}$. (a) What is the temperature increase of the wire? (b) What is the initial length of the wire, assuming it is a cylinder and its radius is 0.080 cm? (c) By what percentage does the volume increase because of the temperature increase? (d) Do you think the engineer should be considering this expansion in the design?

**(a) $q = mc\Delta T$ = (mass, g) (specific heat, J/g · °C) (temperature change, °C)**

$$\Delta T = \frac{q}{mc} = 580 \text{ J} \div [(65 \text{ g})(0.385 \text{ J g}^{-1} \text{ °C}^{-1})] = 23\text{°C}$$

**(b) The volume is found from the density:** $V = \frac{m}{d} = 65 \text{ g} \div 8.94 \text{ g/cm}^3 = 7.3 \text{ cm}^3$

$$V = \pi r^2 \ell;\ \ell = V/\pi r^2 = 7.3 \text{ cm}^3 \div [(3.1416)(0.08 \text{ cm})^2] = 3.6 \times 10^2 \text{ cm}$$

**(c) The thermal expansion gives the increase in radius per meter of wire per kelvin of temperature increase.**

$$\text{Expansion} = [16.6 \times 10^{-6} \frac{\text{m}}{\text{m} \cdot \text{K}}] \times [3.6 \text{ m}] \times [23.2 \text{ K}] = 1.4 \times 10^{-3} \text{ m} = 0.14 \text{ cm}$$

**New radius = 0.08 cm + 0.14 cm = 0.22 cm.**

$$\text{New volume} = V = \pi r^2 \ell = (3.1416)(0.22 \text{ cm})^2(3.6 \times 10^2 \text{ cm}) = 55 \text{ cm}^3$$

$$\% \text{ increase} = \frac{55 \text{ cm}^3 - 7.3 \text{ cm}^3}{7.3 \text{ cm}^3} \times 100 = 650 \%$$

**(d) Yes, the engineer should consider this expansion. If not accounted for, it could cause structural failure of whatever contains the wire.**

Problem 9.89 A substance has the following properties:

$\Delta H_{fusion}$ = 10.0 kJ/mol $\Delta H_{vaporization}$ = 20.0 kJ/mol
$C_p$(solid) = 30 J/mol/K $C_p$(liquid) = 60 J/mol/K $C_p$(gas) = 30 J/mol/K

(a) Which of the four graphs below would be most consistent with these data?

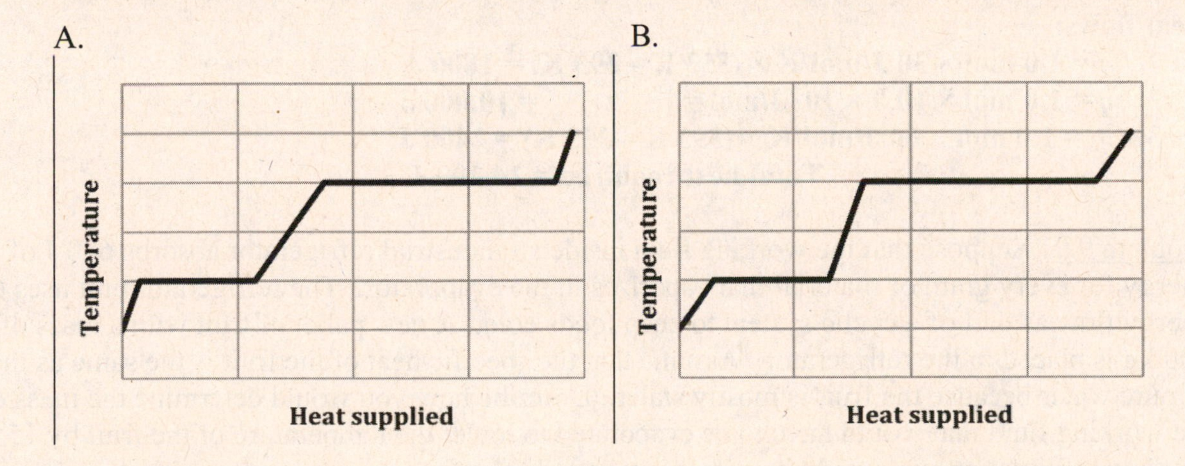

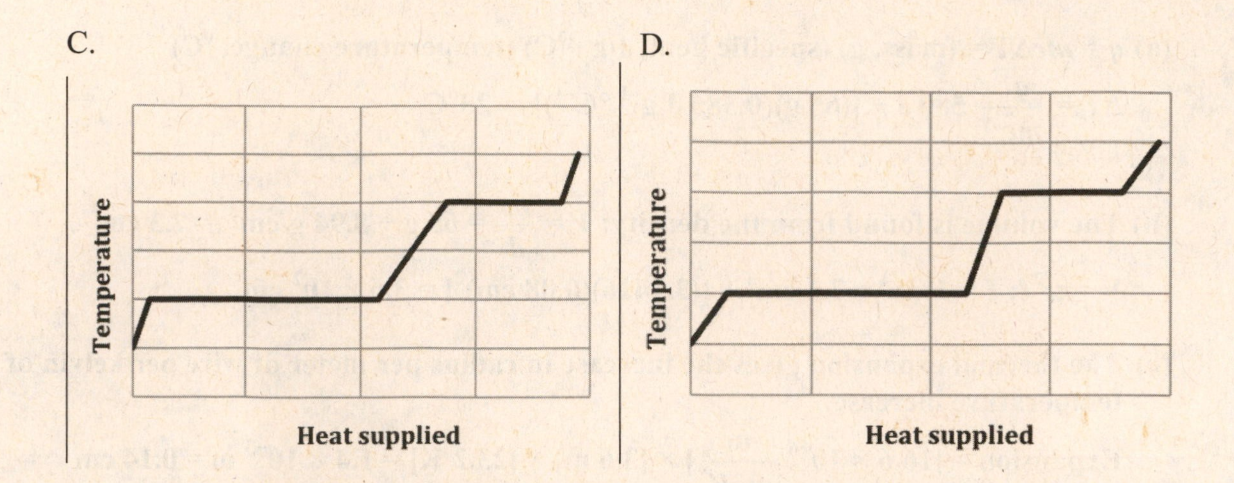


**Graph A. The heat flow from vaporization (second horizontal line) is two times the heat flow from fusion (first horizontal line). The heat supplied to raise the temperature of the solid and gas (first and third diagonal lines are the same as indicated by the same values of molar heat capacity and those heat flows are half that required to heat the liquid.**

(b) If the melting point of the substance is 80°C, how much heat would be required to convert 1.0 mol of it from solid at 20°C into liquid at 120°C?

**There are a total of three heat flows to calculate:**

1. **Heat required to raise the temperature of the solid to the melting point: $q = nC_p\Delta T$**
2. **Heat required to melt the solid: $q = n\Delta H_{fusion}$**
3. **Heat required to raise the temperature of the liquid to 120°: $q = nC_p\Delta T$**

**Heat flows:**

1. **$q$ = 1.0 mol × 30 J/mol/K × (353 K – 293 K) = 1800 J**
2. **$q$ = 1.0 mol × 10.0 × $10^3$ J/mol = = 10,000 J**
3. **$q$ = 1.0 mol × 60 J/mol/K × (393 K – 353 K) = 2400 J**

**Total heat required = 14,200 J**

Problem 9.91 Suppose that the working fluid inside an industrial refrigerator absorbs 680 J of energy for every gram of material that vaporizes in the evaporator. The refrigerator unit uses this energy flow as part of a cyclic system to keep foods cold. A new pallet of fruit with a mass of 500 kg is placed in the refrigerator. Assume that the specific heat of the fruit is the same as that of pure water because the fruit is mostly water. Describe how you would determine the mass of the working fluid that would have to be evaporated to lower the temperature of the fruit by 15°C. List any information you would have to measure or look up.

**We could first calculate the heat that would have to be removed from the fruit using the mass, temperature change, and the specific heat of water; the specific heat would have to be looked up ($q = mc\Delta T$).**

**Using that heat flow, we can calculate the mass of working fluid required using the value of 680 J absorbed per gram fluid vaporized.**

Problem 9.93 The chemical reaction $BBr_3(g) + BCl_3(g) \rightarrow BBr_2Cl(g) + BCl_2Br(g)$ has an enthalpy change very close to zero. Using Lewis structures of the molecules, all of which have a central boron atom, provide a molecular-level description of why $\Delta H^o$ for this reaction might be very small.

$BBr_3 + BCl_3 \longrightarrow BBr_2Cl + BCl_2Br$ (Lewis structures: each molecule has a central B bonded to three halogen atoms)

**The total number and type of bonds in the reactants and products is the same (3 B–Br, 3 B–Cl bonds). Since there is no difference between the bond energies of the reactant and products, we would expect to $\Delta H^o$ be very small.**

Problem 9.95 Silicon nitride, $Si_3N_4$, has physical, chemical, and mechanical properties that make it a useful industrial material. For a particular engineering project, it is crucial that you know the heat of formation of this substance. A clever experiment allows direct determination of the $\Delta H^o$ of the following reaction:

$$3\ CO_2(g) + Si_3N_4(s) \rightarrow 3\ SiO_2(s) + 2\ N_2(g) + 3\ C(s)$$

Based on the fact that you know the enthalpy change in this reaction, state what additional data might be looked up or measured to determine $\Delta H^o_f$ for silicon nitride.

**Applying Hess's law, the net change in enthalpy for the reaction, $\Delta H^o_{rxn}$, is equal to the sum of the heats of formation of the product (each times its coefficient) minus the sum of the heats of formation of the reactants (each times its coefficient). In equation form:**

$$\Delta H^o_{rxn} = \sum n\Delta H_f^o{}_{products} - \sum n\Delta H_f^o{}_{reactants}$$

**In applying this formula to our equation, we already know $\Delta H^o_{rxn}$; by looking up the heat of formation of $CO_2$, $SiO_2$, $N_2$, and C, we can solve algebraically for $\Delta H^o_f$ of silicon nitride.**

Problem 9.97 One reason why the energy density of a fuel is important is that to move a vehicle one must also move its unburned fuel. Octane is a major component of gasoline. It burns according to the reaction

$$2\ C_8H_{18}(\ell) + 25\ O_2(g) \rightarrow 16\ CO_2(g) + 18\ H_2O(g) \qquad \Delta H^\circ = -1.10 \times 10^4\ \text{kJ}$$

Starting from this thermochemical equation, describe the energy density, in kJ/g, for octane. Be sure to indicate what you would need to calculate or look up to complete this problem.

**The enthalpy change gives us the heat flow per mole of reaction (or per 2 moles of octane); we need to convert that to heat flow per gram octane. Using the molar mass of octane, we can convert from moles to grams.**

Problem 9.99 In passive solar heating, the goal is to absorb heat from the sun during the day and release it during the night. Which material would be better for this application: one with a high heat capacity or one with a low heat capacity? Explain.

**A material with a higher heat capacity (or specific heat) would be better because it would be able to absorb more heat with a relatively smaller increase in temperature:**

**$q = mc\Delta T$ = (mass, g) (specific heat, J/g · °C) (temperature change, °C) or**

$$\Delta T = \frac{q}{mc}$$

**A material that does not vary greatly in temperature is preferable because it is easier to handle (it doesn't get too hot).**

Problem 9.101 ■ At the beginning of 2011, the United State had 25.2 billion barrels of proven oil resources. One barrel of oil can produce about 19.5 gallons of gasoline. Assume that the gallon of gasoline is pure octane, with a density of 0.692 g/mL. If all 25.2 billion barrels of oil were converted to gasoline and burned, how much energy would be released?

**We need to use the heat of combustion of octane from Problem 9.95: $\Delta H^\circ = -1.10 \times 10^4$ kJ/mol rxn.**

$$\frac{1.10\times10^4\ \text{kJ}}{\text{mol rxn}} \times \frac{1\ \text{mol rxn}}{2\ \text{mol octane}} \times \frac{1\ \text{mol}\ C_8H_{18}}{114.23\ \text{g}\ C_8H_{18}} \times \frac{0.692\ \text{g}}{\text{mL}} \times \frac{1000\ \text{mL}}{1\ \text{L}} \times \frac{1\ \text{L}}{1.056\ \text{quarts}} \times$$

$$\frac{4\ \text{quarts}}{1\ \text{gallon}} \times \frac{19.5\ \text{gallon of octane}}{1\ \text{barrel of oil}} \times (25.2 \times 10^9\ \text{barrels of oil}) = 6.20 \times 10^{16}\ \text{kJ}$$

Problem 9.103 ■ Suppose that the car in the previous problem has a fuel efficiency of 24 mpg. How much energy is released in burning a gallon of gasoline (assuming that all of it is octane)? Based on this calculation and the work required to move the car those 24 miles, what percentage of the energy released in the combustion is wasted (doesn't directly contribute to the work of moving the car)? How do the assumptions you make in carrying out these calculations affect the value you obtain?

**Referring to Problem 9.101:**

$$\frac{\mathbf{1.10 \times 10^4\ kJ}}{\mathbf{mol\ rxn}} \times \frac{\mathbf{1\ mol\ rxn}}{\mathbf{2\ mol\ octane}} \times \frac{\mathbf{1\ mol\ C_8H_{18}}}{\mathbf{114.23\ g\ C_8H_{18}}} \times \frac{\mathbf{0.692\ g}}{\mathbf{mL}} \times \frac{\mathbf{1000\ mL}}{\mathbf{1\ L}} \times \frac{\mathbf{1\ L}}{\mathbf{1.056\ quarts}} \times$$

$$\frac{\mathbf{4\ quarts}}{\mathbf{1\ gallon}} = \mathbf{1.26 \times 10^5\ kJ\ released\ per\ gallon\ of\ octane\ burned.}$$

**From Problem 9.100, the work required to move a 930-kg car 24 miles is $8.7 \times 10^4$ kJ.**

**The percentage of wasted energy (energy from the burning gasoline not used to move the car) is:**

$$\frac{\mathbf{(1.26 \times 10^5\ kJ - 8.7 \times 10^4\ kJ)}}{\mathbf{1.26 \times 10^5\ kJ}} \times \mathbf{100 = 31\%\ wasted\ energy.}$$

**We have assumed that there are no frictional losses, wind resistance, mechanical energy transfer losses, etc.; in reality these are significant and will cause the efficiency to be much lower.**

## CHAPTER OBJECTIVE QUIZ

This quiz will test your understanding of the basic text chapter objectives and give you additional practice problems. You should work this quiz after completing the end-of-chapter questions. The solutions to these questions are found at the end of this chapter.

1. TRUE or FALSE: The process of converting the potential of coal or natural gas into electricity has only minimal losses and inefficiency.

2. Every month, a business uses an average of 2500 kilowatt-hours (kWh) of electricity for lighting. Assuming they use incandescent bulbs now, how much money could they save per month switching the fluorescent bulbs? (Consult Table 9.1 and assume 1 kWh of electricity costs $0.12.)

3. The heat flow for a system is +600 J and 300 J of work is performed by the system. Which statement is FALSE?
   a. Heat is absorbed.
   b. The system is exothermic.
   c. $\Delta E = +300$ J.
   d. The sign of work is negative.
   e. $\Delta E = q + w$

4. What is the specific heat of a metal when a 75.13 g sample increases temperature by 11.3 °C when 785 J are added?

5. Which of the following statements about the first law of thermodynamics is INCORRECT?
   a. All energy change in a chemical reaction is in the form of heat.
   b. $\Delta E_{\text{universe}} = 0$.
   c. $\Delta E = \Delta E_{\text{system}} + \Delta E_{\text{surroundings}}$.
   d. Energy cannot be created nor destroyed in ordinary chemical reactions.
   e. The combined amount of matter and energy in the universe is constant.

6. In a coffee cup calorimeter, 100 mL of 0.10 M $HClO_4$ is mixed with 100 mL of 0.10 M KOH. The initial temperature of both solutions is 23.2°C. The resulting mixing temperature after the reaction is 26.3°C. The calorimeter constant is 27.8 J/°C. Write the balanced chemical equation and calculate how much heat is released by the experiment.

7. How much heat must be supplied to change 100.0 g of water at 30.0°C to steam at 110.0°C? The specific heats of water and steam are 4.18 J/g · °C and 2.03 J/g · °C, respectively. The molar heat of vaporization of water ($\Delta H_{\text{vap}}$) is 40.7 kJ/mol.

8. Given the following reaction,

   $2\ CH_4(g) + 4\ O_2(g) \rightarrow 2\ CO_2(g) + 4\ H_2O(\ell)\quad \Delta H = -1780.8$ kJ/mol rxn,

   Is the reaction exothermic or endothermic? What would be the value for the following reaction?

   $CO_2(g) + 2\ H_2O(\ell) \rightarrow CH_4(g) + 2\ O_2(g)$?

9. A fuel has a molar mass of 156.2 g/mol. If 75.6 g of this fuel burns, releasing 1035 kJ of heat, what is the molar heat of combustion?

10. Using Hess's law and the following data, calculate the heat of reaction for $A + B \rightarrow C$.

| | | $\Delta H$, kJ/mol |
|---|---|---|
| 1. | $2\ D \rightarrow B$ | 50 |
| 2. | $2\ A + D \rightarrow C$ | −100 |
| 3. | $D \rightarrow A$ | 75 |

11. Write a formation reaction for ethyl alcohol, $C_2H_5OH(\ell)$.

12. Using Hess's law and heats of formation, calculate the heat of reaction ($\Delta H^\circ_{\text{rxn}}$) at 298 K for:

| | 2 A(g) | + 2 B(g) | → 3 D(g) |
|---|---|---|---|
| $\Delta H^\circ_f$, kJ/mol | −235 | 31 | −177 |

13. How much heat is released when 16.3 g of ethanol burns?

$C_2H_5OH(\ell) + 3\ O_2(g) \rightarrow 2\ CO_2(g) + 3\ H_2O(\ell) \quad \Delta H = -1366.8$ kJ/mol rxn

14. When diamond oxidizes (burns), heat is released:

$C(s)_{diamond} + O_2(g) \rightarrow CO_2(g) \quad \Delta H = -395.4$ kJ/mol rxn

If 15400 kJ are released while burning diamond, how many grams of diamond are consumed?

15. Which of the following is NOT a correct statement regarding the United States electrical grid system?
    a. Electricity travels through transmission lines at very high voltages to save on energy losses.
    b. Computers are critical to monitoring power usage through the grid.
    c. The electrical grid is immune to cyber-attack.
    d. Local utilities must make choices on a daily basis as to how much electrical output they'll need to meet their customers' needs and from where to buy that electricity.
    e. Distribution substations reduce the voltage of electricity.

## ANSWERS TO THE CHAPTER OBJECTIVE QUIZ

1. **FALSE**

2. **The current electric costs for the business are: 2500 kWh × $\frac{\$0.12}{\text{kWh}}$ = $300. From table 9.1, we see that the efficiency of a fluorescent bulb is about four times greater than that of an incandescent bulb (20% vs. 5%). This would result in one-fourth the electricity being used for lighting. The electric cost would be reduced by one-fourth: $300 ÷ 4 = $75. This would be a yearly saving of $2700 to the business.**

3. **b**

4. **Heat flow** $= q = mc\Delta T, \quad c = \frac{q}{m\Delta T} = \frac{(785\ \text{J})}{(75.13\ \text{g})(11.3\ ^\circ\text{C})} = 0.925\ \text{J/g}\cdot{}^\circ\text{C}$

5. **a**

6. $HClO_4 + KOH \rightarrow KClO_4 + H_2O$

$q = C_{calorimeter} \times \Delta T = 27.8\ \text{J/}^\circ\text{C} \times (26.3^\circ\text{C} - 23.2^\circ\text{C}) = 86.2\ \text{J}$

7. **We must calculate three heat flows:**
   - a. **Heat the water form 30.0°C to 100.0°C**
   - b. **Vaporize the water at 100.0°C**
   - c. **Heat the steam from from 100.0°C to 110.0°C**

$q_i = mc_{water}\Delta T = (100.0\ \text{g})(4.18\ \text{J/g}\cdot{}^\circ\text{C})(100.0^\circ\text{C} - 30.0^\circ\text{C}) = 2.93 \times 10^4\ \text{J} = 29.3\ \text{kJ}$

$$q_{ii} = n\,\Delta H_{vap} = 100.0\ \text{g}\ H_2O \times \frac{1\ \text{mol}\ H_2O}{18.0\ \text{g}\ H_2O} \times \frac{40.7\ \text{kJ}}{\text{mol}\ H_2O} = 226\ \text{kJ}$$

$q_{iii} = mc_{ice}\Delta T = (100.0\ \text{g})(2.03\ \text{J/g}\cdot{}^\circ\text{C})(110.0^\circ\text{C} - 100.0^\circ\text{C}) = 2.03 \times 10^3\ \text{J} = 2.03\ \text{kJ}$

**Total heat flow:** $q_T = q_i + q_{ii} + q_{iii} = 29.3\ \text{kJ} + 226\ \text{kJ} + 2.03\ \text{kJ} = 257\ \text{kJ}$

8. **The negative sign of the enthalpy change indicates the heat is released; the reaction is exothermic.**
   **Reversing the equation will cause the heat flow to go in the opposite direction, and dividing the reaction by two will divide the enthalpy change by two.** $\Delta H = +890.4$ **kJ/mol rxn.**

9. **The molar heat of combustion,** $\Delta H_{comb}$**, is equal to the total heat released divided by the moles of fuel burning.**

$$75.6 \text{ g} \times \frac{1 \text{ mol fuel}}{156.2 \text{ g fuel}} = 0.484 \text{ mol} \qquad \Delta H_{comb} = \frac{1035 \text{ kJ}}{0.484 \text{ mol fuel}} = 2140 \text{ kJ/mol}$$

**10. The three reactions must be combined in such a way that they add to give the desired reaction. Then the enthalpy changes will be combined in the same way to find the enthalpy change of the desired reaction.**

| | | $\Delta H$, kJ/mol |
|---|---|---|
| Reverse reaction 1: | B → 2 D | –(50) |
| Reaction 2: | 2 A + D → C | –100 |
| Reaction 3: | D → A | 75 |
| Overall Reaction: | A + B → C | $\Delta H_{rxn}$ = –75 kJ/mol |

**11. A formation reaction shows one mole of a compound formed from its elements in their standard states.**

$2\ C(s)_{graphite} + 3\ H_2(g) + \frac{1}{2}\ O_2(g) \rightarrow C_2H_5OH(\ell)$

**12. Using Formation reactions, we can express the heat of reaction, $\Delta H^o_{rxn}$, in terms of heats of formation, $\Delta H^o_f$:**

$$\Delta H^o_{rxn} = \sum n\Delta H_f^o{}_{products} - \sum n\Delta H_f^o{}_{reactants}$$

**where n = the stoichiometric coefficient in the equation.**

$\Delta H^o_{rxn} = [3\ (\Delta H^o_{f,D})] - [2\ (\Delta H^o_{f,A}) + 2\ (\Delta H^o_{f,A})]$

$\Delta H^o_{rxn} = [3\ (-177 \text{ kJ/mol})] - [2\ (-235 \text{ kJ/mol}) + 2\ (31 \text{ kJ/mol})] = -123 \text{ kJ/mol rxn}$

**13.** $16.3 \text{ g } C_2H_5OH \times \frac{1 \text{ mol } C_2H_5OH}{46.068 \text{ g } C_2H_5OH} \times \frac{1 \text{ mol rxn}}{1 \text{ mol } C_2H_5OH} \times \frac{1366.8 \text{ kJ}}{\text{mol rxn}} = 483.6 \text{ kJ released}$

**14.** $15400 \text{ kJ} \times \frac{1 \text{ mol rxn}}{395.4 \text{ kJ}} \times \frac{1 \text{ mol C}}{1 \text{ mol rxn}} \times \frac{12.011 \text{ g C}}{1 \text{ mol C}} = 468 \text{ g C (diamond) burned.}$

**15. c**

# CHAPTER 10

## *Study Goals*

The study goals outline specific concepts to be mastered in this section of the text chapter. Related problems at the end of the text chapter will also be noted. Working the questions noted should aid in mastery of each study goal and also highlight any areas that you may need additional help in.

**Section 10.1 *INSIGHT INTO* Recycling of Plastics**

1. Outline the plastics recycling process and list several limitations in the effectiveness of plastics recycling. *Work Problems 1–6.*

**Section 10.2 Spontaneity**

2. Define a spontaneous process and relate spontaneity to heat flow. *Work Problems 7–18, 81, 82, 89, 90.*

**Section 10.3 Entropy**

3. Discuss the Carnot cycle and describe how entropy relates to molecular disorder. *Work Problem 19.*
4. Recognize how entropy can be calculated using the statistics of probabilities. *Work Problems 20–22.*
5. Predict how entropy is affected in phase changes, temperature changes, volume changes, mixing, and chemical reactions. *Work Problems 23–34.*

**Section 10.4 The Second Law of Thermodynamics**

6. State the second law of thermodynamics in terms of the entropy change of the universe, system, and surroundings. *Work Problems 35–45, 91.*

**Section 10.5 The Third Law of Thermodynamics**

7. State the third law of thermodynamics and understand its implications in calculating entropy changes in chemical reactions. *Work Problems 46, 48, 52–55.*
8. Calculate $\Delta S^{o}_{rxn}$ using tabulated values of absolute entropies. *Work Problems 47, 49−51.*

**Section 10.6 Gibbs Free Energy**

9. Define Gibbs free energy and relate $\Delta G_{system}$ to $\Delta S_{universe}$. *Work Problems 56−58, 62, 63.*
10. Use the sign of $\Delta G_{system}$ to determine the spontaneous direction of chemical reactions. *Work Problems 59–61, 64–67.*

**Section 10.7 Free Energy and Chemical Reactions**

11. Calculate $\Delta G^{o}_{rxn}$ using tabulated thermodynamic data ($\Delta H^{o}_{f}$, $S^{o}$, $\Delta G^{o}_{f}$). *Work Problems 68–74, 76, 84–87, 92–104.*
12. Using $\Delta H^{o}_{rxn}$ and $\Delta S^{o}_{rxn}$, determine the temperature conditions for which a chemical reaction is spontaneous. *Work Problems 75, 83, 88, 91.*

**Section 10.8 *INSIGHT INTO* The Economics of Recycling**

13. Recognize the economic realities of recycling plastics. *Work Problems 77–80.*

## *Solutions to the Odd-Numbered Problems*

Problem 10.1 "Reduce, reuse, recycle" is a common slogan among environmentalists, and the order of the three words indicates their perceived relative benefits. Why is recycling the least desirable of these three approaches to waste reduction?

**Reducing or reusing materials requires no extra energy to accomplish. Recycling, while usually beneficial for the environment, requires energy to convert the recycled waste into a usable material again.**

Problem 10.3 List some consumer products made from recycled PET.

**Drinking cups, bottled-water bottles, and 2-liter soda bottles are made with poly(ethylene terephthalate) or PET. It is also used as carpet fibers.**

Problem 10.5 Use the web to research a company that specializes in the recycling of plastics. Does the material on their website emphasize environmental, scientific, or economic concerns? Write a brief essay on the company's positions, explaining how they fit with the ideas expressed in this chapter.

**Answers will vary according to the website found.**

Problem 10.7 ■ On the basis of your experience, predict which of the following reactions are spontaneous.

(a) $CO_2(s) \rightarrow CO_2(g)$ at 25°C

(b) $NaCl(s) \rightarrow NaCl(\ell)$ at 25°C

(c) $2\ NaCl(s) \rightarrow 2\ Na(s) + Cl_2(g)$

(d) $CO_2(g) \rightarrow C(s) + O_2(g)$

**Consider physically what is happening in the chemical equation and determine if the process occurs spontaneously.**

**(a) Spontaneous. Solid carbon dioxide readily sublimes at room temperature.**

**(b) Non-spontaneous. Table salt does not melt at 25°C.**

**(c) Non-spontaneous. Table salt does not decompose at 25°C.**

**(d) Non-spontaneous. Carbon dioxide does not decompose at 25°C.**

**The only spontaneous process is (a).**

Problem 10.9 If the combustion of butane is spontaneous, how can you carry a butane lighter safely in your pocket or purse?

**The reaction of butane and oxygen is spontaneous but not at room temperature. This reaction only becomes spontaneous at a much higher temperature. This is why a source of ignition (the spark from the flint) is required to initiate the reaction. Once the reaction begins, it continues because the exothermic reaction releases enough energy to sustain the reaction.**

Problem 10.11 Identify each of the processes listed as spontaneous or nonspontaneous. For each nonspontaneous process, describe the corresponding spontaneous process in the opposite direction.

(a) Oxygen molecules dissociate to form oxygen atoms.

**This is nonspontaneous. Oxygen atoms will spontaneously combine to form diatomic molecular oxygen, the most stable form of oxygen.**

(b) A tray of water is placed in the sun on a warm day and freezes.

**This is nonspontaneous. Water will not freeze at the temperature of a warm day; rather it will evaporate into vapor.**

(c) A solution of salt water forms a layer of acid on top of a layer of base.

**This is not spontaneous. The spontaneous reaction would be the acid and base reacting to form the salt and water.**

(d) Silver nitrate is added to a solution of sodium chloride and a precipitate forms.

**This is spontaneous.**

(e) Sulfuric acid sitting in a beaker turns into water by giving off gaseous $SO_3$.

**This is nonspontaneous. Water and $SO_3$ will spontaneously react to form $H_2SO_4$, sulfuric acid.**

Problem 10.13 Are any of the following exothermic processes not spontaneous under any circumstances?

**An exothermic process will be spontaneous at some conditions but not necessarily any conditions.**

**(a)** Snow forms from liquid water.

**This is spontaneous when the temperature is below 0°C.**

**(b)** Liquid water condenses from water vapor.

**This is spontaneous at temperatures below 100°C.**

**(c)** Fossil fuels burn to form carbon dioxide and water.

**This is always spontaneous.**

**(d)** Monomers react to form a polymer.

**This reaction is generally spontaneous at lower temperatures.**

Problem 10.15 When a fossil fuel burns, is that fossil fuel the system? Explain your answer.

**The burning of a fossil fuel represents a system. Specifically, the fuel and oxygen as reactants and the carbon dioxide and water as products are the system.**

Problem 10.17

Humpty Dumpty sat on a wall,
Humpty Dumpty had a great fall.
All the King's horses and all the King's men
Couldn't put Humpty together again.

In Lewis Carroll's *Through the Looking Glass*, Alice encounters Humpty Dumpty, a human-sized egg sitting on a wall. Alice, who is familiar with the nursery rhyme, asks anxiously, "Don't you think you'd be safer on the ground? That wall is so narrow." Humpty, an egg with an attitude, growls, "Of course I don't think so. Why if I ever *did* fall off – which there's no chance of – but if I did… the King has promised me – with his very own mouth – (that) they'd pick me up again in a minute, they would!"

Write a paragraph, in the voice of seven-and-a-half-year-old Alice, explaining to Humpty in the context of this chapter **(a)** the probability that Humpty will fall off the wall and **(b)** the probability that the King's horses and men will be able to put him back together again.

**Alice: You probably will fall off the wall because you are so big and round and you don't sit very well on top. I don't think the King will be able to put you together because when you fall you will make a big mess and be in a lot of pieces and no one, not even the King, could put you back together.**

Problem 10.19 What observation about the Carnot engine led Carnot to propose the existence of a new state function?

**In the Carnot cycle, an ideal gas undergoes a series of four processes (see Figure 10.2). Two steps are isothermal (constant temperature) and two are adiabatic (no heat flow). Carnot showed that the quantity $\frac{q}{T}$ summed to zero over the entire cycle. Because the system begins and ends in the same state, this quantity $\frac{q}{T}$ must be a new state function.**

Problem 10.21 How does probability relate to spontaneity?

**The most probable outcome is usually the most random outcome. Spontaneous processes tend to increase randomness of the system, so nature tends to move in the direction of increased randomness, which is most likely to occur.**

Problem 10.23 ■ For each pair of items, tell which has the higher entropy and explain why.

(a) Item 1, a sample of solid $CO_2$ at –78°C, or item 2, $CO_2$ vapor at 0°C
(b) Item 1, solid sugar, or item 2, the same sugar dissolved in a cup of tea
(c) Item 1, a 100-mL sample of pure water and a 100-mL sample of pure alcohol, or item 2, the same samples of water and alcohol after they have been poured together and stirred.

**(a) Item 2 has higher entropy since it is identical to item 1 except that its temperature is higher. Molecules at higher temperature have higher entropy.**

**(b) Item 2, dissolved sugar, has higher entropy than item 1, solid sugar, because solute molecules are more random than those in a solid crystal.**

**(c) Item 2, the mixture of water and alcohol together has higher entropy than item 1, water and alcohol separate. Mixing makes the molecules more random.**

Problem 10.25 If a sample of air were separated into nitrogen and oxygen molecules (ignoring other gases present), what would be the sign of $\Delta S$ for this process? Explain your answer.

**The sign of the entropy change, $\Delta S$, would be negative. This means that entropy (and randomness) is decreasing when air is separated into oxygen and nitrogen molecules. Creating a mixture increases entropy (and randomness) so reversing the process decreases entropy. An analogy would be separating the suits of a card deck (entropy decreases) but shuffling the cards, creating a mixture, increases the entropy of the system.**

Problem 10.27 ■ Without doing a calculation, predict whether the entropy change will be positive or negative when each of the following reactions occurs in the direction it is written.

(a) $CH_3OH(\ell) + 3/2\ O_2(g) \rightarrow CO_2(g) + 2\ H_2O(g)$

(b) $Br_2(\ell) + H_2(g) \rightarrow 2\ HBr(g)$

(c) $Na(s) + 1/2\ F_2(g) \rightarrow NaF(s)$

(d) $CO_2(g) + 2\ H_2(g) \rightarrow CH_3OH(\ell)$

(e) $NH_3(g) \rightarrow N_2(g) + 3\ H_2(g)$

**Recall an increase in number of moles of gas reflects increase in entropy, all other things remaining the same. Using the number of moles of gas products minus the number of moles of gas reactants, we find the entropy change to be:**
**(a) positive ($3 - \frac{3}{2}$ moles) (b) positive (2 –1 mole) (c) negative ($0 - \frac{1}{2}$ mole)**
**(d) negative ( 0 – 3 moles) (e) positive (4 – 1 moles)**

Problem 10.29 In many ways, a leaf is an example of exquisite order. So how can it form spontaneously in nature? What natural process shows that the order found in a leaf is only temporary?

**A process that creates order, such as the formation of a leaf, can be spontaneous if the order (entropy) of the surroundings increases. This occurs when the process is exothermic; the energy released to the surroundings causes the entropy of the surroundings to increase. So overall, order is created when the leaf forms, but disorder is created in the rest of the universe and the net change in entropy of the universe is positive.**
**The process reverses when the leaf falls from the tree and decomposes. Randomness is created again and heat is released to the surroundings.**

Problem 10.31 An explosion brings down an old building, leaving behind a pile of rubble. Does this cause a thermodynamic entropy increase? If so, where? Write a paragraph explaining your reasoning.

**The destruction of the building causes an increase in entropy in the sense that the particles are more random now. Small, random particles of matter are more disordered than a large complicated system like the building. The main increase in thermodynamic entropy comes from the heat flow to the surroundings, increasing the molecular motion in the surroundings.**

Problem 10.33 According to Lambert, leaves lying in the yard and playing cards that are in disarray on a table have not undergone an increase in their thermodynamic entropy. Suggest another reason why leaves and playing cards may not be a good analogy for the entropy of a system containing, for example, only $H_2O$ molecules or only $O_2$ molecules.

**Because leaves or cards are not in random motion on their own, they are not a good analogy for the thermodynamic entropy of a system containing molecules. Molecules such as $H_2O$ or $O_2$ are in constant random motion and will spontaneously move to a more**

**probable (random) state without intervention. Leaves or cards can only become more random with assistance.**

<u>Problem 10.35</u> What happens to the entropy of the universe during a spontaneous process?

**The entropy of the universe always increases in a spontaneous process: $\Delta S_{universe} > 0$. The entropy change for the <u>system</u> can be either positive or negative in a spontaneous process but the overall sum of the entropy changes of both the system and the surroundings must be positive: $\Delta S_{universe} = \Delta S_{system} + \Delta S_{surroundings} > 0$.**

<u>Problem 10.37</u> One statement of the second law of thermodynamics is that heat cannot be turned completely into work. Another is that the entropy of the universe always increases. How are these two statements related?

**Work causes a decrease in randomness, which decreases entropy. If heat cannot be converted completely into work, then some heat must be transferred to the surroundings. The heat gained by the surroundings increases the entropy of the universe, offsetting the decrease in entropy of the system. This ensures that the net change in entropy of the universe is positive.**

<u>Problem 10.39</u> How does the second law of thermodynamics explain a spontaneous change in a system that becomes more ordered when that process is exothermic?

**See Problem 10.35. The entropy change for a system can be either positive or negative in a spontaneous system but the overall sum of the entropy changes of both the system and the surroundings must be positive: $\Delta S_{universe} = \Delta S_{system} + \Delta S_{surroundings} > 0$. If the system has a negative entropy change but is exothermic, the energy released to the surroundings causes an increase in entropy in the surroundings, $\Delta S_{surroundings} = \frac{q_{surr}}{T}$. The increase in the entropy of the surroundings must be larger than the decrease in entropy of the system for the entropy change of the universe to be positive. An example of this type of system is when water spontaneously changes to ice in a freezer at –10°C. The change from liquid to a solid represents a decrease in entropy but the process is exothermic.**

<u>Problem 10.41</u> When a reaction is exothermic, how does that influence $\Delta S$ of the system? Of the surroundings?

**An exothermic reaction releases heat. This process has no significant effect on the entropy of the system. The heat released is absorbed by the surroundings, which increases the molecular motion and randomness of the surroundings. This will have a very significant effect (increase) on the entropy of the surroundings.**

<u>Problem 10.43</u> Which reaction occurs with the greater increase in entropy? Explain your reasoning.

**(a)** $2\,NO(g) \rightarrow N_2(g) + O_2(g)$
**(b)** $Br_2(g) + Cl_2(g) \rightarrow 2\,BrCl(g)$

**In reaction (a), there is not much change in entropy; in fact the entropy would most likely decrease slightly. The reason is NO is a diatomic molecule made of two different atoms and it is forming two diatomic molecules formed from two of the same atoms. Heteronuclear diatomic molecules usually have higher entropy than homonuclear.**
**In reaction (b), a heteronuclear molecule is being formed from homonuclear diatomic elements; that usually results in a decrease in entropy as a more complicated substance is formed. Reaction (b) will have a greater increase in entropy.**

Problem 10.45 Limestone is predominantly $CaCO_3$, which can undergo the reaction $CaCO_3(s) \rightarrow CaO(s) + CO_2(g)$. We know from experience that this reaction is not spontaneous, yet ΔS for the reaction is positive. How can the second law of thermodynamics explain that this reaction is not spontaneous?

**The entropy change for the reaction is positive because the reactant is a solid but one of the products is a gas. Gases are much more disordered than solids and this results in a large increase in entropy. The reaction is still not spontaneous at 25°C because it is a very endothermic reaction. The large amount of heat that leaves the surroundings and enters the system causes the entropy of the surroundings to decrease. Even though $\Delta S_{system}$ is positive, $\Delta S_{surroundings}$ is negative and the the sum of $\Delta S_{system} + \Delta S_{surroundings} < 0$. The second law of thermodynamics states that $\Delta S_{universe} > 0$ for spontaneous processes, thus the reaction is not spontaneous, at that temperature.**

Problem 10.47 ■ Use tabulated thermodynamic data to calculate the entropy change of each of the reactions listed below.

**The net change in entropy for a reaction, $\Delta S_{rxn}$, can be calculated by taking the sum of the absolute entropies of the products (each times its coefficient) minus the sum of the absolute entropies of the reactants (each times its coefficient). In equation form:**
$\Delta S^o_{rxn} = \sum nS^o_{products} - \sum nS^o_{reactants}$ **where n = the stoichiometric coefficient in the equation. Absolute entropy values are listed in Appendix E of the textbook.**

**(a)** $Fe(s) + 2\,HCl(g) \rightarrow FeCl_2(s) + H_2(g)$

**Applying the above formula to this reaction:**

$\Delta S^o_{rxn} = [1\text{ mol }(S^o_{FeCl2}) + 1\text{ mol }(S^o_{H2})] - [1\text{ mol }(S^o_{Fe}) + 2\text{ mol }(S^o_{HCl})] =$

$\Delta S^o_{rxn} = [1\text{ mol }(117.9\text{ J/mol·K}) + 1\text{ mol }(130.6\text{ J/mol·K})] - [1\text{ mol }(27.3\text{ J/mol·K}) + 2\text{ mol }(186.8\text{ J/mol·K})] =$

$\Delta S^o_{rxn} = -152.4\text{ J/K}$

(b) $3 NO_2(g) + H_2O(\ell) \rightarrow 2 HNO_3(\ell) + NO(g)$

**$\Delta S^o_{rxn}$ = [2 mol ($S^o_{HNO3}$) + 1 mol ($S^o_{NO}$)] – [3 mol ($S^o_{NO2}$) + 1 mol ($S^o_{H2O}$)] =**

**$\Delta S^o_{rxn}$ = [2 mol (155.6 J/mol·K) + 1 mol (210.7 J/mol·K)] – [3 mol (240.0 J/mol·K) + 1 mol (69.91 J/mol·K)] =**

**$\Delta S^o_{rxn}$ = –268.0 J/K**

(c) $2 K(s) + Cl_2(g) \rightarrow 2 KCl(s)$

**$\Delta S^o_{rxn}$ = [2 mol ($S^o_{KCl}$)] – [2 mol ($S^o_{K}$) + 1 mol ($S^o_{Cl2}$)] =**

**$\Delta S^o_{rxn}$ = [2 mol (82.6 J/mol·K)] – [2 mol (63.6 J/mol·K) + 1 mol (223.0 J/mol·K)] =**

**$\Delta S^o_{rxn}$ = –185.0 J/K**

(d) $Cl_2(g) + 2 NO(g) \rightarrow 2 NOCl(g)$

**$\Delta S^o_{rxn}$ = [2 mol ($S^o_{NOCl}$)] – [1 mol ($S^o_{Cl2}$) + 2 mol ($S^o_{NO}$)] =**

**$\Delta S^o_{rxn}$ = [2 mol (264 J/mol·K)] – [1 mol (223.0 J/mol·K) + 2 mol (210.7 J/mol·K)] =**

**$\Delta S^o_{rxn}$ = –116.4 J/K**

(e) $SiCl_4(g) \rightarrow Si(s) + 2 Cl_2(g)$

**Applying the above formula to this reaction:**
**$\Delta S^o_{rxn}$ = [1 mol ($S^o_{Si}$) + 2 mol ($S^o_{Cl2}$)] – [1 mol ($S^o_{SiCl4}$)] =**

**$\Delta S^o_{rxn}$ = [1 mol (18.8 J/mol·K) + 2 mol (223.0 J/mol·K)] – [1 mol (330.6 J/mol·K)] =**

**$\Delta S^o_{rxn}$ = 134.2 J/K**

Problem 10.49 Calculate $\Delta S^o$ for the dissolution of magnesium chloride: $MgCl_2(s) \rightarrow Mg^{2+}(aq) + 2 Cl^-(aq)$. Use your understanding of the solvation of ions at the molecular level to explain the sign of $\Delta S^o$.
**We'll use the same formula as in Problem 10.47: $\Delta S^o_{rxn} = \sum nS^o_{products} - \sum nS^o_{reactants}$ where n = the stoichiometric coefficient in the equation. The absolute entropy values we need are listed in Appendix E of the textbook.**

**$\Delta S^o_{rxn}$ = [1 mol ($S^o_{MgCl2}$)] – [1 mol ($S^o_{Mg2+}$) + 2 mol ($S^o_{Cl-}$)] =**

**$\Delta S^o_{rxn}$ = [1 mol (89.5 J/mol·K)] – [1 mol (–138.1 J/mol·K) + 2 mol (56.48 J/mol·K)] =**

**$\Delta S^o_{rxn}$ = 114.6 J/K**

**The positive change in entropy represents an increase in molecular disorder. This is expected because the ionic solid, $MgCl_2$, is very ordered, with cations and anions arranged in a crystal lattice structure. As the salt dissolves and the ions go into solution, the lattice breaks down and the ions become randomly arranged in the solution, surrounded by water molecules. In the solution, there is random movement of ions, and much more molecular disorder, thus the positive change in entropy when dissolution of $MgCl_2$ occurs.**

Problem 10.51 Through photosynthesis, plants build molecules of sugar containing several carbon atoms from carbon dioxide. In the process, entropy is decreased. The reaction of $CO_2$ with formic acid to form oxalic acid provides a simple example of a reaction in which the number of carbon atoms in a compound increases:

$$CO_2(g) + HCOOH(s) \rightarrow H_2C_2O_4(s)$$

**(a)** Calculate the entropy change for this reaction and discuss the sign of $\Delta S^o$.

**Note: The reaction in the text shows $CO_2$ in the aqueous state, but the data in Appendix E only lists $CO_2(g)$, so that is how the problem is solved.**

$$\Delta S^o_{rxn} = \sum nS^o_{products} - \sum nS^o_{reactants}$$

$$\Delta S^o_{rxn} = [1 \text{ mol } (S^o_{H2C2O4})] - [1 \text{ mol } (S^o_{CO2}) + 1 \text{ mol } (S^o_{HCOOH})] =$$

$$\Delta S^o_{rxn} = [1 \text{ mol } (45.61 \text{ J/mol·K})] - [1 \text{ mol } (213.6 \text{ J/mol·K}) + 1 \text{ mol } (163.2 \text{ J/mol·K})] =$$

**$\Delta S^o_{rxn}$ = –331.2 J/K**

**The entropy change of the reaction is negative, indicating that molecular disorder is decreasing or the system becomes more ordered.**

**(b)** How do plants carry out reactions that increase the number of carbons in a sugar, given the changes in entropy for reactions like this?

**To make a reaction of this type spontaneous, the energy change must be exothermic. Heat is released to the surroundings, increasing the entropy of the surroundings, causing the net change in entropy for the universe to be positive.**

Problem 10.53 Look up the values of the standard entropy for the following molecules: $CH_4(g)$, $C_2H_5OH(\ell)$, $H_2C_2O_4(s)$. Rank the compounds in order of increasing entropy, and then explain why this ranking makes sense.

**The values of standard or absolute entropy can be found in Appendix E of the textbook:**

1. **$H_2C_2O_4$(s), $S^o$ = 45.61 J/mol·K**

2. **$C_2H_5OH(\ell)$, $S^o$ = 161 J/mol·K**

3. **$CH_4$(g), $S^o$ = 186.2 J/mol·K**

**Oxalic acid, $H_2C_2O_4$, is a solid where the other two are liquid or gas so it is expected that it has the lowest value of standard entropy. Matter has the most order when it is in the solid state. Ethanol's ($C_2H_5OH$) value is smaller than methane's because it is a liquid as opposed to gas and it is a more complicated molecule. Small, gaseous molecules generally have the highest values of absolute entropy.**

Problem 10.55 A beaker of water at 40°C (on the left in the drawing) and a beaker of ice water at 0°C are placed side by side in an insulated container. After some time has passed, the temperature of the water in the beaker on the left is 30°C and the temperature of the ice water is still 0°C. Describe what is happening in each beaker **(a)** on the molecular level and **(b)** in terms of the second law of thermodynamics.

**(a) On the molecular level, heat flows from the water on the left to the ice water mixture on the right, lowering the temperature of the water on the left. As the temperature of the water decreases, the molecules become more orderly. The temperature of the ice water does not decrease because the heat being added is being used to melt the ice (a phase change takes place at constant temperature). The ice water becomes more disordered as the molecules of water change from solid to liquid.**
**(b) According to the second law of thermodynamics, this process cannot take place spontaneously unless the entropy of the universe increases. While the entropy of the water on the right <u>decreases</u>, the entropy of the ice water <u>increases</u> even more, thus making the net change in entropy of the universe positive.**

Problem 10.57 Under what conditions does $\Delta G$ allow us to predict whether a process is spontaneous?

**The assumptions used in deriving the relationship between the Gibbs free energy change of the system, $\Delta G_{system}$, and the entropy change of the universe, $\Delta S_{universe}$, restrict us to conditions of constant pressure.**

Problem 10.59 ■ Calculate $\Delta G°$ at 45°C for reactions for which

(a) $\Delta H° = 293$ kJ; $\Delta S° = -695$ J/K

(b) $\Delta H° = -1137$ kJ; $\Delta S° = 0.496$ kJ/K

(c) $\Delta H° = -86.6$ kJ; $\Delta S° = -382$ J/K

**$\Delta G° = \Delta H° - T\Delta S°$ (Convert $T$ to kelvin and $\Delta S°$ to kJ/K)**

**(a)** $\Delta G° = 293 \text{ kJ} - (318 \text{ K})(-0.695 \text{ kJ/K})$
$\Delta G° = 514 \text{ kJ}$

**(b)** $\Delta G° = -1137 \text{ kJ} - (318 \text{ K})(0.496 \text{ kJ/K})$
$\Delta G° = -1295 \text{ kJ}$

**(c)** $\Delta G° = -86.6 \text{ kJ} - (318 \text{ K})(-0.382 \text{ kJ/K})$
$\Delta G° = +34.9 \text{ kJ}$

Problem 10.61 ■ The reaction

$$CO_2(g) + H_2(g) \rightarrow CO(g) + H_2O(g)$$

is not spontaneous at room temperature but becomes spontaneous at a much higher temperature. What can you conclude from this about the signs of ΔH° and ΔS°, assuming that the enthalpy and entropy changes are not greatly affected by the temperature change? Explain your reasoning.

**At low (room) temperature, $\Delta G°$ or ($\Delta H° - T\Delta S°$) must be positive, but at higher temperatures, $\Delta G°$ or ($\Delta H° - T\Delta S°$) must be negative. Thus, at the higher temperatures, the $-T\Delta S°$ term must become more negative than $\Delta H°$. Thus, $\Delta S°$ must be positive and so must $\Delta H°$. If either were negative, $\Delta G°$ would not become negative at higher temperatures.**

Problem 10.63 Distinguish between a reversible and an irreversible process.

**A reversible process is one that can take place in the forward and/or reverse direction. In other words, the system can be restored to its original state after some change.**
**An irreversible process can only go in the forward direction and cannot return to its original state.**

Problem 10.65 The combustion of acetylene is used in welder's torches because it produces a very hot flame:

$C_2H_2(g) + 5/2\ O_2(g) \rightarrow 2\ CO_2(g) + H_2O(g)$ $\Delta H^\circ = -1255.5$ kJ

(a) Use the data in Appendix E to calculate $\Delta S^\circ$ for this reaction.

**We'll use the formula, $\Delta S^\circ_{rxn} = \sum nS^\circ_{products} - \sum nS^\circ_{reactants}$**

**$\Delta S^\circ_{rxn} = [2\text{ mol } (S^\circ_{CO2}) + 1\text{ mol } (S^\circ_{H2O})] - [1\text{ mol } (S^\circ_{C2H2}) + \frac{5}{2}\text{ mol } (S^\circ_{O2})] =$**

**$\Delta S^\circ_{rxn} = [2\text{ mol } (213.6\text{ J/mol·K}) + 1\text{ mol } (188.7\text{ J/mol•K})] - [1\text{ mol } (200.8\text{ J/mol·K}) + \frac{5}{2}\text{ mol } (205.0\text{ J/mol·K})] =$**

**$\Delta S^\circ_{rxn} = -97.4$ J/K**

(b) Calculate $\Delta G^\circ$ and show that the reaction is spontaneous at 25°C.

**$\Delta G^\circ = \Delta H^\circ - T\Delta S^\circ = -1255.5\text{ kJ} - 298\text{ K}(-97.4\text{ J/K})(1\text{ kJ} / 1000\text{ J})$**

**$\Delta G^\circ = -1226.5$ kJ**

**The value of Gibbs free energy change is negative, indicating a spontaneous reaction at this temperature.**

(c) Is there any temperature range in which this reaction is not spontaneous?

**Since $\Delta H$ is negative and $\Delta S$ is also negative, this reaction becomes more spontaneous (more negative $\Delta G$) at lower temperatures and less spontaneous at higher temperatures (see Problem 10.61). There is some temperature above which the reaction is no longer spontaneous.**

(d) Do you think you could use Equation 10.4 to calculate such a temperature range reliably? Explain your answer.

**The temperature range can be estimated using Equation 10.4 ($\Delta G = \Delta H - T\Delta S$), by setting the equation equal to zero and solving for the temperature for which $\Delta G = 0$. This temperature represents the temperature where the reaction is neither spontaneous nor nonspontaneous (this is called equilibrium). All temperatures above this temperature will give a positive $\Delta G$, so the reaction is nonspontaneous at this temperature range. There is some error in this calculation as we would have to use the <u>standard state</u> values of enthalpy and entropy change, but the calculation is for some temperature other than**

**standard temperature. The error in this calculation is usually not very great because the values of $\Delta H$ and $\Delta S$ don't vary much with temperature.**

Problem 10.67 Silicon forms a series of compounds that is analogous to the alkanes and has the general formula $Si_nH_{2n+2}$. The first of these compounds is silane, $SiH_4$, which is used in the electronics industry to produce thin ultrapure silicon films. $SiH_4(g)$ is somewhat difficult to work with because it is *pyrophoric* at room temperature – meaning that it bursts into flame spontaneously.

(a) Write an equation for the combustion of $SiH_4(g)$. (The reaction is analogous to hydrocarbon combustion, and $SiO_2$ is a solid under standard conditions. Assume the water produced will be a gas.)

**The reactants are gaseous silane and molecular oxygen and the products are solid $SiO_2$ and liquid water.**

$$SiH_4(g) + 2\ O_2(g) \rightarrow SiO_2(s) + 2\ H_2O(g)$$

(b) Use the data from Appendix E to calculate $\Delta S^o$ for this reaction.

**We'll use the formula, $\Delta S^o_{rxn} = \sum nS^o_{products} - \sum nS^o_{reactants}$**

$\Delta S^o_{rxn} = [1\text{ mol } (S^o_{SiO2}) + 2\text{ mol } (S^o_{H2O})] - [1\text{ mol } (S^o_{SiH4}) + 2\text{ mol } (S^o_{O2})] =$

$\Delta S^o_{rxn} = [1\text{ mol } (41.84\text{ J/mol·K}) + 2\text{ mol } (188.7\text{ J/mol·K})] - [1\text{ mol } (204.5\text{ J/mol·K}) + 2\text{ mol } (205.0\text{ J/mol·K})] =$

$\Delta S^o_{rxn} = -193.5\text{ J/K}$

(c) Calculate $\Delta G^o$ and show that the reaction is spontaneous at 25°C.

**To calculate the $\Delta G^o_{rxn}$, we'll need to first calculate $\Delta H^o_{rxn}$.**

**We'll use the formula, $\Delta H^o_{rxn} = \sum n\Delta H_f{}^o{}_{products} - \sum n\Delta H_f{}^o{}_{reactants}$**

$\Delta H^o_{rxn} = [1\text{ mol } (\Delta H^o_{SiO2}) + 2\text{ mol } (\Delta H^o_{H2O})] - [1\text{ mol } (\Delta H^o_{SiH4}) + 2\text{ mol } (\Delta H^o_{O2})] =$

$\Delta H^o_{rxn} = [1\text{ mol } (-910.9\text{ kJ/mol}) + 2\text{ mol } (-241.8\text{ kJ/mol})] - [1\text{ mol } (34.3\text{ kJ/mol}) + 2\text{ mol } (0)] =$

$\Delta H^o_{rxn} = -1428.5\text{ kJ}$

**Now we can use Equation 10.4, $\Delta G = \Delta H - T\Delta S$:**

$\Delta G^o_{rxn} = \Delta H^o_{rxn} - T\Delta S^o_{rxn}$ **= –1428.5 kJ – (298 K) (–193.5 J/K)(1 kJ / 1000 J)**

$\Delta G^o_{rxn}$ **= – 1370.8 kJ**

**(d)** Compare $\Delta G^o$ for this reaction to the combustion of methane. (See the previous problem.) Are the reactions in these two exercises enthalpy or entropy driven? Explain.

**From Problem 10.66, $\Delta G^o_{rxn}$ = –800.9 kJ/mol, which is not as large a negative value as it is for the silane reaction. The reactions are enthalpy driven. The large negative values of the enthalpy change dominate the calculation of $\Delta G^o_{rxn}$.**

Problem 10.69 ■ Using tabulated thermodynamic data, calculate $\Delta G^o$ for these reactions.

**$\Delta G_{rxn}$, can be calculated by taking the sum of the Gibb's free energies of formation of the products (each times its coefficient) minus the sum of the Gibb's free energies of formation of the reactants (each times its coefficient). In equation form:**

$\Delta G^o_{rxn} = \sum n\Delta G_f^{\,o}{}_{products} - \sum n\Delta G_f^{\,o}{}_{reactants}$ **: where n = the stoichiometric coefficient in the equation. Gibb's free energies of formation are listed in Appendix E of the textbook.**

**(a)** $Fe(s) + 2\ HCl(g) \rightarrow FeCl_2(s) + H_2(g)$

**Applying the above equation:**

$\Delta G^o_{rxn} = [1\ mol\ (\Delta G_f^o{}_{FeCl2}) + 1\ mol\ (\Delta G_f^o{}_{H2})] - [1\ mol\ (\Delta G_f^o{}_{Fe}) + 2\ mol\ (\Delta G_f^o{}_{HCl})]$

**= [1 mol (–302.3 kJ/mol) + 1 mol (0)] – [1 mol (0) + 2 mol (–95.30 kJ/mol)]**

$\Delta G^o_{rxn}$ **= –111.7 kJ**

**(b)** $3\ NO_2(g) + H_2O(\ell) \rightarrow 2\ HNO_3(\ell) + NO(g)$

$\Delta G^o_{rxn} = [2\ mol\ (\Delta G_f^o{}_{HNO3}) + 1\ mol\ (\Delta G_f^o{}_{NO})] - [3\ mol\ (\Delta G_f^o{}_{NO2}) + 1\ mol\ (\Delta G_f^o{}_{H2O})]$

**= [2 mol (–80.79 kJ/mol) + 1 mol (86.57 kJ/mol)] – [3 mol (51.30 kJ/mol) + 1 mol (–237.2 kJ/mol)]**

$\Delta G^o_{rxn}$ **= 8.3 kJ**

**(c)** $2\ K(s) + Cl_2(g) \rightarrow 2\ KCl(s)$

$\Delta G^o_{rxn} = [2\ mol\ (\Delta G_f^o{}_{KCl})] - [2\ mol\ (\Delta G_f^o{}_{K}) + 1\ mol\ (\Delta G_f^o{}_{Cl2})]$

$$= [2 \text{ mol } (-408.8 \text{ kJ/mol})] - [2 \text{ mol } (0) + 1 \text{ mol } (0)]$$

$$\Delta G^o_{rxn} = -817.6 \text{ kJ}$$

**(d)** $Cl_2(g) + 2\,NO(g) \rightarrow 2\,NOCl(g)$

$$\Delta G^o_{rxn} = [2 \text{ mol } (\Delta G_f^o{}_{NOCl})] - [1 \text{ mol } (\Delta G_f^o{}_{Cl2}) + 2 \text{ mol } (\Delta G_f^o{}_{NO})]$$

$$= [2 \text{ mol } (66.36 \text{ kJ/mol})] - [1 \text{ mol } (0) + 2 \text{ mol } (86.57 \text{ kJ/mol})]$$

$$\Delta G^o_{rxn} = -40.42 \text{ kJ}$$

**(e)** $SiCl_4(g) \rightarrow Si(s) + 2\,Cl_2(g)$

$$\Delta G^o_{rxn} = [1 \text{ mol } (\Delta G_f^o{}_{Si}) + 2 \text{ mol } (\Delta G_f^o{}_{Cl2})] - [1 \text{ mol } (\Delta G_f^o{}_{SiCl4})]$$

$$= [1 \text{ mol } (0) + 2 \text{ mol } (0)] - [1 \text{ mol } (-617 \text{ kJ/mol})]$$

$$\Delta G^o_{rxn} = 617 \text{ kJ}$$

Problem 10.71 Calculate $\Delta G^o$ for the dissolution of both sodium chloride and silver chloride using data from Appendix E. Explain how the values you obtain relate to the solubility rules for these substances.

**Using the formula:** $\Delta G^o_{rxn} = \sum n\Delta G_f{}^o{}_{products} - \sum n\Delta G_f{}^o{}_{reactants}$ **:**

$NaCl(s) \rightarrow NaCl(aq)$

$$\Delta G^o_{dissolution} = [1 \text{ mol } (\Delta G_f^o{}_{NaCl(aq)})] - [1 \text{ mol } (\Delta G_f^o{}_{NaCl(s)})]$$

$$= [1 \text{ mol } (-393.0 \text{ kJ/mol})] - [1 \text{ mol } (-384 \text{ kJ/mol})]$$

$$\Delta G^o_{dissolution} = -9.0 \text{ kJ}$$

$AgCl(s) \leftrightarrows Ag^+(aq) + Cl^-(aq)$

$$\Delta G^o_{dissolution} = [1 \text{ mol } (\Delta G_f^o{}_{Ag+}) + 1 \text{ mol } (\Delta G_f^o{}_{Cl-})] - [1 \text{ mol } (\Delta G_f^o{}_{AgCl})]$$

$$= [1 \text{ mol } (77.12 \text{ kJ/mol}) + 1 \text{ mol } (-131.26 \text{ kJ/mol})] - [1 \text{ mol } (-109.8 \text{ kJ/mol})]$$

$$\Delta G^o_{rxn} = 55.7 \text{ kJ}$$

**NaCl is a soluble salt: AgCl is considered insoluble. The negative value of $\Delta G$ for sodium chloride indicates that its dissolution is spontaneous. The positive value of $\Delta G$ for the dissolution of AgCl tells us that its dissolution is not spontaneous.**

Problem 10.73 ■ The normal melting point of benzene, $C_6H_6$, is 5.5°C. For the process of melting, what is the sign of each of the following? (a) $\Delta H°$, (b) $\Delta S°$, (c) $\Delta G°$ at 5.5°C, (d) $\Delta G°$ at 0.0°C, (e) $\Delta G°$ at 25.0°C

**(a) positive (endothermic process)**

**(b) positive (solid → liquid)**

**(c) zero (equilibrium)**

**(d) positive (reactant-favored)**

**(e) negative (product-favored)**

Problem 10.75 ■ Estimate the temperature range over which each of the following reactions is spontaneous.

(a) $2\ Al(s) + 3\ Cl_2(g) \rightarrow 2\ AlCl_3(s)$

(b) $2\ NOCl(g) \rightarrow 2\ NO(g) + Cl_2(g)$

(c) $4\ NO(g) + 6\ H_2O(g) \rightarrow 4\ NH_3(g) + 5\ O_2(g)$

(d) $2\ PH_3(g) \rightarrow 3\ H_2(g) + 2\ P(g)$

**(a) $\Delta H°_{rxn} = [2(-704.2)] - [2(0) + 3(0)] = -1408.4$ kJ**
**$\Delta S°_{rxn} = [2(110.7)] - [2(28.3) + 3(223.0)] = -504.2$ J/K**
**$T_{equil.} = \Delta H°_{rxn} / \Delta S°_{rxn} = -1408400$ J/(−504.2 J/K) = 2793 K or 2520°C**
**This reaction is spontaneous at temperatures below 2520°C (below since $\Delta S°_{rxn}$ is negative).**

**(b) $\Delta H°_{rxn} = [2(90.25) + 1(0)] - [2(52.59)] = +75.32$ kJ**
**$\Delta S°_{rxn} = [2(210.7) + 1(223.0)] - [2(264)] = +116$ J/K**
**$T_{equil.} = \Delta H°_{rxn} / \Delta S°_{rxn} = 75320$ J/(116 J/K) = 649 K or 376°C**
**This reaction is spontaneous at all temperatures above 376°C (since $\Delta S°_{rxn}$ is positive).**

**(c) $\Delta H°_{rxn} = [4(-46.11) + 5(0)] - [4(90.25) + 6(-241.8)] = 905.4$ kJ**
**$\Delta S°_{rxn} = [4(192.3) + 5(205.0)] - [4(210.7) + 6(188.7)] = -180.8$ J/K**
**$T_{equil.} = \Delta H°_{rxn} / \Delta S°_{rxn} = 905400$ J/(−180.8 J/K) = −5008 K or no possible temperature**

**This reaction is nonspontaneous at all temperatures (since both enthalpy and entropy are unfavorable).**

**(d) $\Delta H^o_{rxn}$ = [3(0) + 2(314.6)] − [2(5.4)] = 618.4 kJ**
**$\Delta S^o_{rxn}$ = [3(130.6) + 2(163.1)] − [2(210.1)] = 297.8 J/K**
**$T_{equil.}$ = $\Delta H^o_{rxn}$ / $\Delta S^o_{rxn}$ = 618400 J/(297.8 J/K) = 2077 K or 1804°C**
**This reaction is spontaneous at temperatures above 1804°C (below since $\Delta S^o_{rxn}$ is positive).**

Problem 10.77 During polymerization, the system usually becomes more ordered as monomers link together. Could an endothermic polymerization reaction ever occur spontaneously? Explain.

**No, it would violate the second law of thermodynamics; the entropy of the universe would not increase.**
**In terms of Gibbs free energy, a system that becomes more orderly would have a negative $\Delta S$ and an endothermic system would have a positive $\Delta H$. According to Equation 10.4, $\Delta G = \Delta H - T\Delta S$, such a system would never have negative $\Delta G$ and never be considered spontaneous.**

Problem 10.79 When polymers are recycled, the ends of the long-chain polymer molecules tend to break off, and this process eventually results in a loss of physical properties, rendering the recycled polymer unusable. Explain why the breaking off of the ends of the polymer molecules is favorable from the standpoint of the entropy of the system.

**Generally, processes that increase entropy or molecular disorder tend to be spontaneous. Large complicated molecules are more orderly than smaller simpler ones. The process of breaking of large molecules into smaller one is favored because this increases the entropy of the system.**

Problem 10.81 Diethyl ether is a liquid at normal temperature and pressure, and it boils at 35°C. Given that $\Delta H$ is 26.0 kJ/mol for the vaporization of diethyl ether, find its molar entropy change for vaporization.

**When the ether boils, its vapor is in equilibrium with the liquid, meaning that $\Delta G$ is equal to zero. Setting Equation 10.4 equal to zero, we can solve for the value of $\Delta S$.**

$$\Delta G = 0 = \Delta H - T\Delta S, \qquad \Delta S = \frac{\Delta H}{T} = \frac{26.0\ \text{kJ/mol}}{308\ \text{K}} = 0.0844\ \text{kJ/mol·K or } 84.4\ \text{J/mol·K}$$

Problem 10.83 Gallium metal has a melting point of 29.8°C. Use the information below to calculate the boiling point of gallium in °C.

**When gallium is boiling, the liquid is in equilibrium with the vapor. At equilibrium, the value of $\Delta G$ is zero. Using the data, we can calculate the values of $\Delta H$ and $\Delta S$ for the**

**boiling process. We can then solve for the value of $T$ where $G$ equals zero, $T_{\Delta G=0} = \frac{\Delta H}{\Delta S}$, this is the boiling temperature.** $Ga(\ell) \leftrightarrows Ga(g)$

$$\Delta H = \Delta H^{o}_{Ga(g)} - \Delta H^{o}_{Ga(\ell)} = 271.96 \text{ kJ/mol} - 5.578 \text{ kJ/mol} = 266.4 \text{ kJ/mol}$$

$$\Delta S = S^{o}_{Ga(g)} - S^{o}_{Ga(\ell)} = 169.03 \text{ J/mol·K} - 59.25 \text{ J/mol·K} = 109.78 \text{ J/mol·K}$$

$$T_{\Delta G=0} = \frac{\Delta H}{\Delta S} = \frac{266.4 \text{ kJ/mol}}{0.10978 \text{ kJ/mol} \cdot \text{K}} = 2427 \text{ K} \quad \text{or} \quad 2154^{o}\text{C}$$

**Gallium boils at 2154°C. Gallium has the largest liquid temperature range of all elements.**

Problem 10.85 Iodine is not very soluble in water, but it dissolves readily in a solution containing iodide ions by the following reaction:

$$I_2(aq) + I^{-}(aq) \rightarrow I_3^{-}(aq)$$

The following graph shows the results of a study of the temperature dependence of $\Delta G^{o}$ for this reaction. (The solid line is a best fit to the actual data points.) Notice that the quantity on the *y* axis is $\Delta G^{o}/T$, not just $\Delta G^{o}$. Additional data relevant to this reaction are also given below the graph.

**(a)** Calculate $\Delta G^{o}$ for this reaction at 298 K. (DO NOT read this value off the graph. Use the data given to calculate a more accurate value.)

**Using the formula:** $\Delta G^{o}_{rxn} = \sum n\Delta G_f{}^{o}{}_{products} - \sum n\Delta G_f{}^{o}{}_{reactants}$ **:**

$$\Delta G^{o}_{rxn} = [1 \text{ mol } (\Delta G_f{}^{o}{}_{I3-})] - [1 \text{ mol } (\Delta G_f{}^{o}{}_{I2}) + 1 \text{ mol } (\Delta G_f{}^{o}{}_{I-})]$$

$$= [1 \text{ mol } (-51.4 \text{ kJ/mol})] - [1 \text{ mol } (16.37 \text{ kJ/mol}) + 1 \text{ mol } (-51.57 \text{ kJ/mol})]$$

$$\Delta G^{o}_{rxn} = -16.2 \text{ kJ}$$

**(b)** Determine $\Delta H^{o}$ for this reaction. (Assume $\Delta H^{o}$ is independent of $T$.) (HINT: You will need to use the graph provided to find $\Delta H^{o}$. It may help if you realize that the graph is a straight line and try to write an equation for the line.)

**The slope of this plot is equal to $\Delta H$. If we start with Equation 10.4, $\Delta G = \Delta H - T\Delta S$, and divide by $T$, we get: $\frac{\Delta G}{T} = \frac{\Delta H}{T} - \Delta S$. A graph of $\frac{\Delta G}{T}$ versus $\frac{1}{T}$ will have an intercept of $-\Delta S$ and a slope equal to $\Delta H$.**

**Choosing the first and last data points to calculate the slope:**

$$\text{Slope} = \frac{[-0.0585 - (-0.0500)]\ \text{kJ/K}}{[0.00362 - 0.0031]\ 1/\text{K}} = -17\ \text{kJ}$$

Problem 10.87 ■ Determine whether each of the following statements is true or false. If false, modify to make the statement true.

(a) An exothermic reaction is spontaneous.
(b) When $\Delta G°$ is positive, the reaction cannot occur under any conditions.
(c) $\Delta S°$ is positive for a reaction in which there is an increase in the number of moles.
(d) If $\Delta H°$ and $\Delta S°$ are both negative, $\Delta G°$ will be negative.

**(a) False An exothermic reaction is *usually* spontaneous, but not always. One must also consider entropy changes, which could render an exothermic reaction nonspontaneous.**

**(b) False When $\Delta G°$ is positive, the reaction is not spontaneous under the standard conditions (25°C and 1 atm), but may be spontaneous under different conditions.**

**(c) False $\Delta S°$ is positive for a reaction in which there is an increase in the moles of *gas*.**

**(d) False If $\Delta H°$ and $\Delta S°$ are both negative, $\Delta G°$ will be negative at low temperatures and positive at high temperatures.**

Problem 10.89 ■ Polyethylene has a heat capacity of 2.3027 J $g^{-1}$ $°C^{-1}$. You need to decide if 1.0 ounce of polyethylene can be used to package a material that will be releasing heat when in use. Consumer safety specifications indicate that the maximum allowable temperature for the polyethylene is 45°C; it can be assumed that the plastic is initially at room temperature. (a) What temperature will the polyethylene reach if the product generates 1500 J of heat and all of this energy is absorbed by the plastic package? (b) Is this a realistic estimate of the temperature that the polyethylene packaging would reach? Explain your answer. (c) What is the enthalpy change of the polyethylene? (d) Estimate the entropy change of the polyethylene. (You will need to assume that the temperature of the plastic is constant.)

**(a) The temperature increase can be estimated by $q = mc\Delta T$.**

$$\Delta T = q/mc = \frac{1500\ \text{J}}{1.0\ \text{oz} \times \frac{28.35\ \text{g}}{1\ \text{oz}} \times 2.3027\ \frac{\text{J}}{\text{g} \cdot {}^{\text{o}}\text{C}}} = 23.0^{\text{o}}\text{C}$$

**If the plastic is initially 25°C, then the maximum temperature would be 25°C + 23.0°C = 48°C = 321 K.**

**(b) It is not very realistic as a significant amount of the heat would be transferred to the surrounding air.**

**(c) Assuming constant pressure, the enthalpy change is equal to the heat flow = 1500 J.**

**(d)** $\Delta S = \dfrac{\Delta H}{T} = \dfrac{1500\text{ J}}{321\text{ K}} = 4.7\text{ J/K}$

Problem 10.91 ■ The reaction shown below is involved in the refining of iron. (The table that follows provides all of the thermodynamic data you should need for this problem.)

$$2\,Fe_2O_3(s) + 3\,C(s, \text{graphite}) \rightarrow 4\,Fe(s) + 3\,CO_2(g)$$

| Compound | $\Delta H_f°$ (kJ mol$^{-1}$) | $S°$ (J mol$^{-1}$ K$^{-1}$) | $\Delta G_f°$ (kJ mol$^{-1}$) |
|---|---|---|---|
| $Fe_2O_3(s)$ | –824.2 | ? | –742.2 |
| C(s, graphite) | 0 | 5.740 | 0 |
| Fe(s) | 0 | 27.3 | 0 |
| $CO_2(g)$ | –393.5 | 213.6 | –394.4 |

(a) Find $\Delta H°$ for the reaction.
(b) $\Delta S°$ for the reaction above is 557.98 J/K. Find $S°$ for $Fe_2O_3(s)$.
(c) Calculate $\Delta G°$ for the reaction at the standard temperature of 298 K. (There are two ways that you could do this.)
(d) At what temperatures would this reaction be spontaneous?

**(a) We'll use the formula,** $\Delta H°_{rxn} = \sum n\Delta H_f°{}_{\text{products}} - \sum n\Delta H_f°{}_{\text{reactants}}$

$\Delta H°_{rxn}$ = [3 mol ($\Delta H°_{CO2}$) + 4 mol ($\Delta H°_{Fe}$)] – [2 mol ($\Delta H°_{Fe2O3}$) + 3 mol ($\Delta H°_{C}$)] =

$\Delta H°_{rxn}$ = [3 mol (–393.5 kJ/mol) + 4 mol (0)] – [2 mol (–824.2 kJ/mol) + 3 mol (0)] =

$\Delta H°_{rxn}$ = 467.9 kJ

**(b) We'll use the formula,** $\Delta S°_{rxn} = \sum nS°_{\text{products}} - \sum nS°_{\text{reactants}}$

$\Delta S°_{rxn}$ = [3 mol ($S°_{CO2}$) + 4 mol ($S°_{Fe}$)] – [2 mol ($?S°_{Fe2O3}$) + 3 mol ($S°_{C}$)] = 557.98 J/K

$\Delta S°_{rxn}$ = [3 mol (213.6 J/mol·K) + 4 mol (27.3 J/mol·K)] – [2 mol (?) + 3 mol (5.740 J/mol·K)] = 557.98 J/K

$$?S^{o}_{Fe2O3} = [557.98\ J/K - 640.8\ J/K - 109.2\ J/K + 17.22\ J/K] \div (-2) = 87.4\ J/K$$

**(c) Using the formula:** $\Delta G^{o}_{rxn} = \sum n\Delta G_f^{\,o}{}_{products} - \sum n\Delta G_f^{\,o}{}_{reactants}$ **:**

$$\Delta G^{o}_{rxn} = [3\ mol\ (\Delta G^{o}_{CO2}) + 4\ mol\ (\Delta G^{o}_{Fe})] - [2\ mol\ (\Delta G^{o}_{Fe2O3}) + 3\ mol\ (\Delta G^{o}_{C})]$$

$$= [3\ mol\ (-394.4\ kJ/mol) + 4\ mol\ (0)] - [2\ mol\ (-742.2\ kJ/mol) + 3\ mol\ (0)]$$

$$\Delta G^{o}_{rxn} = 301.2\ kJ$$

or $\Delta G^{o}_{rxn} = \Delta H^{o}_{rxn} - T\Delta S^{o}_{rxn} = 467.9\ kJ - (298\ K)\ (557.98\ J/K)(1\ kJ/1000\ J)$

$$\Delta G^{o}_{rxn} = 301.6\ kJ$$

(d) $T_{\Delta G=0} = \dfrac{\Delta H}{\Delta S} = \dfrac{467.9\ kJ}{0.55798\ kJ/K} = 839\ K$ **or** $566\ ^{o}C$

**The reaction is spontaneous above 566 °C.**

Problem 10.93 The graph below shows $\Delta G°$ as a function of temperature for the synthesis of ammonia from nitrogen and hydrogen.

$$N_2(g) + 3\ H_2(g) \rightarrow 2\ NH_3(g)$$

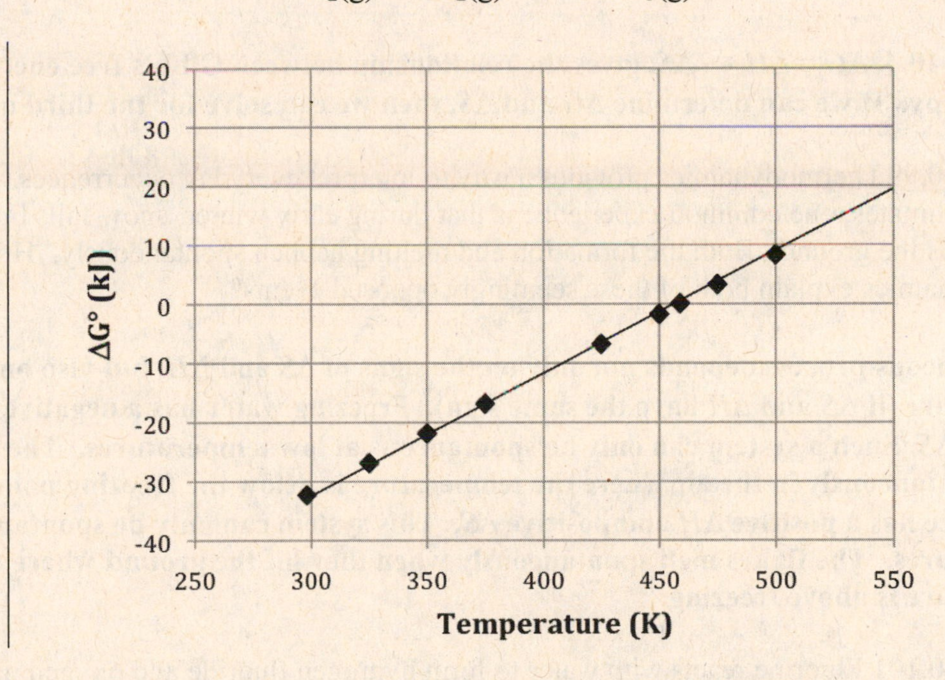


(a). Use the graph to estimate $\Delta S°$ for the ammonia synthesis reaction.

**The graph represents the line:** $\Delta G = \Delta H - T\Delta S$ $\qquad (y = b + mx)$

**Where the slope = $-\Delta S$. Calculating the slope (using data points at 350 K and 500 K):**

$$\text{Slope} = \frac{(9\text{ kJ} - (-22\text{ kJ}))}{(500\text{ K} - 350\text{ K})} = \frac{31\text{ kJ}}{150\text{ K}} = 0.21\text{ kJ/K} \qquad \text{so } \Delta S = -0.21\text{ kJ/K}$$

(b). Given that the standard free energy change of formation for ammonia ($\Delta G_f°$) is −16.50 kJ/mol, estimate $\Delta H°$ for the ammonia synthesis reaction.

**The Gibbs free energy for the reaction is 2 moles × (–16.50 kJ/mol) = – 33.0 kJ.**
$\Delta G° = \Delta H° - T\Delta S°$ **so** $\Delta H° = \Delta G° + T\Delta S° = -33.0\text{ kJ} + 298\text{ K}(-0.21\text{ kJ/K}) =$
$= -96\text{ kJ}$

Problem 10.95 Suppose that you need to know the heat of formation of cyclohexane, $C_6H_{12}$, but the tables you have do not provide the value. You have a sample of the chemical. What could you do to determine the heat of formation?

**The sample can be burned in a calorimeter to determine the heat of combustion. From the values of heat of formation of carbon dioxide and water, the heat of formation of cyclohexane can be calculated.**

Problem 10.97 You have a table of thermodynamic variables that includes standard entropies and free energies of formation. How could you use the information you have to estimate the heat of formation of a substance listed in your table?

**Equation 10.4, $\Delta G = \Delta H - T\Delta S$, gives the relationship between Gibb's free energy, enthalpy and entropy. If we can determine $\Delta G$ and $\Delta S$, then we can solve for the third quantity, $\Delta H$.**

Problem 10.99 Thermodynamics provides a way to interpret everyday occurrences. If you live in northern climates, one common experience is that during early winter, snow falls but then melts when it hits the ground. Both the formation and melting happen spontaneously. How can thermodynamics explain both of these seemingly opposed events?

**A spontaneous process depends not only on the signs of $\Delta S$ and $\Delta H$, but also on temperature (if $\Delta S$ and $\Delta H$ have the same sign). Freezing water has a negative $\Delta H$ and negative $\Delta S$. Such a system can only be spontaneous at low temperatures. The snowflakes form spontaneously in the air where the temperature is below the freezing point of water. Melting ice has a positive $\Delta H$ and positive $\Delta S$. This system can only be spontaneous at high temperatures. The flakes melt spontaneously when they hit the ground where the temperature is above freezing.**

Problem 10.101 Fluorine reacts with water to form hydrogen fluoride and oxygen gas. **(a)** Write a balanced chemical reaction for this reaction. **(b)** Use tabulated data to determine the free energy change for the reaction and comment on its spontaneity. **(c)** Use tabulated data to calculate the enthalpy change of the reaction. **(d)** Determine how much heat flows and in what direction when 34.5 g of fluorine gas is bubbled through excess water.

**(a)** $2\ F_2(g) + 2\ H_2O(\ell) \rightarrow 4\ HF(g) + O_2(g)$

**(b) Using the formula:** $\Delta G^o_{rxn} = \sum n\Delta G_f^{\ o}{}_{products} - \sum n\Delta G_f^{\ o}{}_{reactants}$ **:**

$$\Delta G^o_{rxn} = [4\ mol\ (\Delta G_f^o{}_{HF}) + 1\ mol\ (\Delta G_f^o{}_{O2})] - [2\ mol\ (\Delta G_f^o{}_{F2}) + 2\ mol\ (\Delta G_f^o{}_{H2O})]$$

$$= [4\ mol\ (-273\ kJ/mol) + mol\ (0)] - [2\ mol\ (0) + 2\ mol\ (-237.2\ kJ/mol)]$$

$$\Delta G^o_{rxn} = -\ 618\ kJ$$

**This is a very large negative number, indicating a spontaneous reaction that is strongly favored to go in the forward direction.**

**(c) Using the formula:** $\Delta H^o_{rxn} = \sum n\Delta H_f^{\ o}{}_{products} - \sum n\Delta H_f^{\ o}{}_{reactants}$ **:**

$$\Delta H^o_{rxn} = [4\ mol\ (\Delta H_f^o{}_{HF}) + 1\ mol\ (\Delta H_f^o{}_{O2})] - [2\ mol\ (\Delta H_f^o{}_{F2}) + 2\ mol\ (\Delta H_f^o{}_{H2O})]$$

$$= [4\ mol\ (-271\ kJ/mol) + 1\ mol\ (0)] - [2\ mol\ (0) + 2\ mol\ (-285.8\ kJ/mol)]$$

$$\Delta H^o_{rxn} = -\ 512\ kJ$$

**(d)** $34.5\ g\ F_2 \times \frac{1\ mol\ F_2}{38.00\ g\ F_2} \times \frac{1\ mol\ rxn}{2\ mol\ F_2} \times \frac{-512\ kJ}{mol\ rxn} = -\ 232\ kJ$

**The heat flow is 232 kJ. The negative sign indicates the heat is released from the system.**

<u>Problem 10.103</u> Consider the following thermodynamic data for oxides of manganese. **(a)** What is the correct nomenclature for each oxide? **(b)** Write and balance chemical equations for the conversion of each of these oxides into $Mn_3O_4$. **(c)** Based on the free energy changes of these reactions, which oxide is the most stable at room temperature?

**(a)** **MnO – manganese (II) oxide**
**$MnO_2$ – manganese (IV) oxide**
**$Mn_2O_3$ – manganese (III) oxide**

**(b)** $3\ MnO + \frac{1}{2}\ O_2 \rightarrow Mn_3O_4$
$3\ MnO_2 \rightarrow Mn_3O_4 + O_2$
$3\ Mn_2O_3 \rightarrow 2\ Mn_3O_4 + \frac{1}{2}\ O_2$

**(c)** **We can use the following formula to calculate the Gibb's free energy change for these reactions, using the thermodynamic data provided.**

$$\Delta G^o_{rxn} = \sum n\Delta G_f^{\ o}{}_{products} - \sum n\Delta G_f^{\ o}{}_{reactants}$$

**For MnO:**

$\Delta G^{o}_{rxn} = [1\text{ mol }(\Delta G_f^{o}{}_{Mn3O4})] - [3\text{ mol }(\Delta G_f^{o}{}_{MnO}) + \frac{1}{2}\text{ mol }(\Delta G_f^{o}{}_{O2})]$

$= [1\text{ mol }(-1283\text{ kJ/mol})] - [3\text{ mol }(-362.9\text{ kJ/mol}) + \frac{1}{2}\text{ mol }(0)]$

$\Delta G^{o}_{rxn} = -\ 194.3\text{ kJ}$

**For $MnO_2$:**

$\Delta G^{o}_{rxn} = [1\text{ mol }(\Delta G_f^{o}{}_{Mn3O4}) + 1\text{ mol }(\Delta G_f^{o}{}_{O2})] - [3\text{ mol }(\Delta G_f^{o}{}_{MnO2})]$

$= [1\text{ mol }(-1283\text{ kJ/mol}) + 1\text{ mol }(0)] - [3\text{ mol }(-465.4\text{ kJ/mol})]$

$\Delta G^{o}_{rxn} = 113.2\text{ kJ}$

**For $Mn_2O_3$:**

$\Delta G^{o}_{rxn} = [2\text{ mol }(\Delta G_f^{o}{}_{Mn3O4}) + \frac{1}{2}\text{ mol }(\Delta G_f^{o}{}_{O2})] - [3\text{ mol }(\Delta G_f^{o}{}_{Mn2O3})]$

$= [2\text{ mol }(-1283\text{ kJ/mol}) + \frac{1}{2}\text{ mol }(0)] - [3\text{ mol }(-881.2\text{ kJ/mol})]$

$\Delta G^{o}_{rxn} = 77.6\text{ kJ}$

**Based on the Gibbs free energy changes, $MnO_2$ is the most stable oxide because it has the most positive $\Delta G$ in the reaction to form $Mn_3O_4$. This means the conversion of $MnO_2$ into $Mn_3O_4$ is the least likely to occur.**

## CHAPTER OBJECTIVE QUIZ

This quiz will test your understanding of the basic text chapter objectives and give you additional practice problems. You should work this quiz after completing the end-of-chapter questions. The solutions to these questions are found at the end of this chapter.

1. Which of the following statements about recycling plastics is **not** true?
   a. Recycled plastics are not sorted; they are all melted down together.
   b. Poly (ethylene terephthalate) (PET) and polyethylene (PE) are two of the most commonly recycled plastics.

c. Degradation of the polymer during repeated recycling is a concern.
d. Usually, legal restrictions prohibit recycled plastics from being used for food or beverages.
e. Recycled PET is commonly used as fiberfill for sleeping bags and coats, carpeting, and fleece fabrics.

2. TRUE or FALSE: A spontaneous process will take place under a given set of conditions, without any outside influence.

3. TRUE or FALSE: If a chemical reaction is spontaneous, then it will occur rapidly.

4. Which of the following statements about entropy is incorrect?
   a. Entropy is a thermodynamic state function.
   b. Entropy is defined by q/T in the Carnot cycle.
   c. An increase in entropy means order is increasing in the system.
   d. An increase in disorder is mathematically the most probable outcome.
   e. A larger value of the Boltzmann function represents a greater amount of entropy.

5. Predict the sign of the entropy change, ΔS, for the following reactions:

   a. $2\ H_2O_2(\ell) \rightarrow 2\ H_2O(\ell) + O_2(g)$

   b. $C_{12}H_{22}O_{11}(s) + O_2(g) \rightarrow 12\ CO_2(g) + 11\ H_2O(\ell)$

   c. $Mg(s) + 2\ HCl(aq) \rightarrow MgCl_2(aq) + H_2(g)$

   d. $N_2(g) + 3\ H_2(g) \rightarrow 2\ NH_3(g)$

   e. $Ba(OH)_2(aq) + H_2SO_4(aq) \rightarrow BaSO_4(s) + 2\ H_2O(\ell)$

6. Which of the following statements about the second law of thermodynamics is false?
   a. In any spontaneous process, entropy of the universe must increase.
   b. In any spontaneous process, entropy of the system must increase.
   c. The entropy of the surroundings can increase or decrease.
   d. $\Delta S_{universe} = \Delta S_{system} + \Delta S_{surroundings}$
   e. $\Delta S_{universe} > 0$ for a spontaneous process.

7. Consider 10.0 g of water placed in a freezer at –15.0 °C. What is the entropy change of the surroundings when the water freezes? For water, $\Delta H_{fusion}$ = 338 J/g.

8. Which of the following statements about the third law of thermodynamics is **false**?
   a. All pure substances have zero entropy at zero Kelvin.
   b. Absolute entropy increases with temperature.
   c. Elements in their standard states have zero values of absolute entropy.

d. Cooling matter to absolute zero is not possible, but scientists have come very close.
e. The absolute entropy of a pure substance can be evaluated at standard state conditions.

9. Calculate the entropy change, $\Delta S^\circ_{rxn}$, of the following reactions using thermodynamic data in Appendix E.

   (a) $2\ CO_2(g) + O_2(g) \rightarrow 2\ CO_2(g)$

   (b) $C_3H_8(g) + 5\ O_2(g) \rightarrow 3\ CO_2(g) + 4\ H_2O(\ell)$

   (c) $4\ Fe(s) + 3\ O_2(g) \rightarrow 2\ Fe_2O_3(s)$

10. TRUE or FALSE: In a spontaneous process, $\Delta S_{universe}$ is positive and $\Delta G_{system} = -T(\Delta S_{universe})$.

11. Using thermodynamic data from Appendix E, calculate the Gibbs free energy change, $\Delta G_{system}$, of the following reactions.

    (a) $2\ Fe(s) + 2\ Cl_2(g) \rightarrow 2\ FeCl_2(s)$

    (b) $CH_4(g) + H_2O(\ell) \rightarrow CO(g) + 3\ H_2(g)$

    (c) $S(s,rhombic) + 3\ F_2(g) \rightarrow SF_6(s)$

12. Considering the definition $\Delta G = \Delta H - T\Delta S$, match the following possible combinations of $\Delta H$ and $\Delta S$ (in terms of positive or negative signs).

| | $\Delta H$ | $\Delta S$ | |
|---|---|---|---|
| ____ | + | + | (a) always spontaneous at all temperatures |
| ____ | – | – | (b) only spontaneous at low temperatures |
| ____ | – | + | (c) only spontaneous at high temperatures |

13. Use thermodynamic data from appendix E in the textbook to determine if the following reaction is spontaneous at only high or only low temperatures.

$$N_2(g) + O_2(g) + 180.5\ kJ \rightarrow 2\ NO(g)$$

14. A substance has the following properties:

| | $\Delta H^\circ_f$ (kJ/mol) | $S^\circ$ (J/mol·K) |
|---|---|---|
| X(s) | −52.4 | 109.5 |
| X($\ell$) | −23.4 | 251.1 |

Estimate the normal boiling point of this substance.

15. For bromine, $S^{\circ}_{(\ell)}$ = 152.2 J/mol·K, $S^{\circ}_{(g)}$ = 245.4 J/mol·K and $\Delta H^{\circ}_{f(g)}$ = 30.91 kJ/mol. Verify that the normal boiling point of bromine is 59°C.

16. Use thermodynamic data from Appendix E to calculate the entire temperature range for which the following reaction is spontaneous.

$$C(graphite) + 2\ H_2(g) \rightarrow CH_4(g)$$

# ANSWERS TO THE CHAPTER OBJECTIVE QUIZ

1. a

2. TRUE

3. FALSE

4. c

5. (a) The molecule is breaking down into simpler substances, $\Delta S$ is positive.
   (b) The large sugar molecule (which is a solid) is oxidized to smaller molecules (which are gas and liquid). $\Delta S$ is positive.
   (c) The solid magnesium forms hydrogen gas, $\Delta S$ is positive.
   (d) The compound ammonia is formed from elements; $\Delta S$ is negative.
   (e) A solid precipitates from aqueous solutions; $\Delta S$ is negative.

6. b

7. The entropy change of the surroundings can be defined in terms of the heat flow of the surroundings and its temperature, $\Delta S = \frac{q_{surr}}{T}$. The heat flow of the system (freezing water) can be calculated from the heat of fusion of water.

   $q_{sys} = (-)\ \Delta H_{fusion} \times \text{mass} = -\ 338\ \text{J/g} \times 10.0\ \text{g water} = -\ 3380\ \text{J},$

   $q_{surr} = (-)q_{sys} = +3380\ \text{J}$ Therefore, $\Delta S = \frac{q_{surr}}{T} = \frac{+3380\ \text{J}}{258\ \text{K}} = +13.1\ \text{J/K}$

8. c

9. (a) We'll use the formula, $\Delta S^{o}_{rxn} = \sum nS^{o}_{products} - \sum nS^{o}_{reactants}$

   $\Delta S^{o}_{rxn} = [2\ \text{mol}\ (S^{o}_{CO2})] - [2\ \text{mol}\ (S^{o}_{CO}) + 1\ \text{mol}\ (S^{o}_{O2})] =$

   $\Delta S^{o}_{rxn} = [2\ \text{mol}\ (213.6\ \text{J/mol·K})] - [2\ \text{mol}\ (197.6\ \text{J/mol·K}) + 1\ \text{mol}\ (205.0\ \text{J/mol·K})] =$

   $\Delta S^{o}_{rxn} = -173.0\ \text{J/K}$

   (b) $\Delta S^{o}_{rxn} = [3\ \text{mol}\ (S^{o}_{CO2}) + 4\ \text{mol}\ (S^{o}_{H2O})] - [1\ \text{mol}\ (S^{o}_{C3H8}) + 5\ \text{mol}\ (S^{o}_{O2})] =$

   $\Delta S^{o}_{rxn} = [3\ \text{mol}\ (213.6\ \text{J/mol·K}) + 4\ \text{mol}\ (69.91\ \text{J/mol·K})] - [1\ \text{mol}\ (269.9\ \text{J/mol·K}) + 5\ \text{mol}\ (205.0\ \text{J/mol·K})] =$

**$\Delta S^o_{rxn} = -374.5$ J/K**

**(c) $\Delta S^o_{rxn} = [2 \text{ mol } (S^o_{Fe2O3})] - [4 \text{ mol } (S^o_{Fe}) + 3 \text{ mol } (S^o_{O2})] =$**

**$\Delta S^o_{rxn} = [(2 \text{ mol } (87.4 \text{ J/mol·K})] - [4 \text{ mol } (27.3 \text{ J/mol·K}) + 3 \text{ mol } (205.0 \text{ J/mol·K})] =$**

**$\Delta S^o_{rxn} = -549.4$ J/K**

**10. TRUE**

**11. The fastest way to calculate $\Delta G^o_{rxn}$ is to use:**

$$\Delta G^o_{rxn} = \sum n\Delta G_f^{\ o}{}_{products} - \sum n\Delta G_f^{\ o}{}_{reactants}$$

**where n = the stoichiometric coefficient in the equation.**
**Gibb's free energies of formation are listed in Appendix E of the textbook.**

**$\Delta G^o_{rxn} = [2 \text{ mol } (\Delta G_f^{\ o}{}_{FeCl2})] - [1 \text{ mol } (\Delta G_f^{\ o}{}_{Fe}) + 2 \text{ mol } (\Delta G_f^{\ o}{}_{Cl2})]$**

**$= [2 \text{ mol } (-302.3 \text{ kJ/mol})] - [\ 1 \text{ mol } (0) + 2 \text{ mol } (0)]$**

**$\Delta G^o_{rxn} = -\ 604.6$ kJ**

**(b) $\Delta G^o_{rxn} = [1 \text{ mol } (\Delta G_f^{\ o}{}_{CO}) + 3 \text{ mol } (\Delta G_f^{\ o}{}_{H2})] - [1 \text{ mol } (\Delta G_f^{\ o}{}_{CH4}) + 1 \text{ mol } (\Delta G_f^{\ o}{}_{H2O})]$**

**$= [1 \text{ mol } (-137.2 \text{ kJ/mol}) + 3 \text{ mol } (0)] - [1 \text{ mol } (-50.75 \text{ kJ/mol}) + 1 \text{ mol } (-228.6 \text{ kJ/mol})]$**

**$\Delta G^o_{rxn} = 142.2$ kJ**

**(c) $\Delta G^o_{rxn} = [1 \text{ mol } (\Delta G_f^{\ o}{}_{SF6})] - [1 \text{ mol } (\Delta G_f^{\ o}{}_{S}) + 1 \text{ mol } (\Delta G_f^{\ o}{}_{F2})]$**

**$= [1 \text{ mol } (-1105 \text{ kJ/mol})] - [1 \text{ mol } (0) + 1 \text{ mol } (0)]$**

**$\Delta G^o_{rxn} = -\ 1105$ kJ**

**12.**

| | $\Delta H$ | $\Delta S$ |
|---|---|---|
| c | + | + |
| b | – | – |
| a | – | + |

**13. From the thermochemical equation, we can tell the sign of the enthalpy change is positive. We need to use absolute entropies to determine the sign of the entopy change of the reaction.**

$$\Delta S^o_{rxn} = \sum nS^o_{products} - \sum nS^o_{reactants}$$

$$\Delta S^o_{rxn} = [2 \text{ mol } (S^o_{NO})] - [1 \text{ mol } (S^o_{N2}) + 1 \text{ mol } (S^o_{O2})] =$$

$$\Delta S^o_{rxn} = [2 \text{ mol } (210.7 \text{ J/mol·K})] - [1 \text{ mol } (191.5 \text{ J/mol·K}) + 1 \text{ mol } (205.0 \text{ J/mol·K})] =$$

$$\Delta S^o_{rxn} = +24.9 \text{ J/K}$$

**Since both $\Delta H^o_{rxn}$ and $\Delta S^o_{rxn}$ are positive, this reaction is spontaneous only at relatively high temperature.**

**14. When the substance is melting, the liquid is in equilibrium with the solid. At equilibrium, the value of $\Delta G$ is zero. Using the data, we can calculate the value of $\Delta$H and $\Delta$S for the melting process. We can then solve for the value of *T* where *G* equals zero, $T_{\Delta G=0} = \frac{\Delta H}{\Delta S}$, this is the melting temperature.**

**X(s) ⇆ X(ℓ)**

$$\Delta H = \Delta H^o_{(\ell)} - \Delta H^o_{(s)} = (-23.4) - (-52.4) \text{ kJ/mol} = 29.0 \text{ kJ/mol}$$

$$\Delta S = S^o_{(\ell)} - S^o_{(s)} = (251.1 - 109.5) \text{ J/mol·K} = 141.6 \text{ J/mol·K}$$

$$T_{\Delta G=0} = \frac{\Delta H}{\Delta S} = \frac{29.0 \text{ kJ/mol}}{0.1416 \text{ kJ/mol} \cdot \text{K}} = 204.8 \text{ K} \quad \text{or} \quad -68.2\ ^oC$$

**15. $\Delta G = 0$ when the vapor and liquid are in equilibrium at the boiling point. $T_{\Delta G=0} = \frac{\Delta H}{\Delta S}$ = the boiling point.**

$$\Delta H = \Delta H^o_{Br2(g)} - \Delta H^o_{Br2(\ell)} = (30.91 \text{ kJ/mol} - 0) = 30.91 \text{ kJ/mol}$$

$$\Delta S = S^o_{Br2(g)} - S^o_{Br2(\ell)} = 245.4 \text{ J/mol·K} - 152.2 \text{ J/mol·K} = 93.2 \text{ J/mol·K}$$

$$T_{\Delta G=0} = \frac{\Delta H}{\Delta S} = \frac{30.91 \text{ kJ/mol}}{0.0932 \text{ kJ/mol} \cdot \text{K}} = 332 \text{ K or } 59\ ^oC$$

**16. $\Delta G = \Delta H - T\Delta S$, we can solve for the temperature at which $\Delta G$ is equal to zero. This represents equilibrium. From the enthalpy and entropy data in Appendix E, $\Delta H$ and $\Delta S$ of the reaction can be calculated. Setting $\Delta G = 0$ and solving for T, we can find the entire temperature range for which the reaction is spontaneous:**

$$\Delta G = 0 = \Delta H - T\Delta S,\ \ T_{\Delta G=0} = \frac{\Delta H}{\Delta S}$$

$$\Delta H^{o}_{rxn} = \sum n\Delta H_{f}^{\ o}{}_{products} - \sum n\Delta H_{f}^{\ o}{}_{reactants} = [(\Delta H^{o}_{f,CH4})] - [(\Delta H^{o}_{f,C}) + 2\ (\Delta H^{o}_{f,H2})]$$

$$\Delta H^{o}_{rxn} = [1\text{ mol }(-74.81\text{ kJ/mol})] - [1\text{ mol }(0) + 2\text{ mol }(0)] = -\ 74.81\text{ kJ}$$

$$\Delta S^{o}_{rxn} = \sum nS^{o}{}_{products} - \sum nS^{o}{}_{reactants} = [1\text{ mol }(S^{o}{}_{CH4})] - [1\text{ mol }(S^{o}{}_{C}) + 2\text{ mol }(S^{o}{}_{H2})] =$$

$$\Delta S^{o}_{rxn} = [1\text{ mol }(186.2\text{ J/mol·K})] - [1\text{ mol }(5.74\text{ J/mol·K}) + 2\text{ mol }(130.6\text{ J/mol·K})] = -\ 80.74\text{ J/K}$$

$$T_{\Delta G=0} = \frac{\Delta H}{\Delta S} = \frac{-74.81\text{ kJ}}{-0.08074\text{ kJ/K}} = 926.6\text{ K}$$

**This type of reaction becomes more spontaneous (more negative $\Delta G$) at lower temperatures; the reaction is spontaneous at all temperatures < 926.6 K.**

# CHAPTER 11

## *Study Goals*

The study goals outline specific concepts to be mastered in this section of the text chapter. Related problems at the end of the text chapter will also be noted. Working the questions noted should aid in mastery of each study goal and also highlight any areas that you may need additional help in.

**Section 11.1 *INSIGHT INTO* Ozone Depletion**

1. Describe the formation and destruction of stratospheric ozone in terms of the Chapman cycle. *Work Problems 1–9.*

**Section 11.2 Rates of Chemical Reactions**

2. Express the rate of reaction in terms of the change in the concentration of reactants or products over time (Δ[conc.]/Δtime). *Work Problems 10–21, 37, 38, 93, 97, 102, 108, 109.*

**Section 11.3 Rate Laws and the Concentration Dependence of Rates**

3. Relate the rate of reaction to the concentration of reactants mathematically using the rate law expression. Understand what each term in the rate law represents. *Work Problems 22–29, 31, 104.*
4. Use the method of initial rates to determine the complete rate law from experimental data. *Work Problems 30, 32 – 36.*

**Section 11.4 Integrated Rate Laws**

5. Use the integrated rate law equations to relate the concentration of reactant to time for zero-, first-, and second-order reactions. *Work Problems 39–47, 105.*
6. Determine the reaction order from graphical analysis of experimental data. *Work Problems 48–53, 112.*

**Section 11.5 Temperature and Kinetics**

7. Describe the effect of temperature on reaction rate on a molecular level, including the concept of activation energy and the activated complex. *Work Problems 54–56, 61, 94, 98, 99, 106, 110, 111.*
8. Use the Arrhenius equation to relate rate constants at different temperatures. *Work Problems 57 – 60, 103.*

**Section 11.6 Reaction Mechanisms**

9. Define a reaction mechanism and distinguish between elementary and multi-step overall reactions. *Work Problems 62–68, 100, 107.*
10. Determine the rate law for a reaction by analyzing its mechanism. *Work Problems 69–71.*

### Section 11.7 Catalysis

11. Describe how a catalyst works on a molecular level, distinguishing between heterogeneous and homogeneous catalysts. *Work Problems 72–78, 101.*
12. Give several examples of the importance of catalysts to both the chemical industry and private citizens. *Work Problems 79, 80.*

### Section 11.8 *INSIGHT INTO* Tropospheric Ozone

13. Describe the process of ozone formation in the troposphere, including what aspects of human-activity contribute to its formation. *Work Problems 81–92, 95, 96.*

## *Solutions to the Odd-Numbered Problems*

Problem 11.1 What is an allotrope?

**Allotropes are different forms of an element in the same physical state. Diamond and graphite are allotropes of carbon; molecular oxygen and ozone are allotropes of oxygen.**

Problem 11.3 In what region of the atmosphere is ozone considered a pollutant? In what region is it considered beneficial?

**At the ground level, the troposphere, ozone is a toxic pollutant. In the next higher region of the atmosphere, the stratosphere, ozone is beneficial, absorbing harmful wavelengths of UV radiation as it forms molecular oxygen.**

Problem 11.5 What is the net chemical reaction associated with the Chapman cycle?

**There is no net change occurring in the Chapman cycle; it begins and ends with molecular oxygen.**

Problem 11.7 Which photochemical reaction in the Chapman cycle has the highest energy input via light?

**When the Chapman cycle begins with $O_2$ absorbing a UV photon to produce atomic oxygen, this requires the highest energy photon (one with the shortest wavelength).**

Problem 11.9 Is the ozone hole permanent?

**The level of ozone in the stratosphere is always changing; the ozone hole is not permanent. When ozone decomposes faster than it is formed, a "hole" in the ozone layer may form. When the rate of $O_3$ decomposition decreases, the hole may decrease or disappear. Seasons and weather patterns play a role in the hole shrinking and expanding. Recent evidence suggests efforts to protect the ozone layer are being successful and the hole that forms is shrinking – it is not as large as it used to be.**

Problem 11.11 For each of the following, suggest an appropriate rate unit.

**Rate of "something" means the quantity measured is changing in a given amount of time.**

(a) Heart beating **beats per minute**
(b) Videotape rewinding **feet per second**
(c) Automobile wheels rotating **revolutions (turns) per minute**
(d) Gas evolving in a very fast chemical reaction **mol/L per second**

Problem 11.13 Distinguish between instantaneous rate and average rate. In each of the following situations, is the rate measured the instantaneous rate or the average rate?

**Instantaneous rate is how fast the quantity measure is changing at one instant in time. Average rate looks at the quantity measured over a longer, finite amount of time.**

(a) In a hot dog eating contest, it took the winner only 4 minutes to eat 20 hot dogs, so he ate 5 hot dogs per minute.

**Average rate, the finite length of time is 4 minutes. (He probably ate them faster in the first minute than the last.)**

(b) At minute 1.0, the winner was eating 6 hot dogs per minute, but at minute 3.0, he was down to only 3 hot dogs per minute.

**Instantaneous rate, we know exactly how fast he was eating the hot dog at an exact point in time.**

Problem 11.15 In the description of the candle in Figure 11.4, we mentioned the consumption of oxygen. Assuming the candle wax is a mixture of hydrocarbons with the general formula $C_nH_{2n+2}$, what other variables could be measured besides the concentration of oxygen to determine the rate of the reaction?

**Since the wax is a hydrocarbon, we expect that carbon dioxide and water will be produced. We could measure the increase in concentration of $CO_2$ and water over time to determine the rate of the reaction.**

Problem 11.17 ■ Ammonia can react with oxygen to produce nitric oxide and water.

$$4\ NH_3(g) + 5\ O_2(g) \rightarrow 4\ NO(g) + 6\ H_2O(g)$$

If the rate at which ammonia is consumed in a laboratory experiment is $4.23 \times 10^{-4}$ mol $L^{-1}$ $s^{-1}$, at what rate is oxygen consumed? At what rate is NO produced? At what rate is water vapor produced?

**The rate of consumption of $NH_3$ is related to the rates of the other substances in the equation through stoichoimetry: the molar ratios. Recall that the rate of consumption of a reactant is <u>negative</u> and the rate of formation of a product is <u>positive</u>.**

$$\frac{-4.23\times10^{-4}\ \text{mol NH}_3}{\text{L}\cdot\text{s}}\times\frac{5\ \text{mol O}_2\ \text{consumed}}{4\ \text{mol NH}_3\ \text{consumed}} = -5.29\times10^{-4}\ \text{mol O}_2/\text{L}\cdot\text{s}$$

$$\frac{-4.23\times10^{-4}\ \text{mol NH}_3}{\text{L}\cdot\text{s}}\times\frac{4\ \text{mol NO formed}}{4\ \text{mol NH}_3\ \text{consumed}} = +4.23\times10^{-4}\ \text{mol NO}/\text{L}\cdot\text{s}$$

$$\frac{-4.23\times10^{-4}\ \text{mol NH}_3}{\text{L}\cdot\text{s}}\times\frac{6\ \text{mol H}_2\text{O formed}}{4\ \text{mol NH}_3\ \text{consumed}} = +6.35\times10^{-4}\ \text{mol H}_2\text{O}/\text{L}\cdot\text{s}$$

<u>Problem 11.19</u> A gas, AB, decomposes and the volume of $B_2$ produced is measured as a function of time, *t*. The data obtained are as follows:

| **Time (min)** | 0 | 8.3 | 15.4 | 19.0 |
|---|---|---|---|---|
| **Volume (L)** | 0 | 4.2 | 8.6 | 11.5 |

What is the average rate of production of $B_2$ for the first 8.3 min? For the first 19 min?

**For the first 8.3 min, rate** $= \dfrac{\Delta[\text{volume}]}{\Delta[\text{time}]} = \dfrac{[4.2\ \text{L} - 0\ \text{L}]}{[8.3\ \text{min} - 0\ \text{min}]} = 0.51\ \text{L/min}$

**For the first 19 min, rate** $= \dfrac{\Delta[\text{volume}]}{\Delta[\text{time}]} = \dfrac{[11.5\ \text{L} - 0\ \text{L}]}{[19.0\ \text{min} - 0\ \text{min}]} = 0.605\ \text{L/min}$

<u>Problem 11.21</u> Azomethane, $CH_3NNCH_3$, is not a stable compound, and once generated, it decomposes. The rate of decomposition was measured by monitoring the partial pressure of azomethane, in torr.

| **Time (min)** | 0 | 15 | 30 | 48 | 75 |
|---|---|---|---|---|---|
| **Pressure (torr)** | 36.2 | 30.0 | 24.9 | 19.3 | 13.1 |

Plot the data and determine the instantaneous rate of decomposition of azomethane at $t = 20$ min.

**From the <u>slope</u> of the plot of pressure vs. time, the rate of consumption of azomethane is – 0.34 torr/min.**

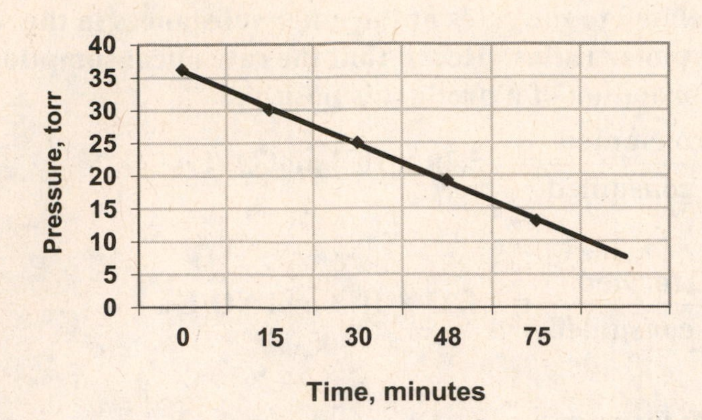


Problem 11.23 ■ A reaction has the experimental rate equation Rate = $k[A]^2$. How will the rate change if the concentration of A is tripled? If the concentration of A is halved?

**Tripling the concentration of A will cause a $(3)^2 = 9\times$ increase in the rate.**
**If the concentration of A is halved the rate will be reduced by a factor of $(\frac{1}{2})^2 = \frac{1}{4}\times$.**

Problem 11.25 ■ For each of the rate laws below, what is the order of the reaction with respect to the hypothetical substances X, Y, and Z? What is the overall order?

**The reaction order is equal to the exponent on the concentration term in the rate law. The overall order is equal to the sum of the individual reaction orders.**

(a) Rate = $k[X][Y][Z]$ **First-order (1) with respect to X, Y, and Z. Overall order = 1 + 1 + 1 = 3; third-order overall.**

(b) Rate = $k[X]^2[Y]^{1/2}[Z]$ **Second-order (2) with respect to X; one-half-order (½) with respect to Y; first-order (1) with respect to Z. Overall order = 2 + ½ + 1 = 3 ½.**

(c) Rate = $k[X]^{1.5}[Y]^{-1}$ **3/2-order with respect to X; negative one-order (−1) with respect to Y; zero-order (0) with respect to Z. Overall order = 3/2 + (−1) = 0 = ½**

(d) Rate = $k[X]/[Y]^2$ **First-order (1) with respect to X; negative 2-order (−2) with respect to Y; zero-order (0) with respect to Z. Overall order = 1 + (−2) + 0 = −1.**

Problem 11.27 Show that if the units of rate are mol $L^{-1}$ $s^{-1}$, then the units of the rate constant in the following second-order reaction are L $mol^{-1}$ $s^{-1}$:

$$H_2(g) + Br_2(g) \rightarrow 2\,HBr \qquad \text{rate} = k\,[H_2][Br_2]$$

**Substituting units in the rate law: rate (mol/L·s) = $k$ (?)[$H_2$ (mol/L)][$Br_2$ (mol/L)]**
**Solving for the units for $k$, $k$ = L/mol·s.**

Problem 11.29 The hypothetical reaction, A + B → C, has the rate law

$$\text{Rate} = k[A]^x[B]^y$$

When [A] is doubled and [B] is held constant, the rate doubles. But the rate increases fourfold when [B] is doubled and [A] is held constant. What are the values of $x$ and $y$?

**When concentration doubles and the rate doubles, with all other things equal, the order is first, $x = 1$.**
**Mathematically, using the rate law:** $2\,(\text{rate}) = [2A]^x$, $2 = 2^x$, $x = 1$

**When concentration doubles and the rate quadruples, with all other things equal, the order is second, $y = 2$.**
**Mathematically using the rate law:** $4\,(\text{rate}) = [2B]^y$, $4 = 2^y$, $y = 2$

Problem 11.31 Give the order with respect to each reactant and the overall order for the hypothetical reaction

$$A + B + C \rightarrow D + E$$

which obeys the rate law Rate = $k\,[A][B]^2$.

**See Problem 11.25. First order (1st) with respect to A, second order (2nd) with respect to B, zero order with respect to C, and third order (3rd) overall. Since no concentration term for C appears in the rate law, the order must be zero.**

Problem 11.33 The following experimental data were obtained for the reaction of $NH_4^+$ and $NO_2^-$ in acid solution.

$$NH_4^+(aq) + NO_2^-(aq) \rightarrow N_2(aq) + 2\,H_2O(\ell)$$

| $[NH_4^+]$ (mol $L^{-1}$) | $[NO_2^-]$ (mol $L^{-1}$) | Rate = $\Delta[N_2] / \Delta t$ (mol $L^{-1}$ $s^{-1}$) |
|---|---|---|
| 0.0092 | 0.098 | $3.33 \times 10^{-7}$ |
| 0.0092 | 0.049 | $1.66 \times 10^{-7}$ |
| 0.0488 | 0.196 | $3.15 \times 10^{-6}$ |
| 0.0249 | 0.196 | $1.80 \times 10^{-6}$ |

Determine the rate law for this reaction, and calculate the rate constant.

**Using the *method of initial rates*, we'll compare two experiments at a time, examining how the rate changes with a given change in concentration. Ideally, we'll try to compare two experiments where only one reactant concentration changes at a time; this simplifies matters.**

**The rate law is:** $\text{Rate} = k\,[NH_4^+]^x[NO_2^-]^y$

**Comparing experiment two to experiment one, the concentration of $NH_4^+$ stays constant and the concentration of $NO_2^-$ doubles (0.098 M/0.049 M = 2). The rate also approximately doubles: $(3.33 \times 10^{-7}\ M\ s^{-1})/(1.66 \times 10^{-7}\ M\ s^{-1}) = 2.0$.**
**This implies that $x = 1$ (see Problem 11.29).**

**Mathematically, using the rate law:** $2\ (Rate) = [2\ NO_2^-]^x$, $2 = 2^x$, $x = 1$

**Next, we compare experiment four to experiment three. The concentration of $NO_2^-$ stays constant and the concentration of $NH_4^+$ doubles (0.0488 M/0.0249 M = 1.96). The rate also approximately doubles: $(3.15 \times 10^{-6}\ M\ s^{-1})/(1.80 \times 10^{-6}\ M\ s^{-1}) = 2.0$.**
**This implies that $y = 1$.**

**Mathematically, using the rate law:** $2\ (Rate) = [2\ NH_4^+]^y$, $2 = 2^y$, $y = 1$

**The rate law now becomes:** $Rate = k\ [NH_4^+]^1[NO_2^-]^1$

**To finish, we can take data for <u>any</u> experiment and substitute into the rate law and solve for *k*.**
**Using experiment one:** $3.33 \times 10^{-6}\ M\ s^{-1} = k\ [0.092\ M][0.098\ M]$,

$$k = \frac{3.33\times10^{-6}\ M\cdot s^{-1}}{[0.0092\ M][0.098\ M]} = 3.7 \times 10^{-4}\ M^{-1}s^{-1}$$

**Putting the complete rate law together:** $Rate = 3.9 \times 10^{-4}\ M^{-1}s^{-1}\ [NH_4^+][NO_2^-]$

<u>Problem 11.35</u> ■ For the reaction

$$2\ NO(g) + 2\ H_2(g) \rightarrow N_2(g) + 2\ H_2O(g)$$

at 1100°C, the following data have been obtained:

| [NO] (mol $L^{-1}$) | [$H_2$] (mol $L^{-1}$) | Rate = $\Delta[N_2] / \Delta t$ (mol $L^{-1}$ $s^{-1}$) |
|---|---|---|
| $5.0 \times 10^{-3}$ | 0.32 | 0.012 |
| $1.0 \times 10^{-2}$ | 0.32 | 0.048 |
| $1.0 \times 10^{-2}$ | 0.64 | 0.096 |

Derive the rate law for the reaction, and determine the value of the rate constant.

**Using the *method of initial rates*, we'll compare two experiments at a time, examining how the rate changes with a given change in concentration. Ideally, we'll try to compare two experiments where only one reactant concentration changes at a time; this simplifies matters.**

**The rate law is:** $Rate = k\ [NO]^x[H_2]^y$

**Comparing experiment one to experiment two, the $[H_2]$ stays constant and the [NO] doubles (0.01 M/0.0050 M = 2). The rate quadruples: $(0.048\ M\ s^{-1})/(0.012\ M\ s^{-1}) = 2$. This implies that $x = 1$.**

**Mathematically, using the rate law:** $4\ (\text{Rate}) = [2\ \text{NO}]^x$, $4 = 2^x$, $x = 2$

**Next, we compare experiment two to experiment three. The [NO] stays constant and the $[H_2]$ doubles (0.64 M/0.32 M = 2.0). The rate doubles: $(0.096\ M\ s^{-1})/(0.048\ M\ s^{-1}) = 2$. This implies that $y = 1$.**

**Mathematically, using the rate law:** $2\ (\text{Rate}) = [2\ H_2]^y$, $2 = 2^y$, $y = 1$

**The rate law now becomes:** $\text{Rate} = k\ [\text{NO}]^2[H_2]^1$

**To finish, we can take data for <u>any</u> experiment and substitute into the rate law and solve for $k$.**
**Using experiment one:** $0.012\ M\ s^{-1} = k\ [0.0050\ M]^2[0.32\ M]$,

$$k = \frac{0.012\ M \cdot s^{-1}}{[0.0050\ M]^2[0.32\ M]} = 1500\ M^{-2}s^{-1}$$

**Putting the complete rate law together:** $\text{Rate} = 1500\ M^{-2}s^{-1}\ [\text{NO}]^2[H_2]$

<u>Problem 11.37</u> In a heterogeneous system such as wood burning in oxygen, the surface area of the solid can be a factor in the rate of the reaction. Increased surface area of the wood means increased collisions with oxygen molecules. To understand how surface area increases when dividing a solid, do the following:

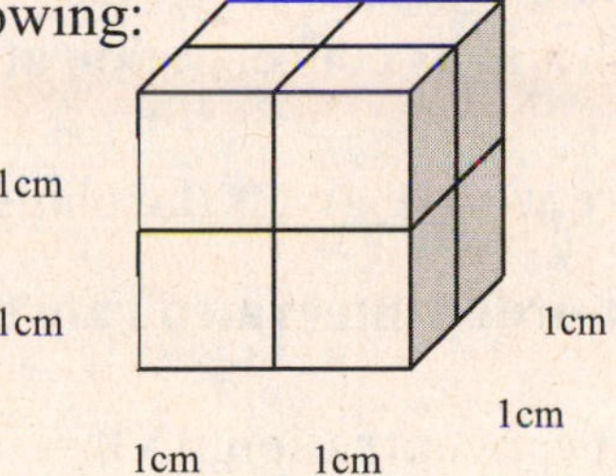


(a) Draw a cube that is 2 cm on a side. What is the total surface area?

**Surface area of a cube is six times the area of a face: 6 × 2 cm × 2 cm = 24 cm$^2$**

(b) Divide each face of the cube into four equal sections. If you were able to take the cube apart on the lines you have just drawn, how many cubes would you have?

**There would now be eight smaller cubes.**

(c) What are the dimensions of the new cubes?

**The length of a face would be half the original; the dimensions would be 1 cm on a side.**

(d) What is the total surface area of all the new cubes? What will happen if you divide the cubes again?

**The surface area of the smaller cube is $6 \times 1$ cm $\times 1$ cm $= 6$ cm$^2$. There are eight of them, so the total is $8 \times 6$ cm$^2$ $= 48$ cm$^2$. If we divide the cubes again, each face will be 0.5 cm and their surface area is 1.5 cm$^2$. There are now 64 ($8 \times 8$) cubes so the total surface area is $64 \times 1.5$ cm$^2$ $= 96$ cm$^2$. Every time we divide the cube, the surface area doubles.**

(e) Now consider burning a log or burning all the toothpicks that could be made from the log. What effect would surface area have on the rate that each burns? Would you rather try to start a campfire with a log or with toothpicks?

**The surface area of the toothpicks would be very much higher than that of the log. The increased surface area will cause the rate of burning for the toothpicks to be much faster than that of the log. Physically, more molecules in the wood can come in contact with oxygen molecules when the surface area is higher. This is why we start a fire with small pieces of kindling instead of large logs.**

Problem 11.39 The decomposition of $N_2O_5$ in solution in carbon tetrachloride is a first-order reaction:

$$2\,N_2O_5 \rightarrow 4\,NO_2 + O_2$$

The rate constant at a given temperature is found to be $5.25 \times 10^{-4}$ s$^{-1}$. If the initial concentration of $N_2O_5$ is 0.200 M, what is its concentration after 10 minutes have passed?

**This is an integrated rate law where we are given the elapsed time and must solve for the final amount of reactant. The first-order integrated rate equation is:** $\ln \frac{[X]_t}{[X]_0} = -kt$

**or $[X]_t = [X]_o e^{-kt}$, where $[X]_t$ = final concentration, $[X]_o$ = initial concentration,**

**t = elapsed time, and $k$ = the rate constant.**

**$[N_2O_5]_o = 0.200$ M, $k = 5.25 \times 10^{-4}$ s$^{-1}$, t = 10 min or 600 s (we have to reconcile the time units with $k$).**
**$[N_2O_5]_t$ is unknown, we will solve for this.**

$$[N_2O_5]_t = 0.200\text{ M}\left(e^{-(5.25\times 10{-4}\text{ s}{-1})(600\text{s})}\right) = 0.200\text{ M}\left(e^{-0.315}\right) = 0.146\text{ M}$$

**The concentration of $N_2O_5$ after 10 minutes of reaction is 0.146 M.**

Problem 11.41 For a drug to be effective in treating an illness, its levels in the bloodstream must be maintained for a period of time. One way to measure the level of a drug in the body is to measure its rate of appearance in the urine. The rate of excretion of penicillin is first order, with a half-life of about 30 min. If a person receives an injection of 25 mg of penicillin at t = 0, how much penicillin remains in the body after 3 hours?

**This is an integrated rate law where we are given the elapsed time and must solve for the final amount of reactant. We must first solve for the rate constant using the half-life, then use the first-order integrated rate law to find the final amount of penicillin after 3 hours.**

$$t_{½} = \frac{\ln 2}{k}, \quad k = \frac{\ln 2}{t_{1/2}} = \frac{0.693}{30\text{ min}} = 0.023\text{ min}^{-1}$$

**The first-order integrated rate equation is:** $\ln \frac{[X]_t}{[X]_0} = -kt$

**or $[X]_t = [X]_o e^{-kt}$, where $[X]_t$ = final concentration, $[X]_o$ = initial concentration,**

**t = elapsed time, and $k$ = the rate constant.**

**Because this formula has the initial and final amounts expressed in a ratio, any units can be used in this formula, not just mol/L.**

**Therefore, $[\text{penicillin}]_o$ = 25 mg, $k$ = 0.023 min$^{-1}$, t = 3 hours or 180 min (we have to reconcile the time units with $k$).**
**$[\text{penicillin}]_t$ is unknown, we will solve for this.**

$$[\text{penicillin}]_t = 25\text{ mg}\left(e^{-(0.023\text{ min}-1)(180\text{ min})}\right) = 25\text{ mg}\left(e^{-4.16}\right) = 0.39\text{ mg}$$

**The amount of penicillin remaining after 3 hours is 0.39 mg.**

Problem 11.43 As with any drug, aspirin (acetylsalicylic acid) must remain in the bloodstream long enough to be effective. Assume that the removal of aspirin from the bloodstream into the urine is a first-order reaction, with a half-life of about 3 hours. The instructions on an aspirin bottle say to take 1 or 2 tablets every 4 hours. If a person takes 2 aspirin tablets, how much aspirin remains in the bloodstream when it is time for the second dose? (A standard tablet of contains 325 mg of aspirin.)

**This is very similar to Problem 11.41. We know the time, initial dose, and half-life and we'll calculate the final amount of aspirin present after 4 hours.**

$$t_{½} = \frac{\ln 2}{k}, \quad k = \frac{\ln 2}{t_{1/2}} = \frac{0.693}{3\text{ hours}} = 0.231\text{ hour}^{-1}$$

**The first-order integrated rate equation is:** $\ln \frac{[X]_t}{[X]_0} = -kt$

**or $[X]_t = [X]_o e^{-kt}$, where $[X]_t$ = final concentration, $[X]_o$ = initial concentration,**

**t = elapsed time, and $k$ = the rate constant.**

**$[\text{aspirin}]_o = 2 \times 325$ mg, $k = 0.231\ \text{hour}^{-1}$, t = 4 hours,**
**$[\text{aspirin}]_t$ is unknown, we will solve for this.**
**$[\text{aspirin}]_t = 650\ \text{mg}\left(e^{-(0.231\ \text{hour}-1)(4\ \text{hours})}\right) = 650\ \text{mg}\left(e^{-0.924}\right) = 258\ \text{mg}$**

**The amount of aspirin remaining after 4 hours is 258 mg.**

Problem 11.45 ■ The initial concentration of the reactant in a first-order reaction A → products is 0.64 mol/L and the half-life is 30.0 s.
(a) Calculate the concentration of the reactant exactly 60 s after initiation of the reaction.
(b) How long would it take for the concentration of the reactant to drop to one-eighth its initial value?
(c) How long would it take for the concentration of the reactant to drop to 0.040 mol/L?

**Use the first-order half-life to get the value of $k$ (Equation 13.7). Then use the integrated rate law to find concentration and time.**

$$k = \frac{\ln 2}{t_{1/2}} = \frac{\ln 2}{30.0\ \text{s}} = 2.3 \times 10^{-2}\ \text{s}^{-1}$$

**(a)** $\ln[A]_t = -kt + \ln[A]_0 = -(2.3 \times 10^{-2}\ \text{s}^{-1})(60\ \text{s}) + \ln(0.64\ \text{mol/L})$

$$\ln[A]_t = -1.83$$

$$[A]_t = e^{-1.83} = 0.16\ \text{mol/L}$$

**The problem could also be solved by recognizing 60 s is two half-lives. Therefore $[A]_t = (\frac{1}{2})^2 \times 0.64\ \text{mol/L} = 0.16\ \text{mol/L}$.**

**(b)** $$[A]_t = \frac{1}{8} \times [A]_0 = \frac{1}{8} \times (0.64\ \text{mol/L}) = 0.080\ \text{mol/L}$$

$$kt = \ln[A]_0 - \ln[A]_t = \ln(0.64\ \text{mol/L}) - \ln(0.080\ \text{mol/L}) = 2.1$$

$$t = \frac{2.1}{k} = \frac{2.1}{2.3 \times 10^{-2}\text{s}^{-1}} = 90\ \text{s}$$

**Again we could solve the problem by recognizing that 1/8 the concentration is three half-lives; 3 × 30 s = 90 s**

**(c)** $[A]_t = 0.040\ \text{mol/L}$

$$kt = \ln[A]_0 - \ln[A]_t = \ln(0.64\ \text{mol/L}) - \ln(0.040\ \text{mol/L}) = 2.8$$

$$t = \frac{2.8}{k} = \frac{2.8}{2.3 \times 10^{-2}\text{s}^{-1}} = 120\ \text{s}$$

**0.040 mol/L is 1/16 the original concentration or four half-lives have occurred; 4 × 30 s = 120 s.**

Problem 11.47 Show that the half-life of a second-order reaction is given by

$$t_{½} = \frac{1}{k[X]_0}$$

In what fundamental way does the half-life of a second-order reaction differ from that of a first-order reaction?

**We start with the integrated rate law equation for second-order kinetics:**

$$\frac{1}{[X]_t} - \frac{1}{[X]_0} = kt$$

**What we want is the time (half-life) required for $[X]_t$ to equal $\frac{1}{2}[X]_0$**

**Substituting,** $\frac{1}{\frac{1}{2}[X]_0} - \frac{1}{[X]_0} = k\, t_{½}$ **, solving for the half-life,** $t_{½} = \frac{1}{k[X]_0}$.

**The half-life of a second-order reaction depends on the initial concentration, but first-order half-life does not. This means a second-order half-life changes over time but a first-order half-life does not.**

Problem 11.49 The rate of photodecomposition of the herbicide picloram in aqueous systems was determined by exposure to sunlight for a number of days. One such experiment produced the following results, (Data from Hedlund, R.T., Youngston, CR. "The Rates of Photodecomposition of Picloram in Aqueous Systems," *Fate of Organic Pesticides in the Aquatic Environment*, Advances in Chemistry Series, #111, American Chemical Society, (1972), 159–172.)

| Exposure Time, t (days) | [Picloram] (mol $L^{-1}$) |
|---|---|
| 0 | $4.14 \times 10^{-6}$ |
| 7 | $3.70 \times 10^{-6}$ |
| 14 | $3.31 \times 10^{-6}$ |
| 21 | $2.94 \times 10^{-6}$ |
| 28 | $2.61 \times 10^{-6}$ |
| 35 | $2.30 \times 10^{-6}$ |
| 42 | $2.05 \times 10^{-6}$ |
| 49 | $1.82 \times 10^{-6}$ |
| 56 | $1.65 \times 10^{-6}$ |

Determine the order of the reaction, the rate constant, and the half-life for the photodecomposition of picloram.

**Using the integrated rate equations as a guide, if the reaction is zero-order, [X] vs. time is linear; if the reaction is first-order, ln [X] vs. time is linear; and if the reaction is second-order, 1/[X] vs. time is linear. We need to examine plots of the data in the form of [picloram] vs. time, ln [picloram] vs. time, and 1/[picloram] vs time to determine reaction order.**

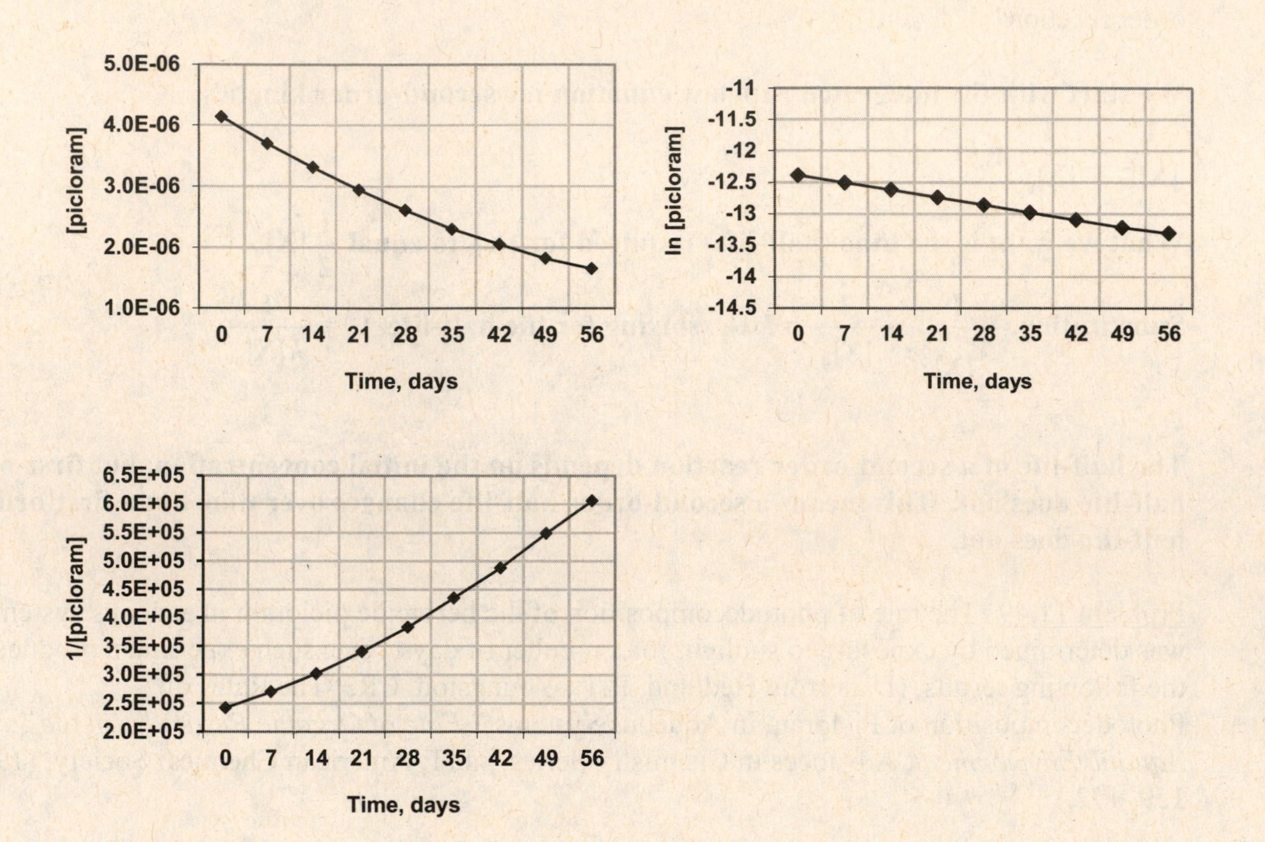


**The slope of the plot ln [picloram] vs. time is linear; therefore the reaction order is <u>one</u>. The slope of the plot is equal to (–*k*). We can approximate the slope using the first and last**

**data points (rise-over-run):** $$\frac{(-13.31-(-12.39))}{(56\,\text{s}-0\,\text{s})} = -0.0164\ \text{days}^{-1} = -k,\quad k = 0.0164\ \text{days}^{-1}$$

**Half-life is related to k:** $$t_{1/2} = \frac{\ln 2}{k} = \frac{0.693}{0.0164\ \text{days}^{-1}} = 42\ \text{days}$$

<u>Problem 11.51</u> ■ Peroxyacetyl nitrate (PAN) has the chemical formula $C_2H_3NO_4$ and is an important lung irritant in photochemical smog. An experiment to determine the decomposition

kinetics of PAN gave the data below. Determine the order of the reaction and calculate the rate constant for the decomposition of PAN.

| Time, t (min) | Partial Pressure of PAN (torr) |
|---|---|
| 0.0 | $2.00 \times 10^{-3}$ |
| 10.0 | $1.61 \times 10^{-3}$ |
| 20.0 | $1.30 \times 10^{-3}$ |
| 30.0 | $1.04 \times 10^{-3}$ |
| 40.0 | $8.41 \times 10^{-4}$ |
| 50.0 | $6.77 \times 10^{-4}$ |
| 60.0 | $5.45 \times 10^{-4}$ |

**We need to examine plots of the data in the form of P vs. time, ln P vs. time and 1/P vs time to determine reaction order.**

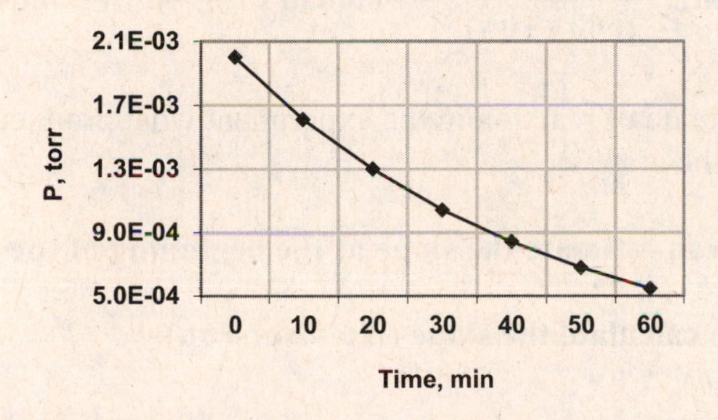


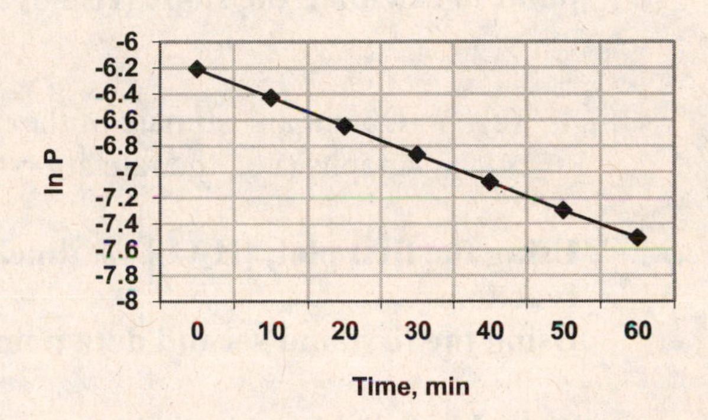


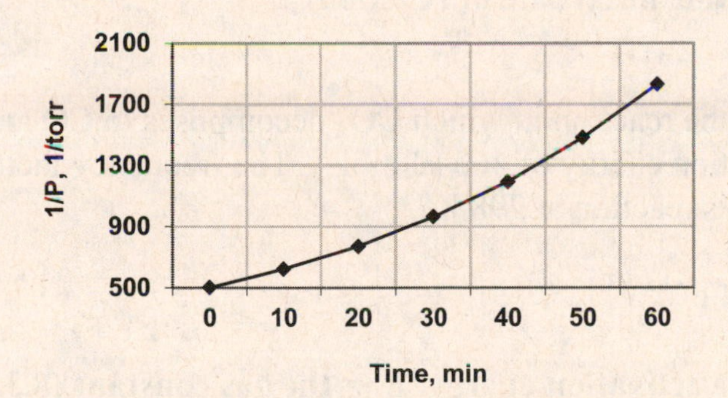


**Since the plot of ln P vs. time is linear, the reaction is first-order. The slope of the plot is equal to –*k*. Using the first and last data point to calculate the slope (rise-over-run):**

$$\textbf{Slope} = \frac{(-7.51-(-6.21))}{(60\,\text{s}-0\,\text{s})} = -0.0217\ \text{s}^{-1} = -k, \qquad k = 0.0217\ \text{s}^{-1}$$

Problem 11.53 Hydrogen peroxide ($H_2O_2$) decomposes into water and oxygen.

$$H_2O_2(aq) \rightarrow H_2O(\ell) + \tfrac{1}{2}\, O_2(g)$$

Ordinarily, this reaction proceeds rather slowly, but in the presence of some iodide ions ($I^-$), the decomposition is much faster. The decomposition in the presence of iodide was studied at 20°C, and the data were plotted in various ways. Use the graphs below, where concentrations are in moles/liter and time is in seconds, to answer the questions that follow.

**(a)** What is the order of reaction for the decomposition of hydrogen peroxide?

**See Problem 11.49. From the graphs, it is observed that the plot of ln [$H_2O_2$] vs. time is linear; therefore the reaction order is one.**

**(b)** Find the numerical value of the rate constant at 20°C, including the correct units.

**Using the plot of ln [$H_2O_2$] vs. time, the slope is equal to –*k*. Using the first and last data point to calculate the slope (rise-over-run): $\frac{(-2.4-(-0.1))}{(600\,s-0\,s)} = -0.0038\ s^{-1} = -k,\ k = 0.0038\ s^{-1}$.**

**(c)** Obtain an estimate of the initial rate of reaction in the experiment what produced the graphs (i.e., the rate at $t = 0$ in the graphs).

**Using the first plot, [$H_2O_2$] vs. time, we can estimate the slope at the beginning of the reaction.**
**Using the first and second data points to calculate the slope (rise-over-run):**

$$\frac{(0.70\,M - 0.89\,M)}{(75\,s - 0\,s)} = -2.5 \times 10^{-3}\ M{\cdot}s^{-1} = \text{the initial rate of reaction}$$

Problem 11.55 The activation energy for the reaction in which $CO_2$ decomposes to CO and a free radical oxygen atom, O, has an activation energy of 460 kJ mol$^{-1}$. The frequency factor is 2 × $10^{11}$ $s^{-1}$. What is the rate constant for this reaction at 298 K?

**We'll use the Arrhenius Equation, $k = A\,e^{-Ea/RT}$**

**where $A$ = the frequency factor, $E_a$ = the activation energy, $R$ = the gas constant (8.314 J/mol·K), and $T$ = temperature in Kelvin.**

$$k = A\,e^{-Ea/RT} = 2 \times 10^{11}\ s^{-1}\ e^{-460,000\ J\ mol{\cdot}1/(8.314\ J/mol{\cdot}K)(298\ K)} = 2 \times 10^{11}\ s^{-1}\ e^{-186} = 5 \times 10^{-70}\ s^{-1}$$

Problem 11.57 The following rate constants were obtained in an experiment in which the decomposition of gaseous $N_2O_5$ was studied as a function of temperature. The products were $NO_2$ and $NO_3$.

| $k$ ($s^{-1}$) | Temperature (K) |
|---|---|
| $3.5 \times 10^{-5}$ | 298 |
| $2.2 \times 10^{-4}$ | 308 |

| | |
|---|---|
| $6.8 \times 10^{-4}$ | 318 |
| $1.6 \times 10^{-3}$ | 328 |

Determine $E_a$ for this reaction in kJ/mol.

**We need the following variation of the Arrhenius equation:** $\ln\left(\frac{k_2}{k_1}\right) = \frac{E_a}{R}\left(\frac{1}{T_1} - \frac{1}{T_2}\right)$

**where $k_2$ = the rate constant at $T_2$, $k_1$ = the rate constant at $T_1$, $E_a$ = the activation energy, $R$ = the gas constant (8.314 J/mol·K), and $T$ = temperature in Kelvin.**

**Substituting the data from 298 K and 328 K,**

$$\ln\left(\frac{1.6\times10^{-3}}{3.5\times10^{-5}}\right) = \frac{E_a}{8.314\ \text{J/mol}\cdot\text{K}}\left(\frac{1}{298\ \text{K}} - \frac{1}{328\ \text{K}}\right)$$

**Solving for $E_a$:** $\ln(46) = \frac{E_a}{8.314\ \text{J/mol}\cdot\text{K}}\left(3.1\times10^{-4}\ \text{K}^{-1}\right)$

**$E_a$ = 103,000 J/mol = 103 kJ/mol**

Problem 11.59 The table below presents rate constants measured at three temperatures for the following reaction, which is involved in the production of nitrogen oxides in internal combustion engines. (Assume the temperatures all have two significant figures.)

$$O(g) + N_2(g) \rightarrow NO(g) + N(g)$$

| $k$ (L mol$^{-1}$ s$^{-1}$) | Temperature (K) |
|---|---|
| $4.4 \times 10^2$ | 2000 |
| $2.5 \times 10^5$ | 3000 |
| $5.9 \times 10^6$ | 4000 |

Determine the activation energy of the reaction in kJ/mol.

**This is very similar to Problem 11.57.**

**We need the following variation of the Arrhenius equation:** $\ln\left(\frac{k_2}{k_1}\right) = \frac{E_a}{R}\left(\frac{1}{T_1} - \frac{1}{T_2}\right)$

**where $k_2$ = the rate constant at $T_2$, $k_1$ = the rate constant at $T_1$, $E_a$ = the activation energy, $R$ = the gas constant (8.314 J/mol·K), and $T$ = temperature in Kelvin.**

**Substituting the data from 2000 K and 4000 K,**

$$\ln\left(\frac{5.9\times10^{6}}{4.4\times10^{2}}\right) = \frac{E_a}{8.314\ \text{J/mol}\cdot\text{K}}\left(\frac{1}{2000\ \text{K}} - \frac{1}{4000\ \text{K}}\right)$$

**Solving for $E_a$:** $\ln\left(1.34 \times 10^4\right) = \dfrac{E_a}{8.314 \text{ J/mol} \cdot \text{K}}\left(2.5 \times 10^{-4} \text{ K}^{-1}\right)$

**$E_a$ = 320,000 J/mol = 320 kJ/mol**

Problem 11.61 Bacteria can cause milk to go sour by generating lactic acid. Devise an experiment that could measure the activation energy for the production of lactic acid by bacteria in milk. Describe how your experiment will provide the information you need to determine this value. What assumptions must be made about this reaction?

**The concentration of lactic acid in the milk must be measured over time, perhaps by using a pH meter to find the $[H^+]$. This assumes that the only source of acidity in the milk is the lactic acid. The data is plotted as in Problem 11.49 to determine the reaction order and the rate constant. This procedure must be repeated at a different temperature, obtaining the rate constant at a different temperature. With two rate constants at two temperatures, the Arrhenius equation,**

$\ln\left(\dfrac{k_2}{k_1}\right) = \dfrac{E_a}{R}\left(\dfrac{1}{T_1} - \dfrac{1}{T_2}\right)$, **can be used to find the activation energy.**

Problem 11.63 Can a reaction mechanism ever be proven correct? Can it be proven incorrect?

**It can be said that a mechanism supports experimental evidence and therefore is possible, but it cannot be proven to be absolutely correct since we can't observe the reaction on a molecular scale. A mechanism can be proven incorrect if it contradicts what is observed experimentally.**

Problem 11.65 Describe how the Chapman cycle is a reaction mechanism. What is the molecularity of each reaction in the Chapman cycle?

**A reaction mechanism is a step-by-step series of small, simple reactions that add up to an overall reaction. The Chapman cycle gives the series of reactions that begin as molecular oxygen is split into atomic oxygen and end as ozone and atomic oxygen reform $O_2$. See figure 11.2.**

**Molecularity refers to the number of molecules involved in the mechanism step and indicates the order of that step. The steps of the Chapman cycle are:**

| | |
|---|---|
| $O_2 \xrightarrow{h\nu} 2\,O$ | **unimolecular** |
| $O_2 + O \longrightarrow O_3$ | **bimolecular** |
| $O_3 \xrightarrow{h\nu} O_2 + O$ | **unimolecular** |
| $O_3 + O \longrightarrow 2\,O_2$ | **bimolecular** |

Problem 11.67 Propane is used as a household fuel in areas where natural gas is not available. The combustion of propane is given by the equation

$$C_3H_8 + 5\ O_2 \rightarrow 3\ CO_2 + 4\ H_2O$$

Does the equation represent a probable mechanism for the reaction? Defend your answer.

**This reaction is not a probable mechanism for this reaction because it represents a very complicated reaction with many molecules. Mechanisms occur as a series of simple reactions because these are much more likely to take place. It is very unlikely that any mechanism step involving more than two molecules will occur.**

Problem 11.69 The following mechanism is proposed for a reaction:

$NO + Br_2 \rightarrow NOBr_2$ (slow)
$NOBr_2 + NO \rightarrow 2\ NOBr$ (fast)

(a) Write the overall equation for the reaction. **The overall reaction is the sum of the mechanism steps. $2\ NO + Br_2 \rightarrow 2\ NOBr$**

(b) What is the rate-determining step? **The slowest step in the mechanism is the rate-determining step. This is the first step in this mechanism: $NO + Br_2 \rightarrow NOBr_2$**

(c) What is the intermediate in this reaction? **An intermediate is a substance in the mechanism that is neither a product nor a reactant. It forms in one step but is consumed in a later step; therefore it does not appear in the overall reaction. $NOBr_2$ is an intermediate in this mechanism.**

(d) What is the molecularity of each step of the reaction?

**Molecularity refers to the number of molecules involved in the mechanism step and indicates the order of that step. Steps one and two are both bimolecular.**

(e) Write the rate expression for each step.
**Step one: $Rate_1 = k_1\ [NO][Br_2]$**
**Step one: $Rate_2 = k_2\ [NOBr_2][NO]$**

Problem 11.71 ■ The reaction of $NO_2(g)$ and $CO(g)$ is thought to occur in two steps:

**Step 1:** $NO_2(g) + NO_2(g) \rightarrow NO(g) + NO_3(g)$ (slow)
**Step 2:** $NO_3(g) + CO(g) \rightarrow NO_2(g) + CO_2(g)$ (fast)

(a) Show that the elementary steps add up to give the overall stoichiometric equation.
(b) What is the molecularity of each step?

(c) For this mechanism to be consistent with kinetic data, what must be the experimental rate equation?
(d) Identify any intermediates in this reaction.

**(a)** $$NO_2(g) + \cancel{NO_2(g)} \rightarrow NO(g) + \cancel{NO_3(g)}$$

$$\cancel{NO_3(g)} + CO(g) \rightarrow \cancel{NO_2(g)} + CO_2(g)$$

---

$$NO_2(g) + CO(g) \rightarrow NO(g) + CO_2(g)$$

**(b)** **Both steps are bimolecular. There are two reactant molecules.**

**(c)** **Rate = $k[NO_2]^2$; the rate-determining step (step 1) shows second-order dependence for $NO_2$**

**(d)** **$NO_3(g)$ is an intermediate. It is neither a reactant nor product in the overall reaction.**

<u>Problem 11.73</u> The word *catalyst* has been adopted outside of science. In sports, a reserve who enters the game might be described as the catalyst for a comeback victory. Discuss how this use of the word is similar to the scientific use of the word. How is it different from the scientific use of the word?

**A catalyst is a substance that is not absolutely necessary for a reaction to take place but can greatly improve the speed of the reaction when used. A reserve player who comes off the bench can similarly improve the play of the team and perhaps help secure the team a victory.**

<u>Problem 11.75</u> What distinguishes homogeneous and heterogeneous catalysis?

**Those terms refer to the phases of the reactants and the catalyst. Homogeneous catalysis involves a catalyst in the same phase as the reactants, whereas heterogeneous catalysis occurs when the catalyst is in a different phase. Gases reacting on solid catalyst surfaces are a very common form of heterogeneous catalysis.**

<u>Problem 11.77</u> Based on the kinetic theory of matter, what would the action of a catalyst do to a reaction that is the reverse of some reaction that we say is catalyzed?

**A catalyst lowers the activation energy of the reaction in <u>both</u> directions equally. The reverse reaction would speed up the same amount as the forward reaction.**

<u>Problem 11.79</u> In Chapter 3, we discussed the conversion of biomass into biofuels. One important area of research associated with biofuels is the identification and development of suitable catalysts to increase the rate at which fuels can be produced. Do a web search to find an article describing biofuel catalysts. Then write one or two sentences describing the reactions being catalyzed, and identify the catalyst as homogeneous or heterogeneous.

**Traditional catalysts for biofuels like biodiesel are base catalysts (NaOH and KOH), acid catalysts ($H_2SO_4$ and $H_3PO_4$), sodium methylate ($NaCH_3O$), and solid or enzyme catalysts. Metal oxide solid catalysts, which are heterogeneous, are newer and still being developed. The majority of biodiesel is currently produced with a homogeneous base catalyst.**

**http://www.smartcatalyst.net/res/Biofuels%20Nov-08%20Catalysis.pdf**

Problem 11.81 Why is smog appropriately labeled *photochemical* smog?

**Smog is a term referring to a combination of several pollutants. These molecules are either formed by absorption of photons from the sun or in some other reaction involving photons. Because photons are required for the chain of reactions producing smog to occur, it is called *photochemical smog*. Principle molecules in smog are NO, $NO_2$, and $O_3$.**

Problem 11.83 What is a VOC? What role do VOCs play in photochemical smog?

**VOC stands for volatile organic chemical. VOCs react with nitrogen oxides to produce other harmful compounds in smog, including ozone, $O_3$.**

Problem 11.85 Many urban centers in North America do not meet air quality standards for ozone levels. What factors associated with living in a city contribute most strongly to the formation of potentially dangerous levels of ozone?

**Formation of $NO_x$ is important and this occurs primarily by automobiles and industry, both usually found in and around a city. Presence of VOCs also contributes to ozone formations and again industry and automobiles (gasoline filling stations) are chiefly responsible for VOCs.**
**Weather patterns can make the situation even worse for a particular city.**

Problem 11.87 Why do ozone concentrations lag in time relative to other pollutants in photochemical smog?

**Ozone forms when VOCs and photochemically active nitrogen oxides in the atmosphere react. Nitrogen oxide is formed when oxygen and nitrogen react during the combustion of gasoline in an automobile. Nitrogen oxide absorbs photons and forms other photochemically active nitrogen oxides. Because it takes several hours for the NO from the exhaust to form the other oxides, there is a time lag between the peak concentration of NO and the peak concentration of $O_3$.**

Problem 11.89 Find a website that tells you whether you live in an area that has unhealthy levels of ozone.

**(a)** What is the URL of the site? Does this site strike you as a reliable source of information? **Answers will vary according to geographic location. A good site is http://airnow.gov/**

**(b)** Does your area have unhealthy air? **Answers will vary according to geographic location.**

Problem 11.91 Western Michigan can experience higher than normal ozone levels in the summer because prevailing southwest winds carry industrial pollution in the form of VOCs from Milwaukee and Chicago. (For maps, visit the EPA Region 5 website at http://www.epa.gov/region5)
**(a)** Rewrite equations 11.10.d-f, using hexane, $C_6H_{14}$, as the VOC pollutant. **(b)** Explain how the $NO_2$ produced in the reactions you wrote in (a) can result in tropospheric ozone production in Western Michigan.

**(a) Refering back to Section 11.8, the reactions are:**

$C_6H_{14} + HO\bullet \rightarrow C_6H_{13}\bullet + H_2O$ **Equation 11.10d**

$C_6H_{13}\bullet + O_2 + M \rightarrow C_6H_{13}O_2\bullet + M$ **Equation 11.10e**

$C_6H_{13}O_2\bullet + NO \rightarrow C_6H_{13}O\bullet + NO_2$ **Equation 11.10f**

**(b) Once $NO_2$ is produced, then a photochemical reaction results in the formation of ozone, $O_3$.**

$NO_2 + h\nu \rightarrow NO + O$ **Equation 11.10a**

$O + O_2 + M \rightarrow O_3 + M$ **Equation 11.10b**

Problem 11.93 According to an article published in 1966 in the *Journal of Chemical Education*, xenon reacts photochemically with fluorine at room temperature (298 K) to produce $XeF_2$. A portion of a table showing ratios of gases in the reaction vessel for four experiments is reproduced below. (Assume that atmospheric pressure is 1 atm.) The reaction is as shown.

$$Xe(g) + F_2(g) \rightarrow XeF_2(s)$$

**(a)** Draw the Lewis structure of $XeF_2$, and predict the molecule's shape.

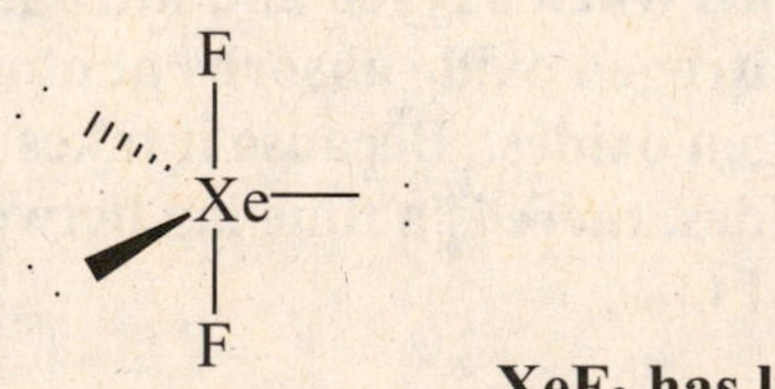


**$XeF_2$ has linear shape.**

**(b)** Calculate the partial pressure of each gas in Experiment 3.

**The relative volumes of each gas can be used to calculate the partial pressures:**

$$\frac{400\text{ mL Xe}}{662\text{ mL Total}} \times 1\text{ atm} \times \frac{760\text{ mm Hg}}{1\text{ atm}} = 459\text{ mm Hg Xe}$$

$$\frac{262\text{ mL F}_2}{662\text{ mL Total}} \times 1\text{ atm} \times \frac{760\text{ mm Hg}}{1\text{ atm}} = 301\text{ mm Hg F}_2$$

(c) Calculate the theoretical yield of $XeF_2$ in Experiment 3 if the amounts shown are combined such that the total pressure remains 1 atm.

**This is a limiting reactant calculation, See Problem 4.29. First, we'll need to use the ideal gas law to calculate the moles:**

$$n_{F2} = \frac{PV}{RT} = \frac{(301\text{ mm Hg})(\frac{1\text{ atm}}{760\text{ mm Hg}})(0.662\text{ L})}{(0.08206\text{ L}\cdot\text{atm/mol}\cdot\text{K})(298\text{ K})} = 0.0107\text{ mol F}_2$$

$$n_{Xe} = \frac{PV}{RT} = \frac{(459\text{ mm Hg})(\frac{1\text{ atm}}{760\text{ mm Hg}})(0.662\text{ L})}{(0.08206\text{ L}\cdot\text{atm/mol}\cdot\text{K})(298\text{ K})} = 0.0164\text{ mol Xe}$$

**Xe and $F_2$ react in a 1:1 ratio, so $F_2$ is the limiting reactant. The theoretical yield of $XeF_2$ can be found from this.**

$$0.0107\text{ mol F}_2 \times \frac{1\text{ mol XeF}_2}{1\text{ mol F}_2} \times \frac{169.29\text{ g XeF}_2}{1\text{ mol XeF}_2} = 1.81\text{ g XeF}_2$$

(d) After the reactants were introduced, the bulb was placed in sunlight. $XeF_2$ began crystallizing on the inner surface of the bulb about 24 hours later. Compare the rate of this reaction to the rate of precipitation of $CaF_2$ from solutions of $Ca^{2+}$(aq) and $F^-$ (aq) in the laboratory.

**The rate of precipitation of $CaF_2$ is very rapid, virtually instantaneous, when concentrated solutions of $Ca^{2+}$ and $F^-$ are mixed. You would need to measure its rate in milliseconds or smaller time units. Since $XeF_2$ crystallized over a period of a day, its rate could be measured in hours, a much slower reaction rate.**

Problem 11.95 On a particular day, the ozone level in Milwaukee exceeded the EPA's 1-hour standard of 0.12 ppm by 10 ppb. How many ozone molecules would be present in 1 liter of air at the detection site?

**We'll assume the pressure is 1 atm and temperature is 298 K. We can calculate the number of molecules in 1 liter of air and using the ozone concentration, find the number of ozone molecules.**

**Moles in 1 liter of air:** $n = \dfrac{PV}{RT} = \dfrac{(1\text{ atm})(1\text{ L})}{(0.08206\text{ L}\cdot\text{atm/mol}\cdot\text{K})(298\text{ K})} = 0.0409\text{ mol air}$

**Molecules in 1 liter of air:** $0.0409\text{ mol air} \times \dfrac{6.022\times10^{23}\text{ molecules}}{1\text{ mol}} = 2.46 \times 10^{22}\text{ molecules}$

**When the EPA standard is exceeded by 10 ppb, the ozone concentration is 130 ppb (0.13 ppm).**

$$2.46 \times 10^{22}\text{ molecules} \times \frac{130\text{ molecules}O_3}{1\times10^{9}\text{ molecules air}} = 3.2 \times 10^{15}\text{ molecules of } O_3 \text{ in one liter of air}$$

Problem 11.97 The following is a thought experiment. Imagine that you put a little water in a test tube and add some NaF crystals. Immediately after you add NaF, you observe that the crystals begin dissolving. The quantity of solid NaF decreases, but before long, it appears that no more NaF is dissolving. The solution is saturated.

(a) The equation for the dissolution of NaF in water is $NaF(s) \rightarrow Na^{+}(aq) + F^{-}(aq)$. As NaF dissolves, what do you think happens to the rate of dissolution? Describe what is happening on the molecular level.

**The rate of dissolution of NaF slows down as the concentration of $Na^{+}$ and $F^{-}$ increase. On a molecular level, the higher the concentration of the ions, the more difficult it is for the solid to dissolve.**

(b) Assume that the reverse reaction, $Na^{+}(aq) + F^{-}(aq) \rightarrow NaF(s)$, also occurs as the crystal dissolves. In other words, both dissolution and precipitation are taking place. When it appears that here is no more change in the quantity of NaF dissolving (the solution is saturated), what has happened to the rates of the forward and reverse reactions? Explain your answer.

**When it appears that no more solid NaF dissolves, what has happened is equilibrium has been established. The rate of dissolution (forward reaction) has become equal to the rate of precipitation (reverse reaction).**

Problem 11.99 The Haber process, which produces ammonia from nitrogen and hydrogen, is one of the world's most important industrial chemical reactions.

$$N_2(g) + 3\,H_2(g) \rightarrow 2\,NH_3(g)$$

For this reaction, both $\Delta H°$ and $\Delta S°$ are negative, meaning that product formation is thermodynamically favored by low temperatures. Yet the industrial reaction is commonly carried out at temperatures between 400 and 600°C. These high temperatures require a significant input of energy; some estimates indicate that the industrial production of ammonia is responsible for 1% of the world's total energy consumption. Suggest a reason why these high temperatures are employed.

**The reaction rate slows down significantly at low temperature. The high temperatures in the Haber process may be necessary to get the reaction rate up to an acceptable level.**

Problem 11.101 ■ Substances that poison a catalyst pose a major concern for many engineering designs, including catalytic converters. One design concept is to add materials that react with potential poisons before they reach the catalyst. Among the commonly encountered catalyst poisons are silicon and phosphorus, which typically form phosphate or silicate ions in the oxidizing environment of an engine. Group 2 elements are added to the catalyst to react with these contaminants before they reach the working portion of the catalytic converter. If estimates show that a catalytic converter will be exposed to 625 g of silicon during its lifetime, what mass of beryllium would need to be included in the design?

**We'll assume that the silicon reacts with beryllium to form $BeSi_2$. Therefore one mole of Be is needed for every two moles of Si.**

$$625 \text{ g Si} \times \frac{1 \text{ mol Si}}{28.09 \text{ g Si}} \times \frac{1 \text{ mol Be}}{2 \text{ mol Si}} \times \frac{9.01 \text{ g Be}}{1 \text{ mol Be}} = 100. \text{ g Be required}$$

Problem 11.103 ■ The flashing of fireflies is the result of a chemical reaction and the rate of flashing can be described by the Arrhenius equation. A certain batch of fireflies was observed to flash at a rate of 16.0 times per minute at 25°C and at a rate of 5.5 times per minute at 15°C. Use these data to find the apparent activation energy for the reaction that causes the flies to flash.

**Let the ratio of flashing at the two temperatures represent the ratio of rate constants:**

$$\frac{k_{25}}{k_{15}} = \frac{16.0 \text{ times/min}}{5.5 \text{ times/min}}$$

**Using the Arrhenius equation,** $$\ln\left(\frac{16.0 \text{ times/min}}{5.5 \text{ times/min}}\right) = \frac{E_a}{8.314 \text{ J/mol} \cdot \text{K}}\left(\frac{1}{288 \text{ K}} - \frac{1}{298 \text{ K}}\right)$$

$$1.07 = \frac{E_a}{8.314 \text{ J/mol} \cdot \text{K}}(0.000117 \text{ K}^{-1})$$

$E_a$ **= 76,000 J/mol or 76 kJ/mol**

Problem 11.105 You measure an archeological specimen and find that the count rate from carbon-14 is $x$ % of that seen in living organisms. How does this information allow you to establish the age of the artifact? What would you have to know or look up?

**If the level of carbon–14 in the artifact is a finite percentage of that seen in living organisms (normal level), then we can use our integrated rate equation (all nuclear decay obeys first-order kinetics) to solve for how long it took that reaction to occur (the age of the artifact). We would need to look up the kinetic order of carbon–14 decay (all nuclear decay are first-order), the half-life of $^{14}C$, and the count rate of $^{14}C$ seen in living organisms.**

Problem 11.107 You are designing a self-heating food package. The chemicals you are using seem to be used up too quickly. What experiments could be run to devise conditions that would allow the heating pack to last longer?

**The rate of reaction needs to be slowed down.**

**An inhibitor is a substance that is opposite of a catalyst. It will slow down the rate of a reaction by increasing the activation energy, but is not actually consumed in the reaction. Several experiments could be performed to find a suitable inhibitor for this reaction. Alternatively, lower reactant concentrations can slow down the rate of reaction, or larger particles sizes if the reaction is heterogeneous.**

Problem 11.109 Fluorine often reacts explosively. What does this fact suggest about fluorine reactions at the molecular level?

**In terms of bond energy, the bonds that F forms with other atoms must be stronger (higher bond energy) than F–F bonds. This means much energy will be released when fluorine reacts.**
**Molecularly speaking, fluorine reactions have very simple mechanisms with very low activation energy.**

Problem 11.111 A free radical is often used as a catalyst for polymerization reactions. Such reactions normally need to be carried out at higher temperatures. Explain this fact from the molecular perspective in terms of the activation energy of the reactants.

**Free radicals are very reactive and don't usually have a stable existence. The high temperature required indicates that a large amount of activation energy is necessary to produce the free radicals. Once the free radicals form, they quickly react with a monomer molecule, creating another radical which continues to react with monomers in a propagating chain.**

# CHAPTER OBJECTIVE QUIZ

This quiz will test your understanding of the basic text chapter objectives and give you additional practice problems. You should work this quiz after completing the end-of-chapter questions. The solutions to these questions are found at the end of this chapter.

1. Which of the following statements about Chapman cycle is false?
   a. Stratospheric $O_3$ is formed when $O_2$ reacts with ultraviolet (UV) radiation.
   b. Hydrocarbons produced by human activity increase the level of $O_3$ in the stratosphere.
   c. Ozone decomposes when it absorbs UV radiation.
   d. Stratospheric $O_3$ has decreased significantly since the 1970's.
   e. If $O_3$ decomposes faster than it forms, the level in the stratosphere will fall.

2. Write three equivalent expressions for the rate of reaction for the following reaction.

$$3\ H_2(g) + N_2(g) \rightarrow 2\ NH_3(g)$$

3. If the rate of decomposition of $N_2O_5$ is –0.53 mol/L·s, what is the rate of formation of *both* products?

$$2\ N_2O_5(g) \rightarrow 2\ NO_2(g) + O_2(g)$$

4. Write a general rate law expression for the following reactions.

(a) $C_4H_8(g) \rightarrow 2\ C_2H_4(g)$

(b) $Cl(g) + H_2(g) \rightarrow HCl(g) + H(g)$

(c) $2\ M(g) + N(g) + 3\ L(g) \rightarrow 2\ X(g) + Z(g)$

5. The following rate law is experimentally determined: Rate = $k$ [A][B]$^2$.

(a) If the [B] is doubled while the [A] remains the same, what is the change in rate?

(b) If the [B] and [A] are both doubled, what is the change in rate?

(c) If the [A] is doubled and the [B] is halved, what is the change in rate?

6. Use the rate law to estimate the initial rate of reaction if a reaction starts with $[N_2]$ = 0.015 M and $[H_2]$ = 0.55 M. Rate = $3.4 \times 10^{-3} M^{-2}s^{-1}\ [N_2][H_2]^2$

7. Use the method of initial rates and the following experimental data to determine the complete rate law for the reaction of $NO_2$ and $F_2$: $2\ NO_2(g) + F_2(g) \rightarrow 2\ NO_2F(g)$

| $[NO_2]$ | $[F_2]$ | Rate = $-\Delta[NO_2]/2\,\Delta t$ |
|---|---|---|

| (mol $L^{-1}$) | (mol $L^{-1}$) | (mol $L^{-1}$ $s^{-1}$) |
|---|---|---|
| $5.0 \times 10^{-3}$ | $5.0 \times 10^{-3}$ | $2.3 \times 10^{-4}$ |
| $5.0 \times 10^{-3}$ | $1.5 \times 10^{-2}$ | $6.9 \times 10^{-4}$ |
| $1.0 \times 10^{-2}$ | $5.0 \times 10^{-3}$ | $4.6 \times 10^{-4}$ |

8. Use the following experimental data to determine the reaction order for the following reaction: $AB_2(g) \rightarrow AB(g) + B(g)$

| Time (s) | Partial Pressure $AB_2$ (atm) |
|---|---|
| 0 | 0.22 |
| 10 | 0.21 |
| 20 | 0.20 |
| 30 | 0.19 |
| 40 | 0.18 |
| 50 | 0.17 |
| 60 | 0.16 |
| 70 | 0.15 |

9. Consider the following first-order reaction: $HOF(g) \rightarrow HF(g) + \frac{1}{2} O_2(g)$ with a half-life of 15.4 seconds.

(a) If a reaction is initiated with 1.55 mol/L HOF, how much remains after 3.25 minutes?

(b) How long would it take for the concentration to be reduced from 0.96 M to $2.5 \times 10^{-3}$ M HOF?

10. The rate constant for the decomposition of phosphine ($PH_3$) is 2.32 L/mol·min at 137 °C.

(a) What is the half-life if the initial concentration is 1.00 M?

(b) How much phosphine is left after 35.8 minutes?

(c) How long would it take for 1.00 M $PH_3$ to be reduced to 0.025 M?

11. If the half-life of a second-order reaction is 34.2 hours, what was the initial concentration of reactant? The rate law is: rate = $0.077\ M^{-1}s^{-1}\ [X]^2$.

12. TRUE or FALSE: The orders for the rate law in an elementary mechanism step are equal to the coefficients in the equation.

13. Use the following mechanism to determine the overall rate law:

| | |
|---|---|
| $2A \rightarrow C$ | (slow) |
| $B + C \rightarrow D$ | (fast) |
| $D + B \rightarrow A + E$ | (fast) |

$A + 2B \rightarrow E$ (overall reaction)

14. Which of the following statements about activation energy ($E_a$) is false?
    a. Activation energy is always positive.
    b. The greater the magnitude of $E_a$, the slower the rate of reaction.
    c. Higher temperature means molecules collide with greater kinetic energy.
    d. The concentration of the activated complex can usually be measured.
    e. The activated complex is a very unstable intermediate between reactants and products.

15. The activation energy of the following reaction is 256 kJ/mol. The rate constant at 55°C is 0.931 $s^{-1}$. What is the value of the rate constant at 255°C?

$$HOF(g) \rightarrow HF(g) + \frac{1}{2} O_2(g)$$

16. Given the following experimental data, what is the activation energy of the reaction?

| Temperature (°C) | $k$ ($s^{-1}$) |
|---|---|
| 40 | 0.035 |
| 50 | 0.130 |
| 60 | 0.278 |

17. Which of the following statements about catalysts is false?
    a. A catalyst will speed up the rate of a reaction.
    b. Catalysts are used in very many commercially important chemical reactions.
    c. Catalytic converters are examples are examples of heterogeneous catalysts.
    d. A catalyst can cause a nonspontaneous reaction take place.
    e. Chlorine radicals catalyzing the destruction of ozone is an example of homogeneous catalysis.

## ANSWERS TO THE CHAPTER OBJECTIVE QUIZ

1. b

2. $\text{Rate} = \dfrac{-1\ \Delta[H_2]}{3\ \Delta[\text{time}]} = \dfrac{-\ \Delta[N_2]}{\Delta[\text{time}]} = \dfrac{1\ \Delta[NH_3]}{2\ \Delta[\text{time}]}$

3. The rates of formation of both products are related to the rate of consumption by the mole ratio. Rates of formation of products are positive values whereas rates of formation of reactants are negative.

$$\frac{-0.53\ \text{mol}\ N_2O_5}{L \cdot s} \times \frac{2\ \text{mol}\ NO_2}{2\ \text{mol}\ N_2O_5} = \frac{0.53\ \text{mol}\ NO_2}{L \cdot s} = \text{rate of formation of } NO_2$$

$$\frac{-0.53\ \text{mol}\ N_2O_5}{L \cdot s} \times \frac{1\ \text{mol}\ O_2}{2\ \text{mol}\ N_2O_5} = \frac{0.27\ \text{mol}\ O_2}{L \cdot s} = \text{rate of formation of } O_2$$

4. For a generic reaction, aA + bB + cC → products, the general rate law expression is rate = $k[A]^x[B]^y[C]^z$.

(a) Rate = $k[C_4H_8]^x$

(b) Rate = $k[Cl]^x[H_2]^y$

(c) Rate = $k[M]^x[N]^y[L]^z$

5. (a) If the [B] is doubled, Rate = $k[A][2 \times B]^2 = 2^2 \times$ (initial rate) = a four-fold increase in rate.

(b) If both [A] and [B] are doubled, Rate = $k[2 \times A][2 \times B]^2$ = $= 2^1 \times 2^2 \times$ (initial rate) = an eight-fold increase in rate.

(c) If the [A] is doubled and the [B] is halved, Rate = $k[2 \times A][\ ½ \times B]^2$ = $= 2^1 \times (½)^2 \times$ (initial rate) = a ½ -fold decrease in rate.

6. Substituting into the rate law, Rate = $3.4 \times 10^{-3} M^{-2} s^{-1}\ [0.015\ M][\ 0.55\ M]^2$
Rate = $1.5 \times 10^{-5}\ M\ s^{-1}$

7. Using the *method of initial rates*, we'll compare two experiments at a time, examining how the rate changes with a given change in concentration. Ideally, we'll try to compare two experiments where only one reactant concentration changes at a time; this simplifies matters.

The general rate law is: Rate = $k[NO_2]^x[F_2]^y$

Comparing experiment one to experiment two, the $[NO_2]$ stays constant and the $[F_2]$ triples (0.015 M/0.0050 M = 3).
The rate also triples: $(6.9 \times 10^{-4}\ M\ s^{-1})/(2.3 \times 10^{-4}\ M\ s^{-1}) = 3$. This implies that $y = 1$.

Mathematically, using the rate law: $3\ (Rate) = [3 \times F_2]^y,\ \ 3 = 3^y,\qquad y = 1$

Next, we compare experiment one to experiment three. The $[F_2]$ stays constant and the $[NO_2]$ doubles (0.010 M/0.005 M = 2).

The rate also doubles: $(4.6 \times 10^{-4}\ Ms^{-1})/(2.3 \times 10^{-4}\ Ms^{-1}) = 4$,
This implies that $x = 1$.

Mathematically, using the rate law: $2\ (Rate) = [2 \times NO_2]^x,\ 2 = 2^x,\qquad x = 1$

The rate law now becomes: $Rate = k[NO_2]^1[F_2]^1$

To finish, we can take data for <u>any</u> experiment and substitute into the rate law and solve for $k$.
Using experiment one: $2.3 \times 10^{-4}\ Ms^{-1} = k[0.0050\ M]^1[0.0050\ M]^1$,

$$k = \frac{2.3 \times 10^{-4}\ M \cdot s^{-1}}{[0.0050\ M][0.0050\ M]} = 9.2\ M^{-1}s^{-1}$$

Putting the complete rate law together: $Rate = 9.2\ M^{-1}s^{-1}\ [NO_2][F_2]$

8. Using the integrated rate equations as a guide, if the reaction is zero-order, [X] vs. time is linear; if the reaction is first-order, ln [X] vs. time is linear; and if the reaction is second-order, 1/[X] vs. time is linear.

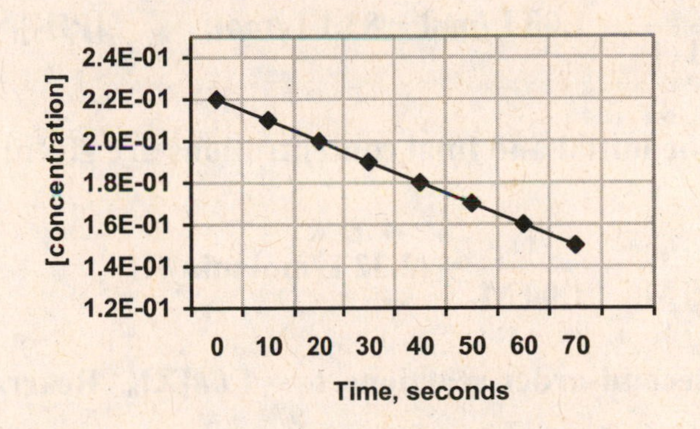


Since the plot of $[AB_2]$ vs. time is linear, the reaction order must be <u>zero</u>.

9. The first-order integrated rate equation is: $\ln \frac{[X]_t}{[X]_0} = -kt$

or $[X]_t = [X]_o e^{-kt}$, where $[X]_t$ = final concentration, $[X]_o$ = initial concentration, t = elapsed time, and $k$ = the rate constant.

For (a), the time is given and the final concentration must be calculated. The rate constant must be found from the half-life.

$$t_{½} = \frac{\ln 2}{k}, \quad k = \frac{\ln 2}{t_{1/2}} = \frac{0.693}{15.4\ \text{s}} = 0.0450\ \text{s}^{-1}; [\text{HOF}]_0 = 1.55\ \text{M}, t = 3.25\ \text{min or } 195\ \text{s}.$$

$$[\text{HOF}]_t = 1.55\ \text{M}\left(e^{-(0.045\ \text{s}-1)(195\ \text{s})}\right) = 1.55\ \text{M}\left(e^{-8.78}\right) = 2.40 \times 10^{-4}\ \text{M HOF}$$

In (b), the initial and final concentration is given and we are solving for time.

$$\ln\left(\frac{2.5\times10^{-3}\ \text{M}}{0.96\ \text{M}}\right) = -(0.0450\ \text{s}^{-1})(t) \qquad t = \frac{-5.95}{-0.045\ \text{s}^{-1}} = 132\ \text{s} = 2.20\ \text{min}$$

10. (a) From the units of the rate constant, we can tell that this is a second-order reaction. The formula for half-life is: $t_{½} = 1 / k[X]_o$

$$t_{½} = \frac{1}{(2.32\ \text{L/mol}\cdot\text{min})(1.00\ \text{M})} = 0.431\ \text{min}$$

(b) The time is given and the final concentration can be found by using the integrated rate equation for second order kinetics: $\frac{1}{[X]_t} - \frac{1}{[X]_0} = kt$

$$\frac{1}{[\text{PH}_3]_t} - \frac{1}{1.00\ \text{M}} = (2.32\ \text{L/mol}\cdot\text{min})(35.8\ \text{min}) = 83.1\ \text{L/mol}$$

$$\frac{1}{[\text{PH}_3]_t} = 1.00\ \text{L/mol} + 83.1\ \text{L/mol}; \qquad [\text{PH}_3]_t = 0.0119\ \text{M}$$

(c) The initial and final concentrations are given; we'll solve for time.

$$\frac{1}{0.025\ \text{M}} - \frac{1}{1.00\ \text{M}} = (2.32\ \text{L/mol}\cdot\text{min})(t) \qquad t = \frac{39\ \text{M}^{-1}}{2.32\ \text{M}^{-1}\text{min}^{-1}} = 16.8\ \text{min}$$

11. For second-order reactions, $t_{½} = 1 / k[X]_o$. Rearranging, $[X]_o = 1/ k\, t_{½}$

$$[X]_o = \frac{1}{(0.077\ \text{M}^{-1}\text{s}^{-1})\left(34.2\ \text{h} \times \frac{3600\ \text{s}}{\text{h}}\right)} = 1.1 \times 10^{-4}\ \text{M}$$

12. TRUE

**13. The slowest step in the mechanism is the rate-determining step. The orders in the rate law for the slow step will be the same as the orders in the overall rate law.**

**The overall rate law is Rate = $k[A]^x[B]^y$**

**The rate law for the slow step is Rate = $k_1\ [A]^2$.**

**Recall that the orders for the rate law in an elementary mechanism step are equal to the coefficients. Comparing, the order with respect to A is second, and the order with respect to B is zero.**

**The overall rate law becomes Rate = $k\ [A]^2[B]^0$**

**14. d**

**15. We need the following variation of the Arrhenius equation to find the rate constant at 255 °C:**

$$\ln\left(\frac{k_2}{k_1}\right) = \frac{E_a}{R}\left(\frac{1}{T_1} - \frac{1}{T_2}\right)$$

**where $k_2$ = the rate constant at $T_2$, $k_1$ = the rate constant at $T_1$, $E_a$ = the activation energy, $R$ = the gas constant (8.314 J/mol·K), and $T$ = temperature in Kelvin.**

**The activation energy and $k_1$ (at 55°C) are given. Substituting the data from 328 K and 528 K,**

$$\ln\left(\frac{k_2}{0.931\ \text{s}^{-1}}\right) = \frac{256{,}000\ \text{J/mol}}{8.314\ \text{J/mol}\cdot\text{K}}\left(\frac{1}{328\ \text{K}} - \frac{1}{528\ \text{K}}\right) = 35.6$$

$$k_2 = 0.931\text{s}^{-1}\ (e^{35.6}) = 2.69 \times 10^{15}\ \text{s}^{-1}$$

**16. We'll use the experimental data in the Arrhenius equation to solve for the activation energy:**

$$\ln\left(\frac{k_2}{k_1}\right) = \frac{E_a}{R}\left(\frac{1}{T_1} - \frac{1}{T_2}\right)$$

**where $k_2$ = the rate constant at $T_2$, $k_1$ = the rate constant at $T_1$, $E_a$ = the activation energy, $R$ = the gas constant (8.314 J/mol·K), and $T$ = temperature in Kelvin. Substituting the data from 313 K (40°C) and 333 K (60°C),**

$$\ln\left(\frac{0.278\ \text{s}^{-1}}{0.035\ \text{s}^{-1}}\right) = \frac{E_a}{8.314\ \text{J/mol}\cdot\text{K}}\left(\frac{1}{313\ \text{K}} - \frac{1}{333\ \text{K}}\right)$$

**Solving for $E_a$:**

$$\ln(7.94) = \frac{E_a}{8.314\ \text{J/mol}\cdot\text{K}}(1.92 \times 10^{-4}\ \text{K}^{-1})$$

$$E_a = 89{,}800\ \text{J/mol} = 89.8\ \text{kJ/mol}$$

**17. d**

# CHAPTER 12

## *Study Goals*

The study goals outline specific concepts to be mastered in this section of the text chapter. Related problems at the end of the text chapter will also be noted. Working the questions noted should aid in mastery of each study goal and also highlight any areas that you may need additional help in.

**Section 12.1 *INSIGHT INTO* Concrete Production and Weathering**

1. Understand the chemical reactions important in the production and weathering of concrete. *Work Problems 1–7.*

**Section 12.2 Chemical Equilibrium**

2. Define dynamic equilibrium and understand the difference between reversible and irreversible reactions. *Work Problems 8–15.*

**Section 12.3 Equilibrium Constants**

3. Use equilibrium concentrations or partial pressures to define an equilibrium constant and recognize the numerical significance of $K$ as it relates to equilibrium position. *Work Problems 21–24, 31, 94, 95.*
4. Write equilibrium constant expressions for homogeneous and heterogeneous reactions. *Work Problems 16–20, 96.*
5. Use experimental equilibrium concentrations to calculate the value of $K$. *Work Problems 32–34, 111.*
6. Manipulate the value of $K$ when a reaction is reversed, reactions are combined, and when reaction stoichoimetry is changed. *Work Problems 25–30, 102.*

**Section 12.4 Equilibrium Concentrations**

7. Calculate equilibrium concentrations given initial concentration and the value of $K$. Use the quadratic formula to perform these calculations. *Work Problems 35–47, 92, 93, 109, 110.*

**Section 12.5 Le Chatelier's Principle**

8. State Le Chatelier's principle and use it to predict equilibrium shifts when changes are made in concentration, pressure, temperature, and catalysts. *Work Problems 48–55, 97–99, 112.*

**Section 12.6 Solubility Equilibria**

9. Apply equilibrium concepts to the solubility of salts and define the solubility product constant. *Work Problems 56, 57, 68–70, 105.*
10. Relate $K_{sp}$ to molar solubility. Calculate $K_{sp}$ given the molar solubility and vice-versa. *Work Problems 58–65, 103.*

11. Perform the above calculations when a common ion is present. *Work Problems 66, 67, 100, 101, 104.*

**Section 12.7 Acids and Bases**

12. Use the Brønsted-Lowry theory to describe the behavior of weak acids and weak bases in aqueous solution. *Work Problems 71–74, 106, 107.*
13. Define pH and relate pH values to acidity. *Work Problem 77*
14. Define an ionization constant for a weak acid or base and use it to calculate the pH of those solutions. *Work Problems 75, 76, 78 – 83.*

**Section 12.8 Free Energy and Chemical Equilibrium**

15. Relate Gibb's free energy change, $\Delta G^o$, to the equilibrium constant, K and calculate K from $\Delta G^o$ and vice-versa. *Work Problems 84 – 87, 108.*

**Section 12.9 *INSIGHT INTO* Bendable Concrete**

16. Describe how engineered cementitious composites or ECC is used to reduce the brittle nature of concrete. *Work Problems 88 – 91.*

## *Solutions to the Odd-Numbered Problems*

Problem 12.1 Identify the first chemical step in the production of Portland cement. How is this reaction related to the chemistry that takes place in the carbonation of concrete?

**Portland cement starts with the production of CaO from limestone, which is mostly calcium carbonate, $CaCO_3$.**

$$CaCO_3 \rightarrow CaO + CO_2$$

**It is almost the reverse process. In carbonation, $CO_2$ from the air diffuses into the concrete. Once there, it can react with calcium hydroxide present in essentially a two-step process represented by the following reactions:**

$$Ca(OH)_2(s) \rightarrow Ca^{2+}(aq) + 2\ OH^-(aq)$$

$$Ca^{2+}(aq) + 2\ OH^-(aq) + CO_2(g) \rightarrow CaCO_3(s) + H_2O(\ell)$$

Problem 12.3 What fraction of the annual release of $CO_2(g)$ into the atmosphere is the result of concrete production? What is the main chemical step that leads to the production of $CO_2$?

**The production of cement for concrete accounts for an estimated 5% of $CO_2$ released into the atmosphere annually.**

**The production of CaO from limestone, which is mostly calcium carbonate, $CaCO_3$, is given by:**

**$CaCO_3 \rightarrow CaO + CO_2$**

Problem 12.5 In Chapter 11, we discussed several factors that can influence chemical kinetics. Use your understanding of those factors to offer an explanation as to why concrete hydration takes days to complete.

**The temperature of the reaction is usually ambient, not very high for a chemical reaction. The reaction also takes place in essentially the solid state where there is little molecular motion (kinetic energy) and therefore activation energy is a rate-limiting factor.**

Problem 12.7 A student finds a piece of old concrete that has recently broken off from the curb alongside a road. Describe what she would expect to observe if she sprayed a solution containing phenolphthalein on all the surfaces of this piece of concrete.

**The inner surfaces would be pink, indicating the presence of hydroxide ions, but the outer surfaces that were exposed to the atmosphere would not be colored. This is due to carbonation, a type of weathering where $CO_2$ from the air diffuses into the concrete. Once there, it can react with calcium hydroxide, neutralizing it and essentially reforming the $CaCO_3$. See Figure 12.2.**

Problem 12.9 In the figure, orange fish are placed in one aquarium and green fish in an adjoining aquarium. The two tanks are separated by a removable partition that is initially closed.

**(a)** Describe what happens in the first few minutes after the partition is opened.
**At first, some of the orange fish will start to swim to the right, and green fish will begin to swim to the left.**

**(b)** What would expect to see several hours later?
**After several hours, there will be a uniform distribution of the different-colored fish in the tank.**

**(c)** How is this system analogous to dynamic chemical equilibrium?
**When the partition is removed, the "system" is not in equilibrium. After several hours, equilibrium is established, but it is dynamic: fish are still moving from side to side but the orange and green fish do so at the same rate. The will be no net change in the distribution of fish after equilibrium is established.**

Problem 12.11 For the system in the preceding problem show the equilibrium condition in terms of the rates of the forward and reverse reactions.

**The equation for this reaction is: $H_2(g) + I_2(g) \leftrightarrows 2\,HI(g)$,**

**At equilibrium, the forward reaction rate equals the reverse reaction rate:**

**$\text{Rate}_{H2(g)+I2(g)\rightarrow 2HI(g)} = \text{Rate}_{2HI(g)\rightarrow H2(g)+I2(g)}$**

Problem 12.13 The graph represents the progress of the reaction, A(g) ⇆ B(g), which starts with only the reactant A present. Label the two lines to show which represents A and which represents B. Discuss what is occurring at each of the times indicated ((a), (b), and (c)).

**The top line represents the concentration of A over time, and the bottom line represents the concentration of B over time.**
**Time (a) represents the beginning of the reaction where there is a high concentration of A and little B. The rate of disappearance of A is large and the rate of formation of B is correspondingly large also.**
**At time (b), the forward reaction has slowed somewhat and the reverse reaction rate is increasing.**
**Finally at time (c), the reverse reaction rate has become equal to the forward rate; there will be no further net change in the amounts of A and B.**

Problem 12.15 In the following equilibrium in lake water, which of the inferences below can be drawn from the equation alone?

$$HCO_3^-(aq) \leftrightarrows H^+(aq) + CO_3^{2-}(aq)$$

**(a)** The rate of the forward reaction equals the rate of the reverse reaction.
**(b)** The equilibrium concentrations of all species are equal.
**(c)** Equilibrium was attained by starting with only $HCO_3^-$ (aq) in solution.

**Only the first statement (a) can be correctly inferred from this equation. Equilibrium is defined by the forward reaction rate equaling the reverse rate.**
**Statement (b) is not correct; the concentration of all species do not have to be equal at equilibrium, in fact usually they are not.**
**Statement (c) is incorrect because equilibrium can be attained by starting with reactants, products, or some of both.**

Problem 12.17 Write equilibrium (mass action) expressions for each of the following reactions:

**The equilibrium constant expression is written in terms of the equilibrium concentrations of products and reactants. Solids and liquids can't be expressed in term of concentration and are omitted (or their values are set = 1). The general expression looks like this:**

$$a\,A + b\,B \leftrightarrows d\,D + e\,E; \quad K = \frac{[D]^d[E]^e}{[B]^b[A]^a}$$

**(a)** $2\,NOBr(g) \leftrightarrows 2\,NO(g) + Br_2(g)$ $\qquad K = \dfrac{[NO]^2[Br_2]}{[NOBr]^2}$

**(b)** $4\,HCl(g) + O_2(g) \leftrightarrows 2\,H_2O(g) + Cl_2(g)$ $\qquad K = \dfrac{[H_2O]^2[Cl_2]}{[HCl]^4[O_2]}$

(c) $SO_2(g) + \frac{1}{2} O_2(g) \leftrightarrows SO_3(g)$ $$K = \frac{[SO_3]}{[SO_2][O_2]^{0.5}}$$

(d) $CH_4(g) + 2\,O_2(g) \leftrightarrows CO_2(g) + 2\,H_2O(g)$ $$K = \frac{[H_2O]^2[CO_2]}{[CH_4][O_2]^2}$$

(e) $C_2H_5OH(g) + 3\,O_2 \leftrightarrows 2\,CO_2(g) + 3\,H_2O(g)$ $$K = \frac{[H_2O]^3[CO_2]^2}{[C_2H_5OH][O_2]^3}$$

Problem 12.19 ■Write equilibrium expressions for each of the following heterogeneous equilibria:

(a) $CaCO_3(s) \leftrightarrows Ca^{2+}(aq) + CO_3^{2-}(aq)$ $$K = \frac{[Ca^{2+}][CO_3^{2-}]}{[1]} = [Ca^{2+}][CO_3^{2-}]$$

(b) $AgCl(s) \leftrightarrows Ag^{+}(aq) + Cl^{-}(aq)$ $$K = \frac{[Ag^{+}][Cl^{-}]}{[1]} = [Ag^{+}][Cl^{-}]$$

(c) $Mg_3(PO_4)_2(s) \leftrightarrows 3\,Mg^{2+}(aq) + 2\,PO_4^{3-}(aq)$ $$K = \frac{[Mg^{2+}]^3[PO_4^{3-}]^2}{[1]} =$$

$$= [Mg^{2+}]^3[PO_4^{3-}]^2$$

(d) $Zn(s) + Cu^{2+}(aq) \leftrightarrows Cu(s) + Zn^{2+}(aq)$ $$K = \frac{[Zn^{2+}][1]}{[Cu^{2+}][1]} = \frac{[Zn^{2+}]}{[Cu^{2+}]}$$

Problem 12.21 For each of the following, are the products or the reactants favored?

**The magnitude of the equilibrium constant shows whether the reaction is reactant favored or product favored:**

**If $K > 1$, it is product favored, If $K < 1$, it is reactant favored.**

(a) $AgCl(s) \leftrightarrows Ag^{+}(aq) + Cl^{-}(aq)$ $K = 1.8 \times 10^{-10}$ **$K < 1$, reactant favored**
(b) $Ca^{2+}(aq) + CO_3^{2-}(aq) \leftrightarrows CaCO_3$ $K = 2.1 \times 10^{8}$ **$K > 1$, product favored**
(c) $N_2O_4 \leftrightarrows 2\,NO_2$ $K = 1.10$ **$K > 1$, product favored (although not by much)**

Problem 12.23 Which of the following is more likely to precipitate the sulfate ion?

(a) $PbSO_4(s) \leftrightarrows Pb^{2+}(aq) + SO_4^{2-}(aq)$ $K = 1.8 \times 10^{-8}$
(b) $CaSO_4(s) \leftrightarrows Ca^{2+}(aq) + SO_4^{2-}(aq)$ $K = 9.1 \times 10^{-6}$

**In these reactions, the sulfate ion precipitates when the reverse reaction is favored. The smaller the value of *K*, the more the reverse reaction is favored. Therefore reaction (a) is more likely to precipitate the sulfate ion.**

Problem 12.25 For each of the following equations, write the equilibrium expression for the reverse reaction:

**For the reverse of these reactions we need to imagine the products as reactants and reactants as products.**

**(a)** $2\,C(s) + O_2(g) \leftrightarrows 2\,CO(g)$ $K = \frac{[1]^2[O_2]}{[CO]^2} = \frac{[O_2]}{[CO]^2}$

**(b)** $AgCl(s) \leftrightarrows Ag^+(aq) + Cl^-(aq)$ $K = \frac{[1]}{[Ag^+][Cl^-]}$

Problem 12.27 Using the following equations, determine each of the following:
**(a)** equilibrium expressions, $K_1$ and $K_2$;
**(b)** the equation for the reaction that is the sum of the two equations; and
**(c)** the equilibrium expression, $K_3$, for the sum of the two equations

**(a)** $CO_3^{2-}(aq) + H^+(aq) \leftrightarrows HCO_3^-(aq)$ $K_1 = ? = \frac{[HCO_3^-]}{[H^+][CO_3^{2-}]}$

$HCO_3^-(aq) + H^+(aq) \leftrightarrows H_2CO_3(aq)$ $K_2 = ? = \frac{[H_2CO_3]}{[H^+][HCO_3^-]}$

**(b) Sum of the reactions:** $CO_3^{2-}(aq) + 2\,H^+(aq) \leftrightarrows H_2CO_3(aq)$

**(c)** $K_3 = \frac{[H_2CO_3]}{[H^+]^2[CO_3^{2-}]}$, **which is the product of $K_1$ and $K_2$.**

Problem 12.29 An engineer is considering the use of bacteria called methanotrophs to remediate the production of small amounts of methane at a mine site. The reaction that takes place can be summarized as $CH_4(g) + 2O_2(g) \leftrightarrows CO_2(g) + H_2O(\ell)$. What is the equilibrium constant expression for this reaction?

**The balanced equation is $CH_4(g) + 2\,O_2(g) \leftrightarrows CO_2(g) + \underline{2}\,H_2O(\ell)$.**

$K = \frac{[CO_2]}{[CH_4][O_2]^2}$ **; $H_2O$ is omitted from the expression because it is in the pure liquid state.**

Problem 12.31 In Exercise 12.30, which reaction has the greater tendency to go to completion as written, reaction 1 or reaction 2?

**The reaction that has the greater tendency to go to completion is the one with the largest value of equilibrium constant.**
**$K_1 = 3.0 \times 10^{-13}$, $K_2 = 1.8 \times 10^7$; $K_2$ is larger; therefore the second reaction has a greater tendency to go in the forward direction.**

Problem 12.33 ■ The following data were collected for a system at equilibrium at 140°C. Calculate the equilibrium constant for the reaction, $3\ H_2(g) + N_2(g) \leftrightarrows 2\ NH_3(g)$, at this temperature.

$[H_2] = 0.10$ mol L$^{-1}$, $[N_2] = 1.10$ mol L$^{-1}$, $[NH_3] = 3.6 \times 10^{-2}$ mol L$^{-1}$

**We need to first write the equilibrium constant expression, and then substitute the values of the equilibrium concentrations to find *K*. (Recall that units are omitted when calculating *K*.)**

$$K = \frac{[NH_3]^2}{[N_2][H_2]^3} = \frac{[0.036]^2}{[1.10][0.10]^3} = 1.2$$

Problem 12.35 ■ Nitrosyl chloride, NOCl, decomposes to NO and $Cl_2$ at high temperatures:

$$2\ NOCl(g) \leftrightarrows 2\ NO(g) + Cl_2(g)$$

Suppose you place 2.00 mol NOCl in a 1.00-L flask and raise the temperature to 462°C. When equilibrium has been established, 0.66 mol NO is present. Calculate the equilibrium constant $K_c$ for the decomposition reaction from these data.

**We'll write the equation and construct a reaction summary table. No products are initially present. Based on how much NO is present at equilibrium, we can determine what the change in concentration is for all species in the reaction.**

**[NOCl] = 2.00 mol NOCl ÷ 1.00 L = 2.00 M initial**
**[NO] = 0.66 mol NO ÷ 1.00 L = 0.66 M at equilibrium**

| | $2\ NOCl(g) \leftrightarrows$ | $2\ NO(g)$ | $+\ Cl_2(g)$ |
|---|---|---|---|
| **initial conc. (M)** | **2.00** | **0** | **0** |
| **change as reaction occurs (M)** | ________ | ________ | ________ |
| **equilibrium conc. (M)** | ________ | **0.66** | ________ |

**Describe the stoichiometric changes in terms of one variable, $X$.**

| | $2\ NOCl(g)$ ⇆ | $2\ NO(g)$ | + $Cl_2(g)$ |
|---|---|---|---|
| **initial conc. (M)** | **2.00** | **0** | **0** |
| **change as reaction occurs (M)** | $-2X$ | $+2X$ | $+X$ |
| **equilibrium conc. (M)** | $2.00 - 2X$ | $0.66 = 2X$ | $X$ |

**Calculate the concentrations of the gases at equilibrium using equilibrium [NO]**

$$[NO] = 0.66\ M = 2X$$

$$X = 0.66\ M \div 2 = 0.33\ M$$

$$2.00 - 2X = 2.00\ M - 0.66\ M = 1.34\ M$$

**Then write the $K_c$ expression and plug the concentrations into the expression to get the value of $K_c$.**

$$K_c = \frac{[NO]^2[Cl_2]}{[NOCl]^2} = \frac{(2X)^2(X)}{(2.00-2X)^2} = \frac{(0.66)^2(0.33)}{(1.34)^2} = 0.080$$

Problem 12.37 A system of 0.100 mole of oxygen gas, $O_2(g)$, is placed in a closed 1-L container and is brought to equilibrium at 600 K.

$$O_2(g) \leftrightarrows 2\ O(g) \qquad K = 2.8 \times 10^{-39}$$

What are the equilibrium concentrations of O and $O_2$?

**This is a basic equilibrium problem: given the $K$ and initial amounts, solve for the equilibrium amounts. A three-step approach is useful for these.**

1. **Fill out a table of concentrations. Let $X$ = unknown mol/L of change taking place in the reaction.**
2. **Write the equilibrium constant expression, set it equal to $K$.**
3. **Substitute the equilibrium concentrations in terms of $X$ into this equation and solve for $X$.**

**In the reaction summary table, $O_2$ decreases by $X$ mol/L and O increases by $2X$:**

| | $O_2(g)$ ⇆ | $2\ O(g)$ |
|---|---|---|
| **Initial** | **0.100 M** | **0.0 M** |
| **Change** | $-X$ | $+2X$ |
| **Equilibrium** | $0.100 - X$ | $2X$ |

**The equilibrium constant expression is:**

$$K = \frac{[O]^2}{[O_2]} = 2.8 \times 10^{-39} \qquad \textbf{Substituting, } K = \frac{[2X]^2}{[0.100 - X]} = 2.8 \times 10^{-39}$$

**Since $K$ is very small, we can assume that $0.100 - X \approx 0.100$. The rule of thumb in using this assumption is when $K < 1 \times 10^{-3}$, the $X$ can be neglected compared to the initial concentration.**

**Our equation for $X$ now becomes:** $K = \frac{[2X]^2}{[0.100]} = 2.8 \times 10^{-39}$

**Solving for $X$,** $4X^2 = (2.8 \times 10^{-39})(0.100)$ $X = 8.4 \times 10^{-21}$ **M**

**Finally, now that we know $X$, the change in mol/L occurring, we can calculate the equilibrium concentrations by returning to our table of concentrations.**

**$[O] = 2X = 2(8.4 \times 10^{-21}$ M$) = 1.7 \times 10^{-20}$ M, $[O_2] = 0.100 - X \approx 0.100$ M**

Problem 12.39 The following reaction establishes equilibrium at 2000 K:

$$N_2(g) + O_2(g) \leftrightarrows 2\,NO \qquad K = 4.1 \times 10^{-4}$$

If the reaction began with 0.100 mol L$^{-1}$ of $N_2$ and 0.100 mol L$^{-1}$ of $O_2$, what were the equilibrium concentrations of all species?

**This is another equilibrium problem, similar to Problem 12.37. We'll use the same three-step approach. As the reaction approaches equilibrium, the NO increases by $2X$ mol/L and $N_2$ and $O_2$ both decrease by $X$ mol/L.**

| | $N_2(g)$ + | $O_2(g)$ ⇆ | 2 NO |
|---|---|---|---|
| **Initial** | **0.100 M** | **0.100 M** | **0.0 M** |
| **Change** | $-X$ | $-X$ | $+2X$ |
| **Equilibrium** | $0.100 - X$ | $0.100 - X$ | $2X$ |

**The equilibrium constant expression is:** $K = \frac{[NO]^2}{[N_2][O_2]}$, **substituting,**

$$K = \frac{[2X]^2}{[0.100 - X][0.100 - X]} = 4.1 \times 10^{-4}$$

**Since we have squares in the numerator and denominator on the left, we can take the square root of both sides to simplify our solution.**

$$K = \sqrt{\frac{[2X]^2}{[0.100 - X][0.100 - X]}} = \sqrt{4.1 \times 10^{-4}}, \qquad \frac{[2X]}{[0.100 - X]} = 2.02 \times 10^{-2}$$ **Solving for $X$,**

**$2X = (2.02 \times 10^{-2})(0.100 - X)$, $\quad 2.0202X = 2.02 \times 10^{-3}$, $\quad X = 1.00 \times 10^{-3}$ M**

**Using $X$ to solve for the equilibrium concentrations,**

**$[NO] = 2X = 2(1.00 \times 10^{-3}\ M) = 2.0 \times 10^{-3}$ M $\qquad [O_2] = [N_2] = 0.100 - X = 0.099$ M**

Problem 12.41 In the reaction in Exercise 12.39, another trial was carried out. The reaction began with an initial concentration $N_2$ equal to the initial concentration of NO. Each had a concentration of 0.100 mol $L^{-1}$. What were the equilibrium concentrations of all species?

**Same approach as Problem 12.39, we'll assume the initial concentration of $O_2$ is zero. To fill out our table of concentrations, we need to consider which species are increasing and which are decreasing. Since $O_2$ starts out at zero, it must increase (by $X$ mol/L), as must $N_2$. NO will decrease by $2X$ mol/L.**

| | $N_2(g)$ + | $O_2(g)$ ⇆ | $2\ NO(g)$ |
|---|---|---|---|
| **Initial** | **0.100 M** | **0.0 M** | **0.100 M** |
| **Change** | **$+X$** | **$+X$** | **$-2X$** |
| **Equilibrium** | **$0.100 + X$** | **$+X$** | **$0.100 - 2X$** |

**The equilibrium constant expression is:** $K = \frac{[NO]^2}{[N_2][O_2]}$, **substituting,**

$$K = \frac{[0.100 - 2X]^2}{[0.100 + X][X]} = 4.1 \times 10^{-4}$$

**This equation will have to be solved using the quadratic formula. We'll rearrange it until it is in the form of a quadratic expression: $aX^2 + bX + c = 0$ then use the quadratic formula:** $X = \frac{-b \pm \sqrt{b^2 - 4ac}}{2a}$ **Rearranging the equation,**

$$(4.1 \times 10^{-4})(X)(0.100 + X) = 0.0100 - 4X + 4X^2$$

$$4.1 \times 10^{-5}X + 4.1 \times 10^{-4}X^2 = 0.0100 - 0.400X + 4X^2$$

**Collecting common terms, $0 = 0.0100 - 0.400041X + 3.99959X^2$**

**$a = 3.99959$, $\quad b = -0.400041$, $\quad c = 0.0100$, $\quad$ using the quadratic formula:**

$$X = \frac{-0.400041 \pm \sqrt{(-0.400041)^2 - 4(3.99959)(0.0100)}}{2(3.99959)} = \frac{0.40041 \pm 0.00701}{7.99916}$$

**$X$ = 0.0509 or 0.0491; $X$ cannot equal 0.0509 because it would give negative concentration.**

**The equilibrium concentrations are:**

**$[NO_2] = 0.100 - 2X = 0.100 - 2(0.0491) = 0.0018$ M**
**$[N_2] = 0.100 + X = 0.1491$ M**
**$[O_2] = X = 0.0491$ M**

Problem 12.43 Again the experiment in Exercise 12.39 was redesigned. This time, 0.15 mol each of $N_2$ and $O_2$ was injected into a 5.0-L container at 2500 K, at which temperature the equilibrium constant is $3.6 \times 10^{-3}$. What was the composition of the reaction mixture at equilibrium?

**Same approach as Problem 12.39. The initial concentration of $N_2$ and $O_2$ is 0.15 mol ÷ 5.0 L = 0.03 M. Since NO starts out at zero, it must increase (by $2X$ mol/L), and $N_2$ and $O_2$ will decrease by $X$ mol/L.**

| | $N_2(g)$ + | $O_2(g)$ ⇆ | 2 NO |
|---|---|---|---|
| Initial | 0.030M | 0.030M | 0.0 M |
| Change | $-X$ | $-X$ | $+2X$ |
| Equilibrium | $0.030 - X$ | $0.030 - X$ | $2X$ |

**The equilibrium constant expression is:** $K = \frac{[NO]^2}{[N_2][O_2]}$, **substituting,**

$$K = \frac{[2X]^2}{[0.030 - X][0.030 - X]} = 3.6 \times 10^{-3}$$

**As in Problem 12.37, we can take the square root of both sides to simplify the solution for $X$.**

$$K = \sqrt{\frac{[2X]^2}{[0.030 - X][0.030 - X]}} = \sqrt{3.6 \times 10^{-3}}, \qquad \frac{[2X]}{[0.030 - X]} = 0.060$$

**Solving for $X$,**

**$2X = (0.060)(0.030 - X)$, $\quad 2.060X = 0.0018$, $\quad X = 8.7 \times 10^{-4}$ M**

**Using $X$ to solve for the equilibrium concentrations,**

**$[NO] = 2X = 2(8.7 \times 10^{-4}\text{ M}) = 1.7 \times 10^{-3}$ M**

**$[O_2] = [N_2] = 0.100 - X = 0.100 - 8.7 \times 10^{-4} = 0.029$ M**

Problem 12.45 A student is simulating the carbonic acid-hydrogen carbonate equilibrium in a lake:

$$H_2CO_3(aq) \leftrightarrows H^+(aq) + HCO_3^-(aq) \qquad K = 4.3 \times 10^{-7}$$

She starts with 0.1000 M carbonic acid. What are the concentrations of all species at equilibrium?

**Another equilibrium problem, the initial concentration of carbonic acid is 0.1000 M and the initial concentration of $H^+$ and $HCO_3^-$ are zero. Carbonic acid will decrease by $X$ mol/L and the products will increase by $X$ mol/L.**

| | $H_2CO_3(aq)$ $\leftrightarrows$ | $H^+(aq)$ + | $HCO_3^-(aq)$ |
|---|---|---|---|
| **Initial** | **0.1000 M** | **0.0 M** | **0.0 M** |
| **Change** | $-X$ | $+X$ | $+X$ |
| **Equilibrium** | $0.1000 - X$ | $X$ | $X$ |

**The equilibrium constant expression is:** $K_a = \dfrac{[HCO_3^-][H^+]}{[H_2CO_3]}$, **substituting,**

$$K_a = \frac{[X][X]}{[0.1000 - X]} = 4.3 \times 10^{-7}$$

**Because of the level of significance of this problem, we can't ignore $X$ relative to the initial concentration and must use the quadratic formula. We'll rearrange it until it is in the form of a quadratic expression:** $aX^2 + bX + c = 0$ **then use the quadratic formula:**

$$X = \frac{-b \pm \sqrt{b^2 - 4ac}}{2a}$$

**Rearranging the equation,**

$$X^2 = 4.3 \times 10^{-8} - 4.3 \times 10^{-7} X$$

**Collecting common terms,** $X^2 + 4.3 \times 10^{-7} X - 4.3 \times 10^{-8} = 0$

$a = 1$, $b = 4.3 \times 10^{-7}$, $c = -4.3 \times 10^{-8}$, **using the quadratic formula:**

$$X = \frac{-4.3\times10^{-7} \pm \sqrt{(4.3\times10^{-7})^2 - 4(1)(-4.3\times10^{-8})}}{2(1)} = \frac{-4.3\times10^{-7} \pm 4.147\times10^{-4}}{2}$$

$X = 2.1 \times 10^{-4}$ **or** $-2.1 \times 10^{-4}$; $X$ **cannot equal** $-2.1 \times 10^{-4}$ **because it would give negative concentration.**

**The equilibrium concentrations are:** $[H^+] = [HCO_3^-] = X = 2.1 \times 10^{-4}$ M

$[H_2CO_3] = 0.1000 - X = 0.0998$ M

Problem 12.47 Because calcium carbonate is a sink for $CO_3^{2-}$ in a lake, the student in Exercise 12.45 decides to go a step further and examine the equilibrium between carbonate ion and $CaCO_3$. The reaction is

$$Ca^{2+}(aq) + CO_3^{2-}(aq) \leftrightarrows CaCO_3(s)$$

The equilibrium constant for this reaction is $2.1 \times 10^8$. If the initial calcium ion concentration is 0.02 M and the carbonate concentration is 0.03 M, what are the equilibrium concentrations of the ions?

**This is a precipitation reaction; the concentrations of ions are too large to exist in equilibrium and a solid phase (a precipitate forms).**

| | $Ca^{2+}(aq)$ | $CO_3^{2-}(aq)$ | $CaCO_3(s)$ |
|---|---|---|---|
| **Initial** | **0.02 M** | **0.03 M** | **---** |
| **Change** | **–0.02** | **–0.02** | **+0.02** |
| **Final** | **0.0 M** | **0.01 M** | **---** |

**Notice that this is not an equilibrium situation. Once the solid forms, however, dissociation begins and an equilibrium is established:** $CaCO_3(s) \leftrightarrows Ca^{2+}(aq) + CO_3^{2-}(aq)$

**This process will produce a small increase in the concentration of both ions:**

| | $CaCO_3(s)$ | $Ca^{2+}(aq)$ | $CO_3^{2-}(aq)$ |
|---|---|---|---|
| **Initial** | **---** | **0.00 M** | **0.01M** |
| **Change** | **$-X$** | **$+X$** | **$+X$** |
| **Equilibrium** | **---** | **$X$** | **$0.01 + X$** |

**The equilibrium expression is:** $K = [Ca^{2+}][CO_3^{2-}]$ **where** $K = \dfrac{1}{2.1 \times 10^8}$ **(because we reversed the equation). Substituting $X$ into the equation:**

$[X][0.01 + X] = 4.8 \times 10^{-9}$, $(0.01 + X) \approx 0.01$ M $\quad X = 4.8 \times 10^{-7}$ M

**The equilibrium concentration of $Ca^{2+} = 4.8 \times 10^{-7}$ M,**

**The equilibrium concentration of $CO_3^{2-} \approx 0.01$ M**

Problem 12.49 In the following equilibrium in a closed system, indicate how the equilibrium is shifted by the indicated stress:

$$HF(g) + H_2O(\ell) \leftrightarrows H_3O^+(aq) + F^-(aq)$$

**The basic idea of Le Chateliér's principle is that whatever change we make to an equilibrium system, the system will respond in an opposite manner as it reestablishes equilibrium.**

**(a)** Additional HF(g) is added to the system.
**More reactant will push the equilibrium more to the products.**

**(b)** Water is added.
**Adding water will have the effect of reducing the concentration of the products. This will push the equilibrium farther to the products.**

**(c)** $Ca(NO_3)$ solution is added, and $CaF_2$ precipitates.
**The precipitation will lower the concentration of $F^-$. Reducing the amount of a product will cause a shift in the equilibrium that favors the products, as the system tries to replace the product.**

**(d)** The volume is reduced.
**A decrease in volume will have the same effect as increase in pressure. This will cause a shift in the reaction direction that has fewer moles of gas. This will be a shift that favors the products.**

**(e)** KOH is added.
**The KOH (a base) will react with the $H_3O^+$. Removal of a product causes a shift toward the products.**

**(f)** A catalyst is added.
**Catalysts do not affect the equilibrium position; both the forward and the reverse reaction rates speed up the same amount.**

Problem 12.51 In each of the reactions below, tell how the equilibrium responds to an increase in pressure.

**An increase in pressure causes the gas molecules to be colliding more often. Le Chatelier's principle states that the system will try to "undo" the change we've made. The system does this by shifting the equilibrium to favor the reaction where fewer gas molecules are produced, thus causing molecules to collide less often.**

**(a)** $CaCO_3(s) \leftrightarrows CaO(s) + CO_2(g)$
**Fewer gas molecules are formed in the reverse direction. The equilibrium will shift more to the reactants.**

**(b)** $N_2O_4(g) \leftrightarrows 2\ NO_2(g)$
**Fewer gas molecules are formed in the reverse direction. The equilibrium will shift more to the reactants.**

**(c)** $HCO_3^-(aq) + H^+(aq) \leftrightarrows H_2CO_3(aq)$

**There are no gas molecules present; a change in pressure will not affect the equilibrium position.**

Problem 12.53 ■ The decomposition of $NH_4HS$:

$$NH_4HS(s) \leftrightarrows NH_3(g) + H_2S(g)$$

is an endothermic process. Using Le Chatelier's principle, explain how increasing the temperature would affect the equilibrium. If more $NH_4HS$ is added to a flask in which this equilibrium exists, how is the equilibrium affected? What if some additional $NH_3$ is placed in the flask? What will happen to the pressure of $NH_3$ if some $H_2S$ is removed from the flask?

**When the temperature is raised, the reaction adjusts to the added heat by consuming reactants and forming more products. The equilibrium shifts to the right. Adding $NH_4HS$, a solid, will have no effect on the equilibrium. Adding $NH_3(g)$, a product, will shift the equilibrium to the left. Removing $H_2S$, a product, will shift the reaction to the right, increasing the $NH_3$ pressure.**

Problem 12.55 The following equilibrium is established in a closed container:

$$C(s) + O_2(g) \leftrightarrows CO_2(g) \qquad \Delta H^\circ = -393 \text{ kJ mol}^{-1}$$

How does equilibrium shift in response to each of the following stresses?

**(a)** The quantity of solid carbon is increased.

**The amount of a solid does not affect equilibrium position, as its concentration does not change. Equilibrium position stays the same.**

**(b)** A small quantity of water is added, and $CO_2$ dissolves in it.

**The concentration/amount of $CO_2$ is decreasing. The equilibrium will shift more to the product side to compensate for this.**

**(c)** The system is cooled.

**The effect of a change in temperature on equilibrium position depends on whether the reaction is exothermic or endothermic. This reaction is exothermic (ΔH is negative). Lowering temperature removes heat, which is a product in this reaction. The equilibrium position is shifted to favor the products.**

**(d)** The volume of the container is increased.

**An increase in volume is the same as a decrease in pressure: the molecules collide less frequently. The reaction will speed up in the direction that creates more molecules to cause them to collide more frequently. Since there is the same number of gas molecules on both sides, there is no change in the equilibrium position.**

Problem 12.57 Write an equilibrium equation and a $K_{sp}$ expression for the dissolution of each of the following in water:

**A solubility product constant expression ($K_{sp}$) is the same as an equilibrium constant expression (see Problem 12.17). These reactions are all of a particular type: dissociation of a slightly soluble salt to yield aqueous ions; that's where the name solubility product constant comes from.**

**(a)** $AgBr(s) \leftrightarrows Ag^+(aq) + Br^-(aq)$ $\quad K_{sp} = \dfrac{[Ag^+][Br^-]}{[1]} = [Ag^+][Br^-]$

**(b)** $Cu(OH)_2(s) \leftrightarrows Cu^{2+}(aq) + 2\ OH^-(aq)$ $\quad K_{sp} = \dfrac{[Cu^{2+}][OH^-]^2}{[1]} = [Cu^{2+}][OH^-]^2$

**(c)** $BaSO_4(s) \leftrightarrows Ba^{2+}(aq) + SO_4^{2-}(aq)$ $\quad K_{sp} = \dfrac{[Ba^{2+}][SO_4^{2-}]}{[1]} = [Ba^{2+}][SO_4^{2-}]$

**(d)** $PbCrO_4(s) \leftrightarrows Pb^{2+}(aq) + CrO_4^{2-}(aq)$ $\quad K_{sp} = \dfrac{[Pb^{2+}][CrO_4^{2-}]}{[1]} = [Pb^{2+}][CrO_4^{2-}]$

**(e)** $Mg_3(PO_4)_2(s) \leftrightarrows 3\ Mg^{2+}(aq) + 2\ PO_4^{3-}(aq)$ $\quad K_{sp} = \dfrac{[Mg^{2+}]^3[PO_4^{3-}]^2}{[1]} =$

$= [Mg^{2+}]^3[PO_4^{3-}]^2$

Problem 12.59 ■ The safe drinking act of 1974 established the maximum permitted concentration of silver ion at 0.05 ppm. What is the concentration of $Ag^+$ in parts per million in a saturated solution of AgCl? (Note: 1 ppm = 1 mg of solute/L of solution.)

**We need to calculate the molarity of silver ion in a saturated solution of AgCl. This is an equilibrium calculation and we need to look up the $K_{sp}$ of AgCl from Appendix H: $K_{sp} = 1.8 \times 10^{-10}$.**

$$AgCl(s) \leftrightarrows Ag^+(aq) + Cl^-(aq)$$

| | AgCl(s) | $Ag^+(aq)$ | $Cl^-(aq)$ |
|---|---|---|---|
| **Initial** | --- | 0.00 M | 0.00 M |
| **Change** | $-X$ | $+X$ | $+X$ |
| **Equilibrium** | --- | $X$ | $X$ |

$K_{sp} = [Ag^+][Cl^-] = 1.8 \times 10^{-10}$, **substituting for $X$,** $[X][X] = 1.8 \times 10^{-10}$,

**Solving for $X$:** $X = \sqrt{1.8 \times 10^{-10}} = 1.3 \times 10^{-5}$ M

**The equilibrium concentration of $Ag^+$ is $1.3 \times 10^{-5}$ mol/L. The last thing we need to do is express this as ppm.**

$$\frac{1.3\times10^{-5}\text{ mol Ag}^+}{\text{L}}\times\frac{107.87\text{ g Ag}^+}{1\text{ mol Ag}^+}\times\frac{1000\text{ mg}}{1\text{ g}}=1.4\text{ mg/L Ag}^+=1.4\text{ ppm Ag}^+$$

Problem 12.61 ■Calculate the solubility of $ZnCO_3$ in (a) water, (b) 0.050 M $Zn(NO_3)_2$, and (c) 0.050 M $K_2CO_3$. $K_{sp}$ of $ZnCO_3 = 3 \times 10^{-8}$

**The $ZnCO_3(s)$ will dissolve $X$ mol/L and both $Zn^{2+}(aq)$ and $CO_3^{2-}$ (aq) will increase by $X$ mol/L. The solubility of $ZnCO_3(s)$ will be $X$ mol/L.**

**In (b) and (c) the common ion effect will cause decreased solubility of $ZnCO_3(s)$.**

| | $ZnCO_3$ (s) | ⇆ $Zn^{2+}$ (aq) | + $CO_3^{2-}$ (aq) |
|---|---|---|---|
| Initial | --- | $[Zn^{2+}]_o$ | $[CO_3^{2-}]_o$ |
| Change | $-X$ | $+X$ | $+X$ |
| Equilibrium | --- | $[Zn^{2+}]_o + X$ | $[CO_3^{2-}]_o + X$ |

**At equilibrium** $K_{sp} = [Zn^{2+}][CO_3^{2-}] = ([Zn^{2+}]_o + X)([CO_3^{2-}]_o + X) = 3 \times 10^{-8}$

**(a) In water, no other source of $Zn^{2+}$ or $CO_3^{2-}$ is present so, before dissociation, $[Zn^{2+}]_o = [CO_3^{2-}]_o = 0$.**

$$(X)(X) = 3 \times 10^{-8} \qquad X = 2 \times 10^{-4}\text{ mol/L}$$

**(b) In $Zn(NO_3)_2$, an external source of $Zn^{2+}$ is present, so $[Zn^{2+}]_o = 0.050$ M. No external sources of $CO_3^{2-}$ are present so $[CO_3^{2-}]_o = 0$.**

$$(0.050 + X)(X) = 3 \times 10^{-8}$$

**Assuming $X$ is small compared to 0.050, ignore its addition.**

$$(0.050)(X) = 3 \times 10^{-8} \quad X = 6 \times 10^{-7}\text{ mol/L}$$

**(*Notice the assumption that X is small is a good one.*)**

**(c) In $K_2CO_3$, an external source of $CO_3^{2-}$ is present. Neglecting the reaction of carbonate as a base, $[CO_3^{2-}]_o = 0.050$ M. No external sources of $Zn^{2+}$ are present so, before dissociation, $[Zn^{2+}]_o = 0$.**

$$(X)(0.050 + X) = 3 \times 10^{-8}$$

**Assuming $X$ is small compared to 0.050, ignore its addition.**

$$(X)(0.050) = 3 \times 10^{-8} \quad X = 6 \times 10^{-7}\text{ mol/L}$$

**(*Notice the assumption that X is small is a good one.*)**

Problem 12.63 The ore cinnabar (HgS) is an important source of mercury. Cinnabar is a red solid whose solubility in water is $5.5 \times 10^{-27}$ mol $L^{-1}$.

(a) Calculate its $K_{sp}$.

**See Problem 12.61. The molar solubility is equal to the *X* mol/L when we solve for equilibrium concentrations.**

$$HgS(s) \leftrightarrows Hg^{2+}(aq) + S^{2-}(aq)$$

| | | | |
|---|---|---|---|
| **Initial** | --- | **0.00 M** | **0.00 M** |
| **Change** | $-X$ | $+X$ | $+X$ |
| **Equilibrium** | --- | $X$ | $X$ |

$K_{sp} = [Hg^{2+}][S^{2-}] = [X][X]$

**Recall *X* = molar solubility = $5.5 \times 10^{-27}$, Therefore,**

$K_{sp} = [X][X] = [5.5 \times 10^{-27}][5.5 \times 10^{-27}] = 3.0 \times 10^{-53}$.

(b) What is the solubility in grams per 100 g of water?

**We can express the molar solubility in terms of g/100 g water:**

$$\frac{5.5 \times 10^{-27} \text{ mol HgS}}{\text{L } H_2O} \times \frac{232.66 \text{ g HgS}}{1 \text{ mol HgS}} \times \frac{1 \text{ L } H_2O}{1000 \text{ g}} \times 100 \text{ g water} =$$

**= $1.3 \times 10^{-25}$ g HgS in 100 g water**

Problem 12.65 ■ From the solubility data given for the following compounds, calculate their solubility product constants.

(a) CuBr, copper(I) bromide, $1.0 \times 10^{-3}$ g/L

(b) AgI, silver iodide, $2.8 \times 10^{-8}$ g/10 mL

(c) $Pb_3(PO_4)_2$, lead(II) phosphate, $6.2 \times 10^{-7}$ g/L

(d) $Ag_2SO_4$, silver sulfate, 5.0 mg/mL

**Molar solubility = mol solute dissolved ÷ L solution**

**(a) Molar solubility of CuBr** $= \frac{1.0 \times 10^{-3} \text{ g}}{\text{L}} \times \frac{1 \text{ mol CuBr}}{143.45 \text{ g}} = 7.0 \times 10^{-6}$ **M**

$$CuBr(s) \leftrightarrows Cu^{+} + Br^{-}$$

| | | | |
|---|---|---|---|
| **Initial** | --- | **0.00 M** | **0.00 M** |
| **Change** | $-7.0 \times 10^{-6}$ | $+7.0 \times 10^{-6}$ | $+7.0 \times 10^{-6}$ |
| **Equilibrium** | --- | $7.0 \times 10^{-6}$ | $7.0 \times 10^{-6}$ |

$K_{sp} = [Cu^{+}][Br^{-}] = (7.0 \times 10^{-6})(7.0 \times 10^{-6}) = 4.9 \times 10^{-11}$

**(b) Molar solubility of AgI =** $\frac{2.8 \times 10^{-6}\ g}{L} \times \frac{1\ mol\ AgI}{234.8\ g} = 1.2 \times 10^{-8}\ M$

| | AgI(s) ⇆ | $Ag^+$ + | $I^-$ |
|---|---|---|---|
| Initial | --- | 0.00 M | 0.00 M |
| Change | $-1.2 \times 10^{-8}$ | $+1.2 \times 10^{-8}$ | $+1.2 \times 10^{-8}$ |
| Equilibrium | --- | $1.2 \times 10^{-8}$ | $1.2 \times 10^{-8}$ |

$K_{sp} = [Ag^+][I^-] = (1.2 \times 10^{-8})(1.2 \times 10^{-8}) = 1.4 \times 10^{-16}$

**(c) Molar solubility of $Pb_3(PO_4)_2$ =** $\frac{6.2 \times 10^{-7}\ g}{L} \times \frac{1\ mol\ Pb_3(PO_4)_2}{811.6\ g} = 7.6 \times 10^{-10}\ M$

| | $Pb_3(PO_4)_2$ (s) ⇆ | 3 $Pb^{2+}$ + | 2 $PO_4^{3-}$ |
|---|---|---|---|
| Initial | --- | 0.00 M | 0.00 M |
| Change | $-7.6 \times 10^{-10}$ | $3(+7.6 \times 10^{-10})$ | $2(+7.6 \times 10^{-10})$ |
| Equilibrium | --- | $2.3 \times 10^{-9}$ | $1.5 \times 10^{-9}$ |

$K_{sp} = [Pb^{2+}]^3[PO_4^{3-}]^2 = (2.3 \times 10^{-9})^3(1.5 \times 10^{-9})^2 = 2.7 \times 10^{-44}$

**(d) Molar solubility of $Ag_2SO_4$ =** $\frac{5.0\ g}{L} \times \frac{1\ mol\ Ag_2SO_4}{311.8\ g} = 1.6 \times 10^{-2}\ M$

| | $Ag_2SO_4$ (s) ⇆ | 2 $Ag^+$ + | $SO_4^{2-}$ |
|---|---|---|---|
| Initial | --- | 0.00 M | 0.00 M |
| Change | $-1.6 \times 10^{-2}$ | $2(+1.6 \times 10^{-2})$ | $+1.6 \times 10^{-2}$ |
| Equilibrium | --- | $3.2 \times 10^{-2}$ | $1.6 \times 10^{-2}$ |

$Ksp = [Ag^+]^2[SO_4^{2-}] = (3.2 \times 10^{-2})^2(1.6 \times 10^{-2}) = 1.6 \times 10^{-5}$

Problem 12.67 ■ Solid $Na_2SO_4$ is added slowly to a solution that is 0.10 M in $Pb(NO_3)_2$ and 0.10 M in $Ba(NO_3)_2$. In what order will solid $PbSO_4$ and $BaSO_4$ form? Calculate the percentage of $Ba^{2+}$ that precipitates just before $PbSO_4$ begins to precipitate.

**$[SO_4^{2-}]$ required to precipitate $PbSO_4$ =** $\frac{K_{sp}\ PbSO_4}{[Pb^{2+}]} = \frac{1.8 \times 10^{-8}}{0.10} = 1.8 \times 10^{-7}$

**$[SO_4^{2-}]$ required to precipitate $BaSO_4$ =** $\frac{K_{sp}\ BaSO_4}{[Ba^{2+}]} = \frac{1.1 \times 10^{-10}}{0.10} = 1.1 \times 10^{-9}$

**$1.1 \times 10^{-9} < 1.8 \times 10^{-7}$, so $BaSO_4$ will precipitate first.**

**(Because these are both 1:1 ionic compounds *and* initial concentrations of $Pb^{2+}$ and $Ba^{2+}$ were equal, this decision could have been made by just comparing $K_{sp}$ values. However, the calculation approach shown here is general.)**

**$PbSO_4$ begins to precipitate when $[SO_4^{2-}] = 1.8 \times 10^{-7}$.**

**At that point,** $$[Ba^{2+}] = \frac{K_{sp}\ BaSO_4}{[SO_4^{2-}]} = \frac{1.1\times10^{-10}}{1.8\times10^{-7}} = 6.1 \times 10^{-4}\text{ M } Ba^{2+} \text{ remaining.}$$

$$\%\ Ba^{2+} \text{ remaining} = \frac{6.1\times10^{-4}}{0.10} \times 100 = 0.61\% \text{ remaining,}$$

**100% − 0.61% = 99.39 % precipitated.**

Problem 12.69 Use the web to look up boiler scale and explain chemically why it is a problem in equipment where water is heated (such as in boilers).

**Answers will vary according to websites found. Calcium carbonate is more soluble in water at higher temperatures. In a boiler, the compound is dissolved in higher concentrations when the water is hot, but will precipitate onto surfaces when the water is cooled. This precipitate is called scale.**

Problem 12.71 Write the formula of the conjugate base of each of the following acids.

**In the Bronsted-Lowry description of acids and bases, conjugate pairs are formed when proton transfer takes place in an acid base reaction. Conjugate acids have one more proton than their conjugate bases.**

| | | |
|---|---|---|
| **(a)** | $HNO_3$ | **$NO_3^-$** |
| **(b)** | $H_2O$ | **$OH^-$** |
| **(c)** | $HSO_4^-$ | **$SO_4^{2-}$** |
| **(d)** | $H_2CO_3$ | **$HCO_3^-$** |
| **(e)** | $H_3O^+$ | **$H_2O$** |

Problem 12.73 ■ For each of the following reactions, indicate the Brønsted-Lowry acids and bases. What are the conjugate acid-base pairs?

(a) $CN^-(aq)$ + $H_2O(\ell)$ ⇆ $HCN(aq)$ + $OH^-(aq)$

**B-L base** **B-L acid** **conj. acid** **conj. base**

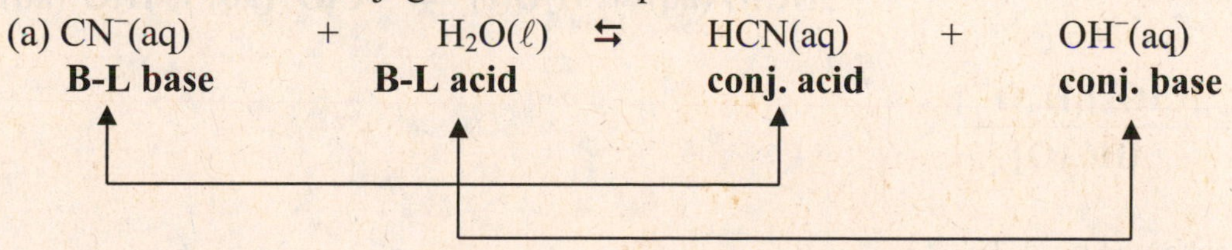

(b) $HCO_3^-(aq)$ + $H_3O^+(aq)$ ⇆ $H_2CO_3(aq)$ + $H_2O(\ell)$

**B-L base** **B-L acid** **conj. acid** **conj. base**

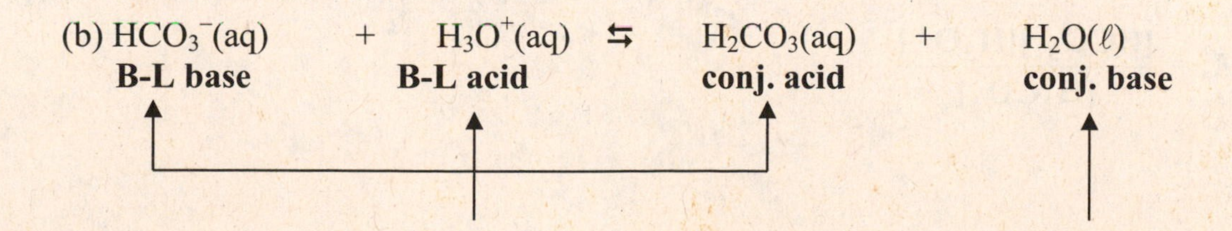

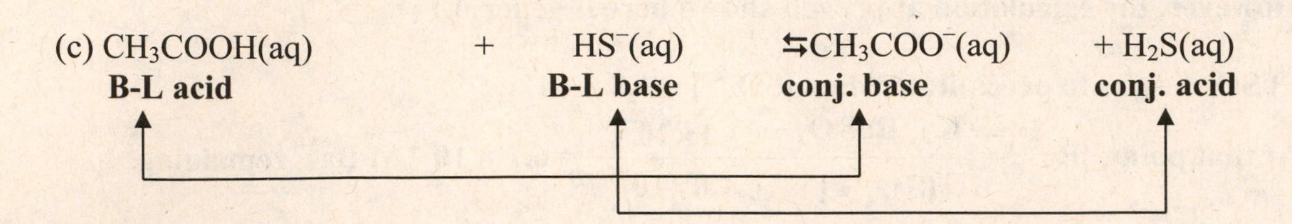


Problem 12.75 Write chemical equations and equilibrium expressions for the reactions of each of the following weak acids with water.

**(a)** acetic acid, $CH_3COOH$

$$CH_3COOH(aq) + H_2O(\ell) \leftrightharpoons CH_3COO^-(aq) + H_3O^+(aq)$$

$$K_a = \frac{[CH_3COO^-][H_3O^+]}{[CH_3COOH]}$$

**(b)** propanoic acid, $C_2H_5COOH$

$$C_2H_5COOH(aq) + H_2O(\ell) \leftrightharpoons C_2H_5COO^-(aq) + H_3O^+(aq)$$

$$K_a = \frac{[C_2H_5COO^-][H_3O^+]}{[C_2H_5COOH]}$$

**(c)** hydrofluoric acid, HF

$$HF(aq) + H_2O(\ell) \leftrightharpoons F^-(aq) + H_3O^+(aq)$$

$$K_a = \frac{[F^-][H_3O^+]}{[HF]}$$

**(d)** chlorous acid, HClO

$$HClO(aq) + H_2O(\ell) \leftrightharpoons ClO^-(aq) + H_3O^+(aq)$$

$$K_a = \frac{[ClO^-][H_3O^+]}{[HClO]}$$

**(e)** carbonic acid, $H_2CO_3$

$$H_2CO_3(aq) + H_2O(\ell) \leftrightharpoons HCO_3^-(aq) + H_3O^+(aq)$$

$$K_a = \frac{[HCO_3^-][H_3O^+]}{[H_2CO_3]}$$

Problem 12.77 Hydrofluoric acid is a weak acid used in the building industry to etch patterns into glass for elegant windows. Because it dissolves glass, it is the only inorganic acid that must be stored in plastic containers. A 0.1 M solution of HF has a pH of 2.1. Calculate $[H_3O^+]$ in this solution.

**The definition of pH is:** $pH = -\log[H_3O^+]$. **Rearranging,** $[H_3O^+] = 10^{-pH}$,

$pH = 2.1$, $[H_3O^+] = 10^{-2.1} = 8 \times 10^{-3}$ **M (one significant digit).**

Problem 12.79 Calculate the pH of 0.10 M solution of propanoic acid and determine its percent ionization.

**To answer this problem, we must solve for the equilibrium concentrations in this solution. The initial concentration is given, we must look up the ionization constant (equilibrium constant) for propanoic acid. From Table 12.5 in the textbook,** $K_a = 1.3 \times 10^{-5}$.

$$C_2H_5COOH(aq) + H_2O(\ell) \leftrightarrows C_2H_5COO^-(aq) + H_3O^+(aq)$$

| Initial | 0.10 M | | 0.00 M | 0.00 M |
|---|---|---|---|---|
| Change | $-X$ | | $+X$ | $+X$ |
| Equilibrium | $0.10 - X$ | | $X$ | $X$ |

$$K_a = \frac{[C_2H_5COO^-][H_3O^+]}{[C_2H_5COOH]} = 1.3 \times 10^{-5}$$, **substituting** $X$ **into the equation:**

$$K_a = \frac{[X][X]}{[0.10 - X]} = 1.3 \times 10^{-5}$$. **Assume that** $0.10 - X \approx 0.10$, $$K_a = \frac{[X][X]}{[0.10]} = 1.3 \times 10^{-5},$$

$X^2 = 1.3 \times 10^{-6}$, $X = 1.1 \times 10^{-3}\ M = [H_3O^+] = [C_2H_5COO^-]$

**The** $pH = -\log[H_3O^+] = -\log[1.1 \times 10^{-3}] = 2.9$.

**Percent ionization is defined as:** $$\%\ \text{ionization} = \frac{[\text{Anion}]_{eq}}{[\text{Acid}]_{init}} \times 100.$$

$$\%\ \text{ionization} = \frac{[C_2H_5COO^-]_{eq}}{[C_2H_5COOH]_{init}} \times 100 = \frac{1.1 \times 10^{-3}\ M}{0.10\ M} \times 100 = 1.1\ \%$$

Problem 12.81 Acrylic acid is used in the polymer industry in the production of acrylates. Its $K_a$ is $5.6 \times 10^{-5}$. What is the pH of a 0.11 M solution of acrylic acid, $CH_2CHCOOH$?

**This is very similar to Problem 12.79.**

$$CH_2CHCOOH + H_2O \leftrightarrows CH_2CHCOO^- + H_3O^+$$

| Initial | 0.11 M | | 0.00 M | 0.00 M |
|---|---|---|---|---|
| Change | $-X$ | | $+X$ | $+X$ |
| Equilibrium | $0.11 - X$ | | $X$ | $X$ |

$$K_a = \frac{[CH_2CHCOO^-][H_3O^+]}{[CH_2CHCOOH]} = 5.6 \times 10^{-5},$$ **substituting $X$ into the equation:**

$$K_a = \frac{[X][X]}{[0.11 - X]} = 5.6 \times 10^{-5}$$ **Assume that $0.11 - X \approx 0.11$.** $$K_a = \frac{[X][X]}{[0.11]} = 5.6 \times 10^{-5},$$

$X^2 = 6.2 \times 10^{-6}$, $X = 2.5 \times 10^{-3}$ M $= [H_3O^+] = [CH_2CHCOO^-]$

**The pH = $-\log [H_3O^+] = -\log [2.5 \times 10^{-3}] = 2.6$.**

Problem 12.83 Morphine, an opiate derived from the opium poppy (genus *Papaver*), has the molecular formula $C_7H_{19}NO_3$. It is a weakly basic amine, with a $K_b$ of $1.6 \times 10^{-6}$. What is the pH of a 0.0045 M solution of morphine?

**We'll solve this problem just like the previous two, with the exception of the reaction will be that of a base, and the equilibrium concentrations will give us $OH^-$ not $H_3O^+$.**

$$C_7H_{19}NO_3 + H_2O \leftrightarrows C_7H_{20}NO_3^+ + OH^-$$

| Initial | 0.0045 M | | 0.00 M | 0.00 M |
|---|---|---|---|---|
| Change | $-X$ | | $+X$ | $+X$ |
| Equilibrium | $0.0045 - X$ | | $X$ | $X$ |

$$K_b = \frac{[C_7H_{20}NO_3^+][OH^-]}{[C_7H_{19}NO_3]} = 1.6 \times 10^{-6},$$ **substituting $X$ into the equation:**

$$K_b = \frac{[X][X]}{[0.0045 - X]} = 1.6 \times 10^{-6}.$$

**Assume that $0.0045 - X \approx 0.0045$,** $$K_b = \frac{[X][X]}{[0.0045]} = 1.6 \times 10^{-6},$$

$X^2 = 7.2 \times 10^{-9}$, $X = 8.5 \times 10^{-5}$ M $= [OH^-] = [C_7H_{20}NO_3^+]$.

**We need to calculate the $H_3O^+$ to find pH.**

$[H_3O^+][OH^-] = K_w = 1.0 \times 10^{-14}$, $$[H_3O^+] = \frac{[K_w]}{[OH^-]} = \frac{1.0 \times 10^{-14}}{8.5 \times 10^{-5}} = 1.2 \times 10^{-10} \text{ M}.$$

**The pH = $-\log [H_3O^+] = -\log [1.2 \times 10^{-10}] = 9.9$.**

Problem 12.85 For the reaction, $PbCl_5(g) \leftrightarrows PbCl_3(g) + Cl_2(g)$, the measured equilibrium constant is $7.7 \times 10^{-3}$ at 25°C. **(a)** Based on this information, predict if $\Delta G^o$ is positive or negative. Explain your answer. **(b)** Calculate $\Delta G^o$ from this information.

**(a) The relationship between $\Delta G$ and $K$ is $\Delta G = -RT(\ln K)$. If $K$ is smaller than one, then the natural log will be negative and $\Delta G$ will be positive.**

**(b) Because we are looking for $\Delta G^o$, $T = 298$ K.**
**$\Delta G = -RT(\ln K) = -(8.314 \text{ J/mol·K})(298 \text{ K})(\ln 0.0077) = 12{,}000$ J/mol**

Problem 12.87 A group of students working on chemistry homework calculates the equilibrium constants for three reactions. Reaction 1 has $\Delta G^o = -50$ kJ, reaction 2 has $\Delta G^o = -30$ kJ and reaction 3 has $\Delta G^o = -10$ kJ. The students are surprised when their answers for each reaction are very close to $K = 1$. Suggest what error the students must have made and calculate the actual values of $K$ for these reactions.

**It is probably an error in conversion of energy units. The Gibb's free energy is given in units of kJ, but the value of $R$, the gas constant, is usually 8.314 J/mol·K. If that error is made, not much difference is noted in the value of $K$ calculated. $\Delta G$ should be expressed in J/mol if $R = 8.314$ J/mol·K.**

Problem 12.89 Ductile materials, like metal wires, are quite different from brittle materials like concrete. If you have a piece of wire, and no wire cutters, how could you break it into two pieces? How does this differ from breaking a piece of brittle material like glass?

**By bending the wire repeatedly (cyclic loadings) in the place where you want to break it, the metal will eventually fatigue and break. In metal fatigue, micro-cracks form and eventually lead to failure. In a brittle material, macroscopic cracks form and lead to rapid failure.**

Problem 12.91 The monomer of polyvinyl alcohol was describe in Example problem 7.6. (a) Depict a small section of the resulting polymer as a line diagram.

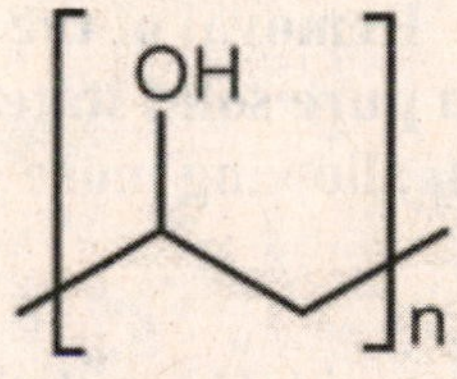


(b) Based on your understanding of intermolecular forces, how might different polymer molecules that are adjacent to each other in a PVA fiber interact with each other?

**The presence of the hydroxyl group (–OH) means hydrogen bonding forces will exist between adjacent PVA molecules. This will result in relatively strong attractive forces between these molecules.**

Problem 12.93 Solid $CaCO_3$ is placed in a closed container and heated to 800°C. What is the equilibrium concentration of $CO_2$ in the following equilibrium, for which $K = 2.5 \times 10^{-3}$?

$$CaCO_3(s) \leftrightarrows CaO(s) + CO_2(g)$$

**This is an equilibrium problem. We'll use the three-step method introduced in Problem 12.37. Calcium carbonate and calcium oxide are solids and don't affect our equilibrium. Carbon dioxide begins at 0.00 M and increases by *X* mol/L.**

| | $CaCO_3(s)$ ⇆ | $CaO(s)$ + | $CO_2(g)$ |
|---|---|---|---|
| **Initial** | --- | --- | **0.00 M** |
| **Change** | $-X$ | $+X$ | $+X$ |
| **Equilibrium** | --- | --- | $X$ |

**The equilibrium constant expression is:** $K = \dfrac{[CO_2][1]}{[1]} = 2.5 \times 10^{-3}$.

**Substituting for *X*:** $[X] = 2.5 \times 10^{-3}$ **M**

**The equilibrium concentration of carbon dioxide is $[CO_2] = 2.5 \times 10^{-3}$ M.**

Problem 12.95 An engineer working on a design to extract petroleum from a deep thermal reservoir wishes to capture toxic hydrogen sulfide gases present by reaction with aqueous iron(II)nitrate to form solid iron(II)sulfide. (a) Write the chemical equation for this process, assuming that it reaches equilibrium. (b) What is the equilibrium constant expression for this system? (c) How can the process be manipulated so that it does not reach equilibrium, allowing the continuous removal of hydrogen sulfide?

**(a)** $Fe(NO_3)_2(aq) + H_2S(g) \leftrightarrows FeS(s) + 2\ HNO_3(aq)$

**(b)** $K = \dfrac{[1][HNO_3]^2}{[Fe(NO_3)_2]P_{H2S}}$

**(c) By removing a product, the forward reaction will be favored. Removal of the solid FeS, however, would not affect the equilibrium because it is a pure solid state. The nitric acid could be neutralized to remove it from the solution, allowing more gaseous $H_2S$ to react.**

Problem 12.97 Methanol, $CH_3OH$, can be produced by the reaction of CO with $H_2$, with the liberation of heat. All species in the reaction are gaseous. What effect will each of the following have on the equilibrium concentration of CO?

**It would probably be helpful to write an equation:** $CO(g) + 2\ H_2(g) \leftrightarrows CH_3OH(g)$

**Recall that the basic idea of Le Chatelier's Principle is that the equilibrium system will respond in the opposite manner to whatever change we make.**

(a) pressure is increased **Will shift the equilibrium toward the side with fewer molecules of gas: shifts toward the products.**

(b) volume of the reaction container is decreased **Will shift the equilibrium toward the side with fewer molecules of gas: shifts towards the products.**

(c) heat is added **Since heat is liberated, this is an exothermic reaction. Heat is a product so adding heat causes a shift to the reactants.**

(d) the concentration of CO is increased **Increasing the amount of a reactant causes a shift towards the products.**

(e) some methanol is removed from the container **Decreasing the amount of a product causes a shift towards the products.**

(f) $H_2$ is added **Increasing the amount of a reactant will cause a shift towards the products.**

Problem 12.99 Using the kinetic-molecular theory, explain why an increase in pressure produces more $N_2O_4$ in the following system:

$$N_2O_4(g) \leftrightarrows 2\,NO_2(g)$$

**An increase in pressure means that molecules are colliding more often, so this is the change we need to consider with respect to Le Chatelier's principle. The system wants to "undo" this change, causing the molecules to collide less frequently. The only way this can be accomplished is for fewer molecules of gas to be present. The reaction speeds up in the direction that produces fewer molecules of gas, the reverse direction in our example. This occurs until there are fewer molecules present and equilibrium is re-established, producing more $N_2O_4$ and less $NO_2$ at the new equilibrium position.**

Problem 12.101 ■ To prepare community wastewater to be released to a lake, phosphate ions (among others) must be removed. $Al_2(SO_4)_3$ can be used for this purpose, precipitating $AlPO_4$, for which $K_{sp}$ at 25°C is $1.3 \times 10^{-20}$. What mass of $Al_2(SO_4)_3$ must be added per gallon of waste water to assure that the concentration of $PO_4^{3-}$ in the water released to the lake will be no higher than $1.0 \times 10^{-12}$ M?

**We need to calculate what concentration of $Al^{3+}$ would be needed to be in equilibrium with $1.0 \times 10^{-12}$ M $PO_4^{3-}$.**

$$AlPO_4\,(s) \leftrightarrows Al^{3+}\,(aq) + PO_4^{3-}\,(aq) \qquad K_{sp} = [Al^{3+}][PO_4^{3-}] = 1.3 \times 10^{-20}$$

$$[Al^{3+}] = \frac{1.3\times10^{-20}}{[PO_4^{\ 3-}]} = \frac{1.3\times10^{-20}}{1.0\times10^{-12}} = 1.3 \times 10^{-8} \text{ M } Al^{3+}$$

**To maintain no more than 1.0 × $10^{-12}$ M $PO_4^{3-}$, the concentration of $Al^{3+}$ must be no less than 1.3 × $10^{-8}$ M.**

**Now we need to find the mass of $Al_2(SO_4)_3$ needed per gallon of water to achieve this concentration:**

$$\frac{1.3\times10^{-8}\text{ mol Al}^{3+}}{1\text{ L}} \times \frac{1\text{ mol Al}_2(\text{SO}_4)_3}{2\text{ mol Al}^{3+}} \times \frac{342.15\text{ g Al}_2(\text{SO}_4)_3}{1\text{ mol Al}_2(\text{SO}_4)_3} \times \frac{0.9463\text{ L}}{1\text{ quart}} \times \frac{4\text{ quarts}}{1\text{ gallon}} =$$

**= 8.4 × $10^{-6}$ g $Al_2(SO_4)_3$ per gallon**

Problem 12.103 Copper(II) iodate has a solubility of 0.136 g per 100 g of water. Calculate its molar solubility in water and its $K_{sp}$.

**See Problem 12.65. We'll use the solubility to determine the molar solubility. From this we can solve for the equilibrium concentrations and calculate $K_{sp}$.**

$$\frac{0.136\text{ g Cu(IO}_3)_2}{100\text{ g H}_2\text{O}} \times \frac{1\text{ mol Cu(IO}_3)_2}{413.34\text{ g Cu(IO}_3)_2} \times \frac{1000\text{ g}}{\text{L H}_2\text{O}} = 3.29 \times 10^{-3}\text{ mol/L Cu(IO}_3)_2$$

**This is the molar solubility.**

**$Cu(IO_3)_2$ (s) ⇆ $Cu^{2+}$(aq) + 2 $IO_3^-$ (aq)**

| | | | |
|---|---|---|---|
| **Initial** | **---** | **0.00 M** | **0.00 M** |
| **Change** | **$-X$** | **$+X$** | **$+2X$** |
| **Equilibrium** | **---** | **$X$** | **$2X$** |

**$K_{sp}$ = $[Cu^{2+}][IO_3^-]^2 = [X][2X]^2$.  Recall $X$ = molar solubility = 3.29 × $10^{-3}$, Therefore,**

$$K_{sp} = [X][2X]^2 = [3.29 \times 10^{-3}][2(3.29 \times 10^{-3})]^2 = 1.42 \times 10^{-7}$$

Problem 12.105 You have three white solids. What experiment could you carry out to rank them in order of increasing solubility?

**Start with equal masses of each solid in separate containers. Start adding equal volumes of water to each, maybe drop-wise or milliliter increments. The first solid to completely dissolve is the most soluble, the last one to completely dissolve the least soluble. You could also estimate the solubility of each solid using the mass and volume of water required to dissolve it.**

Problem 12.107 You find a bottle of acid in the laboratory and it is labeled $H_2C_2O_4$. There is no concentration given and you don't have the reagents to carry out a titration. What other measurement could be made and what information would you need to look up to determine the concentration of the acid?

**You could measure the pH of the solution (a pH meter will give a very accurate number). From the pH, the concentration of $H_3O^+$ can be found. By looking up the ionization constant ($K_a$) of the acid, you can perform an equilibrium calculation to find the concentration of the acid.**

Problem 12.109 A mixture of 1.00 mole of $PCl_3$ and 2.00 moles of $PBr_3$ is placed in a 1.00-L reaction vessel. At the temperature of the experiment, $K = 2.0$ for the following reaction:

$$PCl_3 + PBr_3 \leftrightarrows PCl_2Br + PClBr_2$$

(a) What is the equilibrium concentration of each species in the vessel?

**We are given $K$ and the initial concentrations (divide moles given by 1.00 liters), we can solve for equilibrium concentrations. The reactants will decrease by $X$ mol/L and the products will increase by $X$ mol/L.**

| | $PCl_3$ + | $PBr_3$ ⇆ | $PCl_2Br$ + | $PClBr_2$ |
|---|---|---|---|---|
| **Initial** | **1.00 M** | **2.00 M** | **0.00 M** | **0.00 M** |
| **Change** | $-X$ | $-X$ | $+X$ | $+X$ |
| **Equilibrium** | **1.00 –** $X$ | **2.00 –** $X$ | $X$ | $X$ |

**The equilibrium constant expression is:** $K = \dfrac{[PCl_2Br][PClBr_2]}{[PCl_3][PBr_3]}$**, substituting,**

$K = \dfrac{[X]^2}{[1.00 - X][2.00 - X]} = 2.0.$ **This is another problem where we need to use the quadratic formula:** $X = \dfrac{-b \pm \sqrt{b^2 - 4ac}}{2a}$

**Rearranging our equation:** $X^2 = 2.0\,(2 - 3X + X^2) = 4.0 - 6X + 2X^2$

$0 = 4.0 - 6X + X^2, \quad X = \dfrac{-(-6) \pm \sqrt{(-6)^2 - 4(1)(4)}}{2(1)} = \dfrac{6 \pm \sqrt{20}}{2} = 0.76 \text{ or } 5.2$

**$X$ must be 0.76 or else we would have negative concentration.**

**The equilibrium concentrations are:** $[PCl_2Br] = [PClBr_2] = X = 0.76$ **M**
$[PCl_3] = [1.00 - X] = 1.00 - 0.76 = 0.24$ **M**

$$[PCl_3] = [2.00 - X] = 2.00 - 0.76 = 1.24\ \text{M}$$

(b) Because the number of bonds is the same on both side of the equation, this reaction should be close to thermoneutrality, meaning that it is neither exothermic nor endothermic. What drives this reaction?

**Since the change in energy is not driving the reaction to occur, it must be the other factor that influences spontaneous change: entropy.**

Problem 12.111 ■ The vapor pressure of water at 80.0°C is 0.467 atm. Find the value of $K_c$ for the process

$$H_2O(\ell) \leftrightarrows H_2O(g)$$

at this temperature.

**Write the $K_p$ expression for the reaction and plug in the known value. Then use the relationship between $K_p$ and $K_c$ (Equation 12.4), with $R = 0.08206\ \frac{L \cdot atm}{mol \cdot K}$, and $\Delta n$ = change in the number of moles of gas in the reaction, to get $K_c$.**

$$H_2O(\ell) \leftrightarrows H_2O(g) \qquad K_p = P_{H2O(g)} = 0.467\ \text{atm}$$

$$T = 80.\ °C + 273.15 = 353\ K$$

$$\Delta n = 1\ \text{mol } H_2O \text{ gas product} - 0 \text{ mol gas reactants} = 1$$

$$K_p = K_c(RT)^{\Delta n} \qquad\qquad K_p\,(RT)^{-\Delta n} = K_c$$

$$K_c = (0.467\ \text{atm}) \times \left\{\left(0.08206\frac{L \cdot atm}{mol \cdot K}\right) \times (353K)\right\}^{-1} = 0.0161$$

## CHAPTER OBJECTIVE QUIZ

This quiz will test your understanding of the basic text chapter objectives and give you additional practice problems. You should work this quiz after completing the end-of-chapter questions. The solutions to these questions are found at the end of this chapter.

1. TRUE or FALSE: Portland cement contains CaO, whose formation from $CaCO_3$ releases tremendous amounts of carbon dioxide.

2. TRUE or FALSE: *Admixtures* are components added to concrete to manipulate the concrete into having desired properties.

3. Which of the following statements about dynamic equilibrium is **false**?
   a. When equilibrium is established, the reaction has stopped.
   b. When equilibrium is established, the forward reaction rate and the reverse reaction rate are equal.
   c. Once equilibrium is established, the concentration of products will not change.
   d. After equilibrium is established, the reactants continue to form products.
   e. A reversible reaction can begin with the reverse reaction rate faster than the forward rate.

4. Write the equilibrium constant expressions for the following reactions:

   (a) $2\ NO(g) + O_2(g) \leftrightarrows 2\ NO_2(g)$

   (b) $Al(OH)_3(s) + OH^-(aq) \leftrightarrows Al(OH)_4^-(aq)$

   (c) $HOCl(aq) + H_2O(\ell) \leftrightarrows H_3O^+(aq) + OCl^-(aq)$

5. Identify whether the following equilibrium constants represent an equilibrium position favoring the reactants or products.
   (a) $K = 2.3 \times 10^4$ (b) $K = 0.11$ (c) $K = 63.1$ (d) $K = 7.7 \times 10^{-6}$

6. After equilibrium is established in this reaction, the following concentrations are measured:

   $2\ A(g) + C_2(g) \leftrightarrows A_2C_2(g)$

   $[A] = 2.3$ M, $[C_2] = 1.5$ M, $[A_2C_2] = 0.32$ M

   What is the value of the equilibrium constant? What are the units of the equilibrium constant?

7. For the following reaction at 500 K, $K = 23.4$.

   $H_2(g) + Br_2(g) \leftrightarrows 2\ HBr(g)$
   If 2.2 M HBr is placed in a 1.0 L reactor, at 500 K, what are the equilibrium concentrations of all species?

8. When 0.57 M $PCl_5$ is placed in a 5.0 L reactor at 350 K, what will be the equilibrium concentrations of all species?
   $PCl_5(g) \leftrightarrows PCl_3(g) + Cl_2(g)$ $K = 0.14$ @ 350K

9. What is the molar solubility of $CaF_2$? $K_{sp} = 3.9 \times 10^{-11}$

10. What is the $K_{sp}$ of $BaSO_4$? One liter of water contains 0.0025 g dissolved $BaSO_4$.

11. Which of the following statements about acids and bases is **false**?
    a. An acid donates a proton ($H^+$) to a base.
    b. An acid donates a proton to a base.
    c. pH is defined as pH = $-\log [H_3O^+]$
    d. Higher pH values represent more acidic solutions.
    e. The conjugate acid of $NH_3$ is $NH_4^+$.

12. What is the concentration of $[H_3O^+]$ if the pH of a solution is 11.2?

13. What is the pH of 0.0567 M $HNO_3$?

14. Write an equilibrium constant expression for the ionization of (a) formic acid, HCOOH, (b) methylamine, $CH_3NH_2$.

15. Calculate the pH of (a) 0.080 M formic acid, (b) 0.080 M methylamine.

16. Predict whether the following changes will shift the reactions equilibrium position to the reactants, products, or no change.

$$N_2(g) + 3\ H_2(g) \leftrightarrows 2\ NH_3(g) \quad \Delta H = -92\ kJ/mol$$

(a) Add more $N_2$ (b) Add more $H_2$ (c) Add more $NH_3$
(d) Remove $H_2$ (d) Increase temperature (f) Increase pressure
(g) Add a catalyst

17. When the following reaction is in equilibrium at 408 K, the concentrations are:

[A] = 1.1 M, [B] = 2.1 M, [C] = 4.3 M $A(g) + (B) \leftrightarrows C(g)$

If 0.50 M of C is added, what will be the new equilibrium concentrations?

18. Use thermodynamic data to calculate the equilibrium constant for:
$C(s) + H_2O(g) \leftrightarrows CO(g) + H_2(g)$ @ 298 K

19. ECC concrete ("bendable concrete") has a strain capacity of only about 5%. Why is this number significant?

## ANSWERS TO THE CHAPTER OBJECTIVE QUIZ

1. TRUE

2. TRUE

3. a

4. (a) $K = \dfrac{[NO_2]^2}{[NO]^2[O_2]}$ (b) $K = \dfrac{[Al(OH)_4^-]}{[1][OH^-]}$ (recall solids are assigned concentration = 1)

   (c) $K_1 = \dfrac{[H_3O^+][OCl^-]}{[HClO][1]}$ (recall liquids are assigned concentration = 1)

5. When $K > 1$, the equilibrium position favors the products; when $K < 1$ it favors the reactants.
   (a) $K = 2.3 \times 10^4$, product favored (b) $K = 0.11$, reactant favored
   (c) $K = 63.1$, product favored (d) $K = 7.7 \times 10^{-6}$, reactant favored

6. The equilibrium constant is defined in terms of equilibrium concentration. The molarity values are used in the calculation but the units are omitted. Thus, the equilibrium constant is unitless.

   $$K = \frac{[A_2C_2]}{[A]^2[C_2]} = \frac{[0.32]}{[2.3]^2[1.5]} = 0.040$$

7. The initial concentrations of both reactants zero. Since $H_2$ and $Br_2$ start out at zero, they must increase (by $X$ mol/L), HBr will decrease by $2X$ mol/L.

   $H_2(g) + Br_2(g) \leftrightarrows 2\,HBr(g)$

| | $H_2(g)$ | $Br_2(g)$ | $2\,HBr(g)$ |
|---|---|---|---|
| Initial | 0.0 M | 0.0 M | 2.2 M |
| Change | $+X$ | $+X$ | $-2X$ |
| Equilibrium | $X$ | $X$ | $2.2 - 2X$ |

The equilibrium constant expression is: $K = \dfrac{[HBr]^2}{[H_2][Br_2]}$, substituting,

$$K = \frac{[2.2 - 2X]^2}{[X][X]} = 23.4.$$

Since we have squares in the numerator and denominator on the left, we can take the square root of both sides to simplify our solution.

$$K = \sqrt{\frac{[2.2-2X]^2}{[X]^2}} = \sqrt{23.4}, \quad \frac{[2.2-2X]}{[X]} = 4.84$$

**Solving for $X$,**

**$2.2 - 2X = 4.84\,X$, $\quad 2.2 = 6.84\,X$, $\quad X = 0.32$ M.**

**Using X to solve for the equilibrium concentrations,**

**$[HBr] = 2.2 - 2X = 2.2 - 2(0.32\text{ M}) = 1.56$ M $\quad [H_2] = [O_2] = X = 0.32$ M**

8. **The initial concentrations of both reactants are zero. The concentration of $PCl_5$ will decrease by $X$ mol/L, the concentrations of $PCl_3$ and $Cl_2$ will increase by $X$ mol/L.**

| | $PCl_5(g)$ ⇆ | $PCl_3(g)$ + | $Cl_2(g)$ |
|---|---|---|---|
| Initial | 0.57 M | 0.0 M | 2.2 M |
| Change | $-X$ | $+X$ | $+X$ |
| Equilibrium | $0.57 - X$ | $X$ | $X$ |

**The equilibrium constant expression is:** $K = \dfrac{[PCl_3][Cl_2]}{[PCl_5]}$,

**substituting,** $K = \dfrac{[X][X]}{[0.57 - X]} = 0.14.$

**This equation will have to be solved using the quadratic formula. We'll rearrange it until it is in the form of a quadratic expression: $aX^2 + bX + c = 0$, then use the quadratic formula:** $X = \dfrac{-b \pm \sqrt{b^2 - 4ac}}{2a}$. **Rearranging the equation,**

$(X)(X) = (0.57 - X)(0.14) = 0.0798 - 0.14\,X.$

**Collecting common terms,** $X^2 + 0.14\,X - 0.0798 = 0.$

**a = 1, b = 0.14, c = – 0.0798,**

**Using the quadratic formula:** $X = \dfrac{-0.14 \pm \sqrt{(0.14)^2 - 4(1)(-0.0798)}}{2(1)} =$

$\dfrac{0.14 \pm 0.582}{2}$

**$X = 0.221$ or $-0.361$; $X$ cannot equal $-0.361$ because it would give negative concentration.**

**The equilibrium concentrations are: $[PCl_3] = [Cl_2] = X = 0.221$ M,**
**$[PCl_5] = 0.57 - X = 0.57 - 0.221 = 0.349$ M.**

**9. Recall that the molar solubility is equal to the $X$ mol/L when we solve for equilibrium concentrations.**

**(a)** $CaF_2(s) \leftrightarrows Ca^{2+}(aq) + 2\,F^-(aq)$

| | $CaF_2(s)$ | $Ca^{2+}(aq)$ | $2\,F^-(aq)$ |
|---|---|---|---|
| Initial | --- | 0.00 M | 0.00 M |
| Change | $-X$ | $+X$ | $+2X$ |
| Equilibrium | --- | $X$ | $2X$ |

$K_{sp} = [Ca^{2+}][F^-]^2 = 3.9 \times 10^{-11}$, **substituting for $X$,** $[X][2X]^2 = 3.9 \times 10^{-11}$.

**Solving for $X$:** $4X^3 = 3.9 \times 10^{-11}$ $X = \sqrt[3]{9.75 \times 10^{-12}} = 2.1 \times 10^{-4}$ M

**The molar solubility in pure water is $2.1 \times 10^{-4}$ M.**

**10. We'll use the solubility to determine the molar solubility. From this we can solve for the equilibrium concentrations and calculate $K_{sp}$.**

$$\frac{0.0025\text{ g BaSO}_4}{1\text{ L H}_2\text{O}} \times \frac{1\text{ mol BaSO}_4}{233.4\text{ g BaSO}_4} = 1.1 \times 10^{-5}\text{ mol/L BaSO}_4$$

$BaSO_4(s) \leftrightarrows Ba^{2+}(aq) + SO_4^{2-}(aq)$

| | $BaSO_4(s)$ | $Ba^{2+}(aq)$ | $SO_4^{2-}(aq)$ |
|---|---|---|---|
| Initial | --- | 0.00 M | 0.00 M |
| Change | $-X$ | $+X$ | $+X$ |
| Equilibrium | --- | $X$ | $X$ |

$K_{sp} = [Ba^{2+}][SO_4^{2-}] = [X][X]$ **Recall X = molar solubility = $1.1 \times 10^{-5}$ M, therefore,**

$K_{sp} = [X][X] = [1.1 \times 10^{-5}][1.1 \times 10^{-5}] = 1.1 \times 10^{-10}$.

**11. d**

**12. Using the definition of pH, the equation can be rearranged to solve for $[H_3O^+]$:**
**pH = $-\log[H_3O^+]$, $[H_3O^+] = 10^{-pH} = 10^{-11.2} = 6.3 \times 10^{-12}$ M.**

**13. Nitric acid is a *strong* acid as opposed to a *weak* one. This means that it will be 100% ionized in solution. So, 0.0567 M $HNO_3$ will produce 0.0567 M $[H^+]$.**
**pH = $-\log[0.0567] = 1.246$**

**14. Weak acids donate a proton, $H^+$, to a base (water when alone in solution according to the Brønsted-Lowry theory. This produces a hydronium ion and an anion.**

**(a) HCOOH:** $HCOOH(aq) + H_2O(\ell) \leftrightharpoons HCOO^-(aq) + H_3O^+(aq)$

$$K_a = \frac{[HCOO^-][H_3O^+]}{[HCOOH]}$$

**Weak bases accept a proton from an acid (water when alone in solution) according to the Brønsted-Lowry theory. This produces a hydroxide ion and a cation.**

**(b) $CH_3NH_2$:** $CH_3NH_2(aq) + H_2O(\ell) \leftrightharpoons CH_3NH_3^+(aq) + OH^-(aq)$

$$K_b = \frac{[CH_3NH_3^+][OH^-]}{[CH_3NH_2]}$$

**15. (a) To answer this problem, we must solve for the equilibrium concentrations in this solution. The initial concentration is given; we must look up the ionization constant (equilibrium constant) for formic acid. From Table 12.6 in the textbook, $K_a = 1.8 \times 10^{-4}$.**

| | $HCOOH(aq)$ | $+ H_2O(\ell) \leftrightharpoons$ | $HCOO^-(aq)$ | $+ H_3O^+(aq)$ |
|---|---|---|---|---|
| Initial | 0.08 M | | 0.00 M | 0.00 M |
| Change | $-X$ | | $+X$ | $+X$ |
| Equilibrium | $0.08 - X$ | | $X$ | $X$ |

$K_a = \dfrac{[HCOO^-][H_3O^+]}{[HCOOH]} = 1.8 \times 10^{-4}$, **substituting $X$ into the equation:**

$K_a = \dfrac{[X][X]}{[0.08 - X]} = 1.8 \times 10^{-4}$. **Assume that $0.08 - X \approx 0.08$,** $K_a = \dfrac{[X][X]}{[0.08]} = 1.8 \times 10^{-4}$,

$X^2 = 1.4 \times 10^{-5}$, $X = 3.8 \times 10^{-3}$ M $= [H_3O^+] = [HCOO^-]$.

**The pH = $-\log [H_3O^+] = -\log [3.8 \times 10^{-3}] = 2.4$**

**(b) We'll solve this problem just like the previous one, with the exception of the reaction will be that of a base, and the equilibrium concentrations will give us $OH^-$ not $H_3O^+$. From Appendix G in the textbook, $K_b = 5.0 \times 10^{-4}$.**

| | $CH_3NH_2$ | $+ H_2O \leftrightharpoons$ | $CH_3NH_3^+$ | $+ OH^-$ |
|---|---|---|---|---|
| Initial | 0.080 M | | 0.00 M | 0.00 M |
| Change | $-X$ | | $+X$ | $+X$ |
| Equilibrium | $0.080 - X$ | | $X$ | $X$ |

$K_b = \dfrac{[CH_3NH_3^+][OH^-]}{[CH_3NH_2]} = 5.0 \times 10^{-4}$, **substituting $X$ into the equation:**

$$K_b = \frac{[X][X]}{[0.080 - X]} = 5.0 \times 10^{-4}.$$ Assume that $0.080 - X \approx 0.080$,

$$K_b = \frac{[X][X]}{[0.080]} = 5.0 \times 10^{-4}, \quad X^2 = 4.0 \times 10^{-5}, \quad X = 6.3 \times 10^{-3}\ \text{M} = [OH^-] = [CH_3NH_3^+].$$

We need to calculate the $H_3O^+$ to find pH.

$$[H_3O^+][OH^-] = K_w = 1.0 \times 10^{-14}, \quad [H_3O^+] = \frac{[K_w]}{[OH^-]} = \frac{1.0 \times 10^{-14}}{6.3 \times 10^{-3}} = 1.6 \times 10^{-12}\ \text{M}$$

The pH = $-\log [H_3O^+] = -\log [1.6 \times 10^{-12}] = 11.8$.

16. (a) Adding more reactants causes an equilibrium shift to the products.
(b) Adding more reactants causes an equilibrium shift to the products.
(c) Adding more products causes an equilibrium shift to the reactants.
(d) Removing reactants causes an equilibrium shift to the reactants.
(e) Increasing temperature for an exothermic reaction cause a shift to the reactants.
(f) Increasing pressure will cause a shift in the direction of fewer moles of gas, which in this case is the direction of the products.
(h) Adding a catalyst will not change the equilibrium position. The reaction rates in both the forward and reverse directions speed up equally.

17. The first step in solving this problem is calculating the equilibrium constant using the original equilibrium concentrations.

$$K = \frac{[C]}{[A][B]} = \frac{[4.3]}{[1.1][2.1]} = 1.9.$$

The system is in equilibrium until the additional 0.50 M C is added; causing a shift to the reactants as the system re-establishes equilibrium. $X$ mol/L additional A and B will form and $X$ mol/L C will react. The value of $K$ doesn't change.

| | A(g) + | B(g) ⇆ | C(g) |
|---|---|---|---|
| Original Equilibrium | 1.1 M | 2.1 M | 4.3 M |
| Initial | 1.1 | 2.1 | 4.8 |
| Change | $+X$ | $+X$ | $-X$ |
| New Equilibrium | $1.1 + X$ | $2.1 + X$ | $4.8 - X$ |

Substituting into the equilibrium constant expression, $$K = \frac{[4.8 - X]}{[1.1 + X][2.1 + X]} = 1.9$$

This equation will have to be solved using the quadratic formula. We'll rearrange it until it is in the form of a quadratic expression: $aX^2 + bX + c = 0$, then use the quadratic formula: $X = \frac{-b \pm \sqrt{b^2 - 4ac}}{2a}$. Rearranging the equation,

$(4.8 - X) = 1.9(1.1 + X)(2.1 + X) = 1.9(2.31 + 3.2X + X^2) = 4.39 + 6.08X + 1.9X^2$

Collecting common terms, $0 = 1.9X^2 + 7.08X - 0.41$,

a = 1.9, b = 7.08, c = – 0.41,

Using the quadratic formula: $X = \frac{-7.08 \pm \sqrt{(7.08)^2 - 4(1.9)(-0.41)}}{2(1.9)} =$

$\frac{7.08 \pm 7.30}{3.8}$

X = 0.058 or – 3.8; X cannot equal – 3.8 because it would give negative concentration.

The equilibrium concentrations are:

$[A] = 1.1 + X = 1.1 + 0.058 = 1.16$ M

$[B] = 2.1 + X = 2.1 + 0.058 = 2.16$ M

$[C] = 4.8 - X = 4.8 - 0.058 = 4.74$ M

18. From the thermodynamic data in Appendix E, the Gibb's free energy change of the reaction can be calculated. From $\Delta G^o_{rxn}$, the equilibrium constant can be found.

$$\Delta G^o_{rxn} = \sum n\Delta G_f^{\,o}{}_{products} - \sum n\Delta G_f^{\,o}{}_{reactants}$$

$$\Delta G^o_{rxn} = [(\Delta G_f^o{}_{CO}) + (\Delta G_f^o{}_{H2})] - [(\Delta G_f^o{}_{C}) + (\Delta G_f^o{}_{H2O})]$$

$$= [(-137.2 \text{ kJ/mol}) + 0] - [0 + (-228.6 \text{ kJ/mol})]$$

$$\Delta G^o_{rxn} = 91.4 \text{ kJ/mol}$$

$$\Delta G = -RT(\ln K), \quad \ln(K) = \frac{-\Delta G}{RT} = \frac{-(91{,}400 \text{ J/mol})}{(8.314 \text{ J/mol}\cdot\text{K})(298 \text{ K})} = -36.9$$

$$K = e^{-36.9} = 9.5 \times 10^{-17}$$

19. The 5% number is significant because it represents such a large increase over traditional concrete (0.01% strain capacity). This puts ECC concrete closer (a little bit) to metals in terms of its strain capacity. In comparison, some metals can achieve up to 100% "strain to rupture".

# CHAPTER 13

## *Study Goals*

The study goals outline specific concepts to be mastered in this section of the text chapter. Related problems at the end of the text chapter will also be noted. Working the questions noted should aid in mastery of each study goal and also highlight any areas that you may need additional help in.

**Section 13.1 *INSIGHT INTO* Corrosion**

1. Understand the type of chemical reaction involved in corrosion. *Work Problem 1.*
2. List several types of corrosion and where they might occur. *Work Problems 2–4.*

**Section 13.2 Oxidation–Reduction Reactions and Galvanic Cells**

3. Review the terms and concepts associated with oxidation-reduction reactions and recognize the relationship between electrochemical cells and oxidation-reduction reactions. *Work Problems 5, 6.*
4. Diagram a typical galvanic cell, identifying all key components. *Work Problems 7–10, 14, 51.*
5. Express the reaction in a galvanic cell using standard cell notation. *Work Problems 11–13, 17.*
6. Compare and contrast galvanic and uniform corrosion. *Work Problems 15, 16, 35, 36.*

**Section 13.3 Cell Potentials**

7. Be familiar with the standard reduction potential table (T13.1 or Appendix I), and all the information that may be obtained from it, including $E^{\circ}_{red}$, $E^{\circ}_{ox}$, strength as oxidizing agent, and strength as reducing agent. *Work Problems 19, 21–24, 28, 33, 34.*
8. Use the standard reduction potential table to calculate standard cell potential for galvanic cells. *Work Problems 18, 20, 25–27, 98.*
9. Describe how galvanic cells are used in the cathodic protection method of corrosion prevention. *Work Problems 52 – 55, 100, 101.*
10. Use the Nernst equation to calculate cell potentials under non-standard conditions. *Work Problems 29–32, 97.*

**Section 13.4 Cell Potentials and Equilibrium**

11. Relate the standard cell potential to Gibb's free energy change and the equilibrium constant. Determine $\Delta G^{\circ}$ and $K$ from the standard cell potential. *Work Problems 37–50, 99.*

**Section 13.5 Batteries**

12. Discuss the design and reactions of several common primary and secondary batteries. *Work Problems 52–63.*

**Section 13.6 Electrolysis**

13. Contrast electrolytic cells with galvanic cells and be able to diagram a typical electrolytic cell. *Work Problems 65, 66.*
14. Describe the passive electrolysis used in the refining of aluminum. *Work Problems 64, 67, 68.*
15. Describe the principles of active electrolysis and electroplating. *Work Problems 69–74.*

**Section 13.7 Electrolysis and Stoichiometry**

16. Relate the electricity used to the amount of product produced in electrolysis. Calculate mass of metal deposited in electroplating given the time and vice-versa. *Work Problems 75–87, 102–105.*

**Section 13.8 *INSIGHT INTO* Batteries in Engineering Design**

17. Describe the basic principles in the design of the lithium-ion battery. *Work Problems 88–96.*

## *Solutions to the Odd-Numbered Problems*

Problem 13.1 When you look at several older cars that are showing signs of rust formation, where do you expect to find the most rust? What does this observation imply about conditions that lead to corrosion?

**Usually the rust begins around an area of exposed metal where the paint has been chipped off or around the wheel well under the car. This is evidence of two requirements for corrosion to occur: oxygen and water must contact the metal. Presence of an electrolyte (salt) will accelerate the process by increasing conductivity. This is why cars from geographic regions where salt is used to remove snow from roads generally rust more quickly than those from warmer climates.**

Problem 13.3 Using the Internet, find three cases where corrosion was the cause of some sort of failure or malfunction of a device or structure.

**Answers will vary according to the websites found.**

**Some of the many examples of corrosion that can be quickly found are: Corrosion failure of stainless steel surgical implants due to adverse tissue reaction, corrosion of wastewater treatment piping due to microbiologically induced corrosion, and corrosion failures of stainless steel valve stems in nuclear power plants due to improper tempering of the 410 stainless steel.**

Problem 13.5 For the following oxidation-reduction reactions, identify the half reactions and label them as oxidation or reduction.

**Oxidation is when atoms lose one or more electrons. Reduction is when atoms gain one or more electrons.**

(a) $Cu(s) + Ni^{2+}(aq) \rightarrow Ni(s) + Cu^{2+}(aq)$ **oxidation: $Cu(s) \rightarrow Cu^{2+}(aq) + 2\ e^-$**

**reduction: $Ni^{2+}(aq) + 2\ e^- \rightarrow Ni(s)$**

(b) $2\ Fe^{3+}(aq) + 3\ Ba(s) \rightarrow 3\ Ba^{2+}(aq) + 2\ Fe(s)$ **oxidation: $Ba(s) \rightarrow Ba^{2+}(aq) + 2\ e^-$**

**reduction: $Fe^{3+}(aq) + 3\ e^- \rightarrow Fe(s)$**

Problem 13.7 What is the role of the salt bridge in the construction a galvanic cell?

**The salt bridge has several functions. It allows the two half-cells to be in electrical contact but prevents mixing of the two solutions. Most importantly, the salt bridge allows charge neutrality to be maintained as the oxidation-reduction reaction takes place.**

Problem 13.9 If a salt bridge contains $KNO_3$ as the electrolyte, which ions diffuse into solution in the anode compartment of the galvanic cell? Explain your answer.

**In the anode compartment, electrons are lost when oxidation occurs. This usually results in cations being formed, creating extra positive charges. The primary role of the salt bridge is to maintain charge balance, so the negative nitrate ion ($NO_3^-$) diffuses into the anode compartment to maintain charge neutrality.**

Problem 13.11 ■ The following oxidation-reduction reactions are used in electrochemical cells. Write them using cell notation.

**In the "shorthand" notation for galvanic cells, we represent the anode reaction on the left and the cathode reaction on the right. We use the symbol ‖ to represent the salt bridge that separates the half-cells. The anode reactant and anode product are shown on the left (separated by a single | ) and the cathode reactant and cathode product are shown on the right (separated by a | ). For example:**

**anode reactant | anode product ‖ cathode reactant | cathode product.**

**Standard cell concentration is 1.0 M; if the concentration is something other than standard, then it is written after the symbol.**

(a) $2\ Ag^+(aq)(0.50\ M) + Ni(s) \rightarrow 2\ Ag(s) + Ni^{2+}(aq)(0.20\ M)$

**$Ni(s) \mid Ni^{2+}(aq)(0.20\ M) \parallel Ag^+(aq)(0.50\ M) \mid Ag(s)$**

(b) $Cu(s) + PtCl_6^{2-}(aq)(0.10\ M) \rightarrow$
$Cu^{2+}(aq)(0.20\ M) + PtCl_4^{2-}(aq)(0.10\ M) + 2\ Cl^-(aq)(0.40\ M)$

**$Cu(s) \mid Cu^{2+}(aq)(0.20\ M) \parallel PtCl_6^{2-}(aq)(0.10\ M), PtCl_4^{2-}(aq)(0.10\ M), Cl^-(aq)(0.40M) \mid Pt(s)$**

(c) $Pb(s) + SO_4^{2-}(aq)(0.30\ M) + 2\ AgCl(s) \rightarrow PbSO_4(s) + 2\ Ag(s) + 2\ Cl^-(aq)(0.20\ M)$

**$Pb(s) \mid PbSO_4(s) \mid SO_4^{2-}(aq)\ )(0.30\ M) \parallel Cl^-(aq)(0.20\ M) \mid AgCl(s) \mid Ag(s)$**

(d) In a galvanic cell, one half-cell contains 0.10 M HCl and a platinum electrode, over which $H_2$ is bubbled at a pressure of 1.0 atm. The other half-cell is composed of a zinc electrode in a 0.125 M solution of $Zn(NO_3)_2$.

**$Zn(s) \mid Zn^{2+}(aq)(0.125\ M) \parallel H^+(aq)(0.10\ M) \mid H_2(g)(1\ atm) \mid Pt(s)$**

Problem 13.13 For the reactions in (a) and (b) in the preceding problem, no anions at all are shown in the cell notation. Explain why this is not a concern.

**Anions are not involved in the oxidation-reduction reactions. They are present, as counter ions, but not changing (they are spectator ions). The reactions must show that charge is balanced but do not have to show both cations and ions.**

Problem 13.15 How does galvanic corrosion differ from uniform corrosion of iron?

**Corrosion of iron involves only one metal; the reduction reaction is usually water and oxygen reduced to hydroxide ion. Galvanic corrosion involves two metals in contact reacting. One of the metals has a more positive reduction potential and it will be the one that is reduced; the other with the more negative reduction potential will be oxidized.**

Problem 13.17 A student who has mercury amalgam fillings in some of her teeth is eating a piece of candy. She accidentally bites down on a piece of the aluminum foil wrapper and experiences a sharp sensation in her mouth. Explain what has happened in terms of electrochemistry.

**Mercury is a metal with a relatively positive reduction potential (for a metal). Aluminum is a metal with a large negative reduction potential, meaning it tends to be oxidized. When the student bit down on the aluminum foil, a galvanic cell was created with the aluminum acting as the anode and the mercury amalgam the cathode. A small electrical current was created, no doubt irritating the nerve of the tooth.**

Problem 13.19 If the reaction at a standard hydrogen electrode is $2\ H^+(aq) + 2\ e^- \rightarrow H_2$, why is there a platinum foil in the system?

**The platinum foil is an inert electrode. It does not actively participate in the reaction but provides a conduit for the flow of electrons and a surface for the reactions to occur.**

Problem 13.21 If the SHE was assigned a value of 3.00 V rather than 0.00 V, what would happen to all of the values listed in the table of standard reduction potentials?

**All the values would increase by 3.00 V but the order in the table would be exactly the same.**

Problem 13.23 In tables of standard reduction potentials that start from large positive values at the top and proceed through 0.0 V to negative values at the bottom, the alkali metals are normally at that bottom of the table. Use your chemical understanding of alkali metals and how they behave in bonding to explain why this is so.

**The alkali metals are very reactive metals; they have only one valence electron and it is easily removed to form a +1 cation. This means that alkali metals tend to be oxidized (lose electrons) easily and this will be reflected in a large negative reduction potential. Metals with large negative reduction potentials have large positive oxidation potentials.**

Problem 13.25 ■ Using values from the tables of standard reduction potentials, calculate the cell potential of the following cells.

**From the cell notation, half reactions can be written (See Problem 13.11). The reduction potential for the cathode reaction is recorded from Appendix I. The sign of the reduction potential of the anode reaction from Appendix I is changed to make it an oxidation potential. The reduction potential and the oxidation potentials of the cell are then added to give the overall cell potential: $E^o_{cell} = E^o_{red} + E^o_{ox}$.**

(a) $Ga(s) \mid Ga^{3+}(aq) \parallel Ag^{+}(aq) \mid Ag(s)$

**cathode: $Ag^{+}(aq) + e^{-} \rightarrow Ag(s)$** $\quad E^o_{red} = 0.7994$ V
**anode: $Ga(s) \rightarrow Ga^{3+}(aq) + 3e^{-}$** $\quad E^o_{ox} = -(-0.53\text{ V})$
$E^o_{cell} = 1.33$ V

(b) $Zn(s) \mid Zn^{2+}(aq) \parallel Cr^{3+}(aq) \mid Cr(s)$

**cathode: $Cr^{3+}(aq) + 3e^{-} \rightarrow Cr(s)$** $\quad E^o_{red} = -0.74$ V
**anode: $Zn(s) \rightarrow Zn^{2+}(aq) + 2e^{-}$** $\quad E^o_{ox} = -(-0.763\text{ V})$
$E^o_{cell} = 0.023\text{ V} \approx 0.02$ V

(c) $Fe(s), S^{2-}(aq) \mid FeS(s) \parallel Sn^{2+}(aq) \mid Sn(s)$

**cathode: $Sn^{2+}(aq) + 2e^{-} \rightarrow Sn(s)$** $\quad E^o_{red} = (-0.14$ V
**anode: $S^{2-}(aq) + Fe(s) \rightarrow FeS(s) + 2e^{-}$** $\quad E^o_{ox} = -(-1.01\text{ V})$
$E^o_{cell} = 0.87$ V

Problem 13.27 ■ One half-cell in a voltaic cell is constructed from a copper wire dipped into a $4.8 \times 10^{-3}$ M solution of $Cu(NO_3)_2$. The other half-cell consists of a zinc electrode in a 0.40 M solution of $Zn(NO_3)_2$. Calculate the cell potential.

**The Nernst equation is used to calculate nonstandard cell potentials. At 298 K, the Nernst equation is:** $E_{cell} = E^o_{cell} - \frac{0.0257 \text{ V} \cdot \text{mol e}^-}{n} \ln [Q]$**, where** $Q$ **is the reaction quotient, and n is the moles of** $e^-$ **transferred in the overall oxidation-reduction reaction.**

$Zn(s) + Cu^{2+}(aq) \rightarrow Zn^{2+}(aq) + Cu(s)$

$$E^o_{cell} = E^o_{cathode} - E^o_{anode} = (0.337 \text{ V}) - (-0.763 \text{ V}) = +1.100 \text{ V}$$

$$E_{cell} = E^o_{cell} - \frac{0.0257 \text{ V} \cdot \text{mol e}^-}{n} \ln \frac{[Zn^{2+}]}{[Cu^{2+}]} = 1.100 \text{ V} - \frac{0.0257 \text{ V} \cdot \text{mol e}^-}{2 \text{ mol e}^-} \ln \frac{[0.40]}{[4.8 \times 10^{-3}]} = 1.043\text{V}$$

Problem 13.29 ■ Use the Nernst equation to calculate the cell potentials of the following cells at 298 K.

(a) $2 Ag^+(aq)(0.50 M) + Ni(s) \rightarrow 2 Ag(s) + Ni^{2+}(aq)(0.20 M)$

(b) $Cu(s) + PtCl_6^{2-}(aq)(0.10 M) \rightarrow Cu^{2+}(aq)(0.20 M) + PtCl_4^{2-}(aq)(0.10 M) + 2 Cl^-(aq)(0.40 M)$

(c) $Pb(s) + SO_4^{2-}(aq)(0.30 M) + 2 AgCl(s) \rightarrow PbSO_4(s) + 2 Ag(s) + 2 Cl^-(aq)(0.20 M)$

**The Nernst equation is used to calculate nonstandard cell potentials. At 298 K, the Nernst equation is:** $E_{cell} = E^o_{cell} - \frac{0.0257 \text{ V} \cdot \text{mol e}^-}{n} \ln [Q]$**, where** $Q$ **is the reaction quotient, and n is the moles of** $e^-$ **transferred in the overall oxidation-reduction reaction.**

**(a)**

| | |
|---|---|
| **cathode:** $2 \{Ag^+(aq) + e^- \rightarrow Ag(s)\}$ | $E^o_{red} = 0.7994$ V |
| **anode:** $Ni(s) \rightarrow Ni^{2+}(aq) + 2e^-$ | $E^o_{ox} = -(-0.25 \text{ V})$ |
| **overall:** $2Ag^+(aq) + Ni(s) \rightarrow 2Ag(s) + Ni^{2+}(aq)$ | $E^o_{cell} = 1.049$ V |

n = 2 mol $e^-$

$$Q = \frac{[Ni^{2+}]^2[1]}{[1][Ag^+]^2} = \frac{[0.20][1]}{[1][0.50]^2} = 0.80$$

**(Recall the "concentrations" of solids are set = 1 in the reaction quotient.)**

$$E_{cell} = E^o_{cell} - \frac{0.0257 \text{ V} \cdot \text{mol e}^-}{n} \ln [Q] = 1.049 \text{ V} - \frac{0.0257 \text{ V} \cdot \text{mol e}^-}{2 \text{ mol e}^-} \ln [0.80] = 1.052 \text{ V}$$

**(b)**

| | |
|---|---|
| **cathode:** $PtCl_6^{2-}(aq) + 2e^- \rightarrow PtCl_4^{2-}(aq) + 2 Cl^-(aq)$ | $E^o_{red} = 0.68$ V |
| **anode:** $Cu(s) \rightarrow Cu^{2+}(aq) + 2e^-$ | $E^o_{ox} = -(0.337 \text{ V})$ |
| **overall:** $PtCl_6^{2-}(aq) + Cu(s) \rightarrow PtCl_4^{2-}(aq) + 2 Cl^-(aq) + Cu^{2+}(aq)$ | $E^o_{cell} = 0.343 \text{ V} \approx 0.34\text{V}$ |

**n = 2 mol e⁻** $$Q = \frac{[PtCl_4^{2-}][Cl^-]^2[Cu^{2+}]}{[PtCl_6^{2-}][1]} = \frac{[0.10][0.40]^2[0.20]}{[0.10][1]} = 0.032$$

$$E_{cell} = E^o_{cell} - \frac{0.0257\ V\cdot mol\ e^-}{n}\ln[Q] = 0.343\ V - \frac{0.0257\ V\cdot mol\ e^-}{2\ mol\ e^-}\ln[0.032] =$$
$$= 0.387\ V \approx 0.39\ V$$

(c) **cathode: 2 {$AgCl(s) + e^- \rightarrow Ag(s) + Cl^-(aq)$}** $E^o_{red} = 0.222$ V
**anode: $SO_4^{2-}(aq) + Pb(s) \rightarrow PbSO_4(s) + 2e^-$** $E^o_{ox} = -(-0.356\ V)$
**overall: $SO_4^{2-}(aq)+Pb(s)+2AgCl(s) \rightarrow PbSO_4(s)+2Ag(s)+2Cl^-(aq)$** $E^o_{cell}=0.578$ V

**n = 2 mol e⁻** $$Q = \frac{[1][Cl^-]^2[1]}{[SO_4^{2-}][1][1]} = \frac{[1][0.20]^2[1]}{[1][0.30][1]} = 0.133$$

$$E_{cell} = E^o_{cell} - \frac{0.0257\ V\cdot mol\ e^-}{n}\ln[Q] = 0.578\ V - \frac{0.0257\ V\cdot mol\ e^-}{2\ mol\ e^-}\ln[0.133] = 0.604\ V$$

Problem 13.31 We noted that a tin-plated steel can corrodes more quickly than an unplated steel can. In cases of galvanic corrosion, one cannot expect standard conditions. Suppose that you want to study the galvanic corrosion of tin-plated steel by constructing a cell with low concentrations of the ions. You have pieces of tin and iron. You also a have a solution of tin(II) chloride that is 0.05 M and one of iron(II) nitrate that is 0.10 M.
(a) Describe the half-reactions you construct for this experiment.
(b) Which half-reaction will be the anode and which the cathode?

**cathode: $Sn^{2+}(aq) + 2e^- \rightarrow Sn(s)$**

**anode: $Fe(s) \rightarrow Fe^{2+}(aq) + 2e^-$**

(c) Based on the solutions you have, calculate the cell potential for your experiment.

**For the cathode reaction, $E^o_{red} = -0.14$ V, for the anode reaction, $E^o_{ox} = -(-0.44$ V).**

**$E^o_{cell} = E^o_{red} + E^o_{ox} = -0.14\ V + 0.44\ V = 0.30\ V$**

Problem 13.33 You are a chemical engineer in a chemical firm that manufactures computer chips. A new accountant who, understands budgets but not chemistry, sends out the following memo:

> "Because of the price of gold, the company should change from gold-plated pins on our computer chips to pins made of copper, which is much cheaper."

Write a memo to your boss explaining why the accountant's choice of copper is a poor one and why gold should continue to be used in pin manufacturing.

**Copper has a less positive reduction potential ($E^o_{red}$ = 0.337 V) compared to gold ($E^o_{red}$ = 1.68 V), meaning gold has a much smaller tendency to oxidize than copper. Using copper pins could lead to poor conductivity due to oxidation on the surface of the pin, leading to poor chip performance in the marketplace and lower profits.**

Problem 13.35 In May 2000, a concrete pedestrian walkway collapsed in North Carolina, injuring more than 100 people. Investigation revealed that $CaCl_2$ had been mixed into the grout that filled the holes around the steel reinforcing cables inside the concrete, resulting in corrosion of the cables, thus weakening the structure. Based on your understanding of corrosion, explain why the use of chloride compounds in steel reinforced concrete is discouraged by the concrete industry.

**Corrosion is a spontaneous electrochemical reaction. Any substance that increases the flow of electrons will increase the rate of this reaction. $CaCl_2$ is an electrolyte; the chloride ions increase conductivity within the concrete and accelerate the corrosion process of the steel reinforcing rods within.**

Problem 13.37 How is the relationship between maximum electrical work and cell potential used to determine the free energy change for electrochemical cells?

**The free energy change is related to the maximum possible useful work in the system (see Section 10.6 of the text).**

**In a galvanic cell, the work done is electrical work:**

$$\textbf{electrical work = joules} = q \times V \textbf{ or } q \times E^o.$$

**The charge, $q$, can be replaced by n$F$, which is generally easier to calculate than charge ($n$ = moles of electrons and $F$ is Faraday's constant.)**

**This gives maximum work, $\Delta G^o = -nFE^o$.**

**The negative sign is added because the cell potential of a spontaneous electrochemical reaction must be positive and Gibb's free energy change of a spontaneous reaction must be negative.**

Problem 13.39 ■ Calculate the free energy change for the following reactions using standard cell potentials for the half-reactions that are involved.

**From the overall reaction, the anode and cathode half-reactions can be written. The oxidation and reduction potentials can be obtained from Appendix I and**

**added to get the standard cell potential. The number of moles of electrons transferred is most easily determined by looking at the half-reactions. With all that information, we'll use Equation 13.5 to calculate the Gibb's free energy change:**

$$\Delta G^o = -nFE^o$$

(a) $Fe(s) + Hg_2^{2+}(aq) \rightarrow Hg(\ell) + Fe^{2+}(aq)$

**anode:** $Fe(s) \rightarrow Fe^{2+}(aq) + 2\,e^-$ $\quad E^o_{ox} = -(-0.44\ V)$
**cathode:** $Hg_2^{2+}(aq) + 2\,e^- \rightarrow 2\,Hg(\ell)$ $\quad \underline{E^o_{red} = 0.789\ V}$
$E^o_{cell} = 1.229\ V \approx 1.23\ V$

$n = 2\ mol\ e^-$ **transferred,** $F = 96{,}485\ J/V{\cdot}mol\ e^-$

$\Delta G^o = -nFE^o = -(2\ mol\ e^-)(96{,}485\ J/V{\cdot}mol\ e^-)(1.23\ V) = -237{,}000\ J$ **or** $-237\ kJ$

(b) $Fe^{3+}(aq) + Ag(s) + Cl^-(aq) \rightarrow Fe^{2+}(aq) + 2\,AgCl(s)$

**anode:** $Ag(s) + Cl^-(aq) \rightarrow AgCl(s) + e^-$ $\quad E^o_{ox} = -(0.222\ V)$
**cathode:** $Fe^{3+}(aq) + e^- \rightarrow Fe^{2+}(aq)$ $\quad \underline{E^o_{red} = 0.771\ V}$
$E^o_{cell} = 0.549\ V$

$n = 1\ mol\ e^-$ **transferred,** $F = 96{,}485\ J/V{\cdot}mol\ e^-$

$\Delta G^o = -nFE^o = -(1\ mol\ e^-)(96{,}485\ J/V{\cdot}mol\ e^-)(0.549\ V) = -53{,}000\ J$ **or** $-53.0\ kJ$

(c) $2\,MnO_4^-(aq) + 5\,Zn(s) + 16\,H_3O^+(aq) \rightarrow 2\,Mn^{2+}(aq) + 5\,Zn^{2+}(aq) + 24\,H_2O(\ell)$

**anode:** $5\,\{Zn(s) \rightarrow Zn^{2+}(aq) + 2e^-\}$ $\quad E^o_{ox} = -(-0.763\ V)$
**cathode:** $2\,\{MnO_4^-(aq) + 8\,H_3O^+(aq) + 5\,e^- \rightarrow Mn^{2+}(aq) + 12\,H_2O(\ell)\}$ $\quad \underline{E^o_{red} = 1.507\ V}$
$E^o_{cell} = 2.270\ V$

$n = 10\ mol\ e^-$ **transferred,** $F = 96{,}485\ J/V{\cdot}mol\ e^-$

$\Delta G^o = -nFE^o = -(10\ mol\ e^-)(96{,}485\ J/V{\cdot}mol\ e^-)(2.270\ V) =$

$= -2{,}190{,}000\ J$ **or** $-2190.\ kJ$

Problem 13.41 Use the potential of the galvanic cell, $Co(s)\,|\,Co^{2+}(aq)\,\|\,Pb^{2+}(aq)\,|\,Pb(s)$, to determine $\Delta G_f^o(Pb^{2+})$, given that $\Delta G_f^o(Co^{2+}) = -54.4\ kJ/mol$.

**The strategy here is to calculate the standard cell potential and then calculate the Gibb's free energy change as in Problem 13.39. Then, an equation can be written for $\Delta G^o_{rxn}$ in terms of Gibb's free energy changes of formation reactions, as in Chapter 10, see Problem 10.69. With that equation, we can solve algebraically for $\Delta G_f^o(Pb^{2+})$.**
**Writing half-reactions,**

**cathode:** $Pb^{2+}(aq) + 2e^- \rightarrow Pb(s)$ $E^o_{red} = (-0.126\ V)$
**anode:** $Co(s) \rightarrow Co^{2+}(aq) + 2e^-$ $E^o_{ox} = -(-0.28\ V)$
**overall reaction:** $Pb^{2+}(aq) + Co(s) \rightarrow Pb(s) + Co^{2+}(aq)$ $E^o_{cell} = 0.154\ V \approx 0.15\ V$

**n = 2 mol e⁻ transferred, F = 96,485 J/V·mol e⁻**

$\Delta G^o = -nFE^o = -(2\text{ mol e}^-)(96{,}485\text{ J/V·mol e}^-)(0.15\text{ V}) =$

$= -29{,}00\text{ J}$ **or** $-29\text{ kJ}$

**Using the formula:** $\Delta G^o_{rxn} = \sum n\Delta G_f{}^o{}_{products} - \sum n\Delta G_f{}^o{}_{reactants}$ **:**

$\Delta G^o_{rxn} = [1\text{ mol }(\Delta G_f{}^o{}_{Pb}) + [1\text{ mol }(\Delta G_f{}^o{}_{Co2+})] - [1\text{ mol }(\Delta G_f{}^o{}_{Pb2+}) + 1\text{ mol }(\Delta G_f{}^o{}_{Co})]$

$= [1\text{ mol }(0) + 1\text{ mol }(-54.4\text{ kJ/mol})] - [1\text{ mol }\Delta G_f{}^o{}_{Pb2+} + 1\text{ mol }(0)]$

$\Delta G^o_{rxn} = -29\text{ kJ} = -54.4\text{ kJ} - 1\text{ mol }(\Delta G_f{}^o{}_{Pb2+})$

$\Delta G_f{}^o{}_{Pb2+} = [-54.4\text{ kJ} + 29\text{ kJ}]/\ 1\text{ mol} = -25\text{ kJ/mol}$

Problem 13.43 ■ Consult a table of standard reduction potentials and determine which of the following reactions are spontaneous under standard electrochemical conditions.

(a) $Mn(s) + 2\ H^+(aq) \rightarrow H_2(g) + Mn^{2+}(aq)$

**cathode:** $2\ H^+(aq) + 2\ e^- \rightarrow H_2(g)$ $E^o_{red} = 0.00\ V$
**anode:** $Mn(s) \rightarrow Mn^{2+}(aq) + 2\ e^-$ $E^o_{ox} = -(-1.18)\ V$
$E^o_{cell} = 1.18\ V$

**$E^o_{cell}$ is positive, so the given reaction would be spontaneous under standard conditions.**

(b) $2\ Al^{3+}(aq) + 3\ H_2(g) \rightarrow 2\ Al(s) + 6\ H^+(aq)$

**cathode:** $2\ Al^{3+}(aq) + 6\ e^- \rightarrow 2\ Al(s)$ $E^o_{red} = -1.66\ V$
**anode:** $3\ H_2(g) \rightarrow 6\ H^+(aq) + 6\ e^-$ $E^o_{ox} = 0.00\ V$
$E^o_{cell} = -1.66\ V$

**$E^o_{cell}$ is negative, so the given reaction would not be spontaneous under standard conditions.**

(c) $2\ Cr(OH)_3(s) + 6\ F^-(aq) \rightarrow 2\ Cr(s) + 6\ OH^-(aq) + 3\ F_2(g)$

**cathode:** $2\ Cr(OH)_3(s) + 6e^- \rightarrow 2\ Cr(s) + 6\ OH^-(aq)$ $E^o_{red} = -1.30\ V$
**anode:** $6\ F^-(aq) \rightarrow 3\ F_2(g) + 6e^-$ $E^o_{ox} = -2.87\ V$
$E^o_{cell} = -4.17\ V$

**$E^o_{cell}$ is negative, so the given reaction would not be spontaneous under standard conditions.**

(c) $Cl_2(g) + 2\ Br^-(aq) \rightarrow Br_2(\ell) + 2\ Cl^-(aq)$

**cathode: $Cl_2(g) + 2\ e^- \rightarrow 2\ Cl^-(aq)$ $\quad E^o_{red} = 1.360$ V**
**anode: $2\ Br^-(aq) \rightarrow Br_2(\ell) + 4\ e^-$ $\quad \underline{E^o_{ox} = -(1.08)\ V}$**
**$E^o_{cell} = 0.28$ V**

**$E^o_{cell}$ is positive, so the given reaction would be spontaneous under standard conditions.**

Problem 13.45 Some calculators cannot display results of an antilog calculation if the power of 10 is greater than 99. This shortcoming can come into play for determining equilibrium constants of redox reactions, which are sometimes quite large. Solve the following expressions for $K$: (a) $\log K = 45.63$, (b) $\log K = 25.00$, (c) $\log K = 20.63$. What is the relationship between the three expressions and the three answers? How can you use this relationship to solve problems that exceed $10^{99}$, even if your calculator will not carry out the calculation directly?

**The antilog function is $10^x$.**

**(a) $\log K = 45.63$, $\quad K = 10^{45.63} = 4.266 \times 10^{45}$**

**(b) $\log K = 25.00$, $\quad K = 10^{25.00} = 1.000 \times 10^{25}$**

**(c) $\log K = 20.63$, $\quad K = 10^{20.63} = 4.266 \times 10^{20}$**

**The product of (b) and (c) equals (a). This means that if a calculator cannot perform an antilog calculation, it can be broken into two smaller calculations and multiplied.**

Problem 13.47 Use the standard reduction potentials for the reactions:
$AgCl(s) + e^- \rightarrow Ag(s) + Cl^-(ag)$ and $Ag^+(aq) + e^- \rightarrow Ag(s)$
to calculate the $K_{sp}$ of silver chloride. How does your answer compare with the value listed in Table 12.4?

**cathode rxn: $AgCl(s) + e^- \rightarrow Ag(s) + Cl^-(ag)$ $\quad E^o_{red} = 0.222$ V**
**anode rxn: $Ag(s) \rightarrow Ag^+(aq) + e^-$ $\quad \underline{E^o_{ox} = -(0.7994\ V)}$**
**overall rxn: $AgCl(s) \rightarrow Ag^+(aq) + Cl^-(ag)$ $\quad E^o_{cell} = -0.5774$ V**

**From the cell potential, the equilibrium constant can be calculated:**

**$\ln K = \dfrac{nFE^o_{cell}}{RT}$; where $n$ = mol $e^-$ transferred, $F$ = Faraday's constant, 96,485 J/V·mol $e^-$, $R$ = gas constant, 8.314 J/mol·K, $T$ = temperature in Kelvin, and $E^o_{cell}$ = the standard cell potential.**

$$\ln K = \frac{(1\ \text{mol e}^-)(96{,}485\ \text{J/V}\cdot\text{mol e}^-)(-0.5774\ \text{V})}{(8.314\ \text{J/mol}\cdot\text{K})(298\ \text{K})} = -22.4,$$

**$K = e^{-22.4} = 1.71 \times 10^{-10}$**
**This is very similar to the value listed in Appendix H ($1.8 \times 10^{-10}$).**

Problem 13.49 ■ Calculate the equilibrium constant for the redox reactions that could occur in the following situations and use that value to explain whether or not any reaction will be observed.
(a) A piece of iron is placed in a 1.0 M solution of $NiCl_2(aq)$.
(b) A copper wire is placed in a 1.0 M solution of $Pb(NO_3)_2(aq)$.

**(a) The cell potential must first be calculated and then *K* can be found. See Problem 13.47.**

| | | |
|---|---|---|
| **cathode rxn:** | $Ni^{2+}(aq) + 2\ e^- \rightarrow Ni(s)$ | $E^o_{red} = -0.25$ V |
| **anode rxn:** | $Fe(s) \rightarrow Fe^{2+}(aq) + 2\ e^-$ | $E^o_{ox} = -(-0.44\ \text{V})$ |
| **overall rxn:** | $Ni^{2+}(aq) + Fe(s) \rightarrow Fe^{2+}(aq) + Ni(s)$ | $E^o_{cell} = 0.19$ V |

$\ln K = \dfrac{nFE^o_{cell}}{RT}$; **where $n$ = mol $e^-$ transferred, $F$ = Faraday's constant, 96,485 J/V·mol $e^-$, $R$ = gas constant, 8.314 J/mol·K, $T$ = temperature in Kelvin, and $E^o_{cell}$ = the standard cell potential.**

$$\ln K = \frac{(2\ \text{mol e}^-)(96{,}485\ \text{J/V}\cdot\text{mol e}^-)(0.19\ \text{V})}{(8.314\ \text{J/mol}\cdot\text{K})(298\ \text{K})} = 15,$$

**$K = e^{15} = 3.3 \times 10^6$, the reaction will occur because *K* is larger than one ($E^o_{cell}$ is positive).**

| **(b)** | | | |
|---|---|---|---|
| | **cathode rxn:** | $Pb^{2+}(aq) + 2\ e^- \rightarrow Pb(s)$ | $E^o_{red} = -0.126$ V |
| | **anode rxn:** | $Cu(s) \rightarrow Cu^{2+}(aq) + 2\ e^-$ | $E^o_{ox} = -(0.337\ \text{V})$ |
| | **overall rxn:** | $Ni^{2+}(aq) + Fe(s) \rightarrow Fe^{2+}(aq) + Ni(s)$ | $E^o_{cell} = -0.463$ V |

$\ln K = \dfrac{nFE^o_{cell}}{RT}$; **where $n$ = mol $e^-$ transferred, $F$ = Faraday's constant, 96,485 J/V·mol $e^-$, $R$ = gas constant, 8.314 J/mol·K, $T$ = temperature in Kelvin, and $E^o_{cell}$ = the standard cell potential.**

$$\ln K = \frac{(2\ \text{mol e}^-)(96{,}485\ \text{J/V}\cdot\text{mol e}^-)(-0.463\ \text{V})}{(8.314\ \text{J/mol}\cdot\text{K})(298\ \text{K})} = -36.1,$$

**$K = e^{-36.1} = 2.2 \times 10^{-16}$**

**The reaction won't occur as *K* is smaller than one ($E^o_{cell}$ is negative).**

Problem 13.51 An engineer is assigned to design an electrochemical cell that will deliver a potential of exactly 1.52 V. Design and sketch a cell to provide this voltage, detailing the solutions, their concentrations, and the electrodes you will need. Write equations for all relevant reactions.

**One possibility would be to construct a cell using a $Sn/S^{2+}$ half-cell for the cathode and a $Al/Al^{3+}$ half-cell for the anode:**

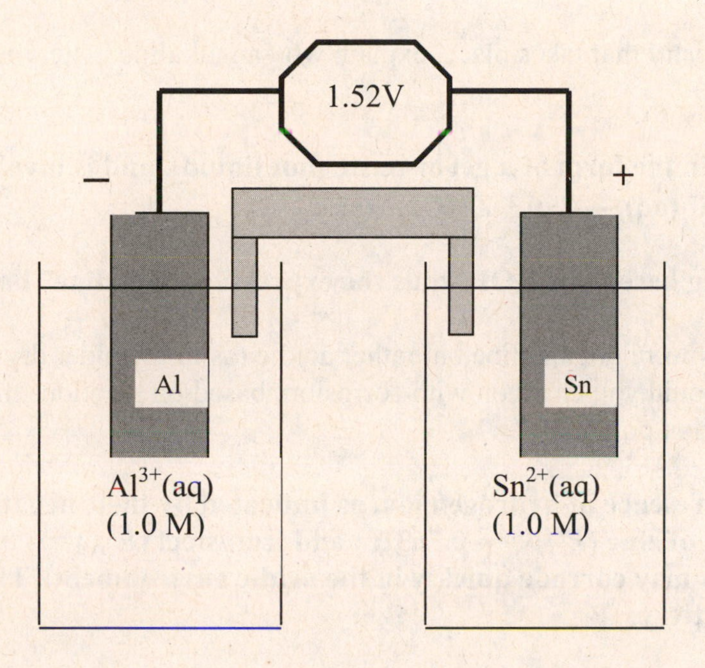


| | | |
|---|---|---|
| **cathode rxn:** | **$3\ \{Sn^{2+}(aq) + 2\ e^- \rightarrow Sn(s)\}$** | **$E^o_{red} = -\ 0.14$ V** |
| **anode rxn:** | **$2\ \{Al(s) \rightarrow Al^{3+}(aq) + 3\ e^-\}$** | **$E^o_{ox} = -\ (-\ 1.66$ V)** |
| **overall rxn:** | **$3\ Sn^{2+}(aq) + 2\ Al(s) \rightarrow 2\ Al^{3+}(aq) + 3\ Sn(s)$** | **$E^o_{cell} = 1.52$ V** |

Problem 13.53 Would nickel make an acceptable sacrificial anode to protect steel? Explain your answer.

**A sacrificial anode is a metal selected to be oxidized preferentially over some other metal such as steel. A galvanic cell is intentionally created where the steel is the cathode and some other metal, with a greater tendency to oxidize, is the anode.**

**Nickel would not make an acceptable sacrificial anode for steel because it has a more positive reduction potential, $E^o_{red} = -\ 0.25$ V, compared to iron's: $E^o_{red} = -\ 0.44$ V. This means that steel would spontaneously oxidize rather than nickel, which is the opposite of what you want to occur.**

Problem 13.55 Sacrificial anodes are sometimes connected to steel using a copper wire. If the anode is completely corroded away and not replaced, so that only the copper wire remains, what could happen where the copper and iron meet? Explain your answer.

**While the sacrificial anode remains, reduction occurs at the steel (the cathode) as oxidation occurs at the anode. When the anode is gone, a galvanic cell will be created where the steel and copper wire meet, with the metal which has the most negative reduction potential being oxidized. Comparing reduction potentials: $E^o_{red} = -0.44$ V for steel (iron) and $E^o_{red} = +0.337$ V for copper. Steel would begin to oxidize via galvanic corrosion where the iron and copper meet.**

Problem 13.57 Based on the chemistry that takes place, explain why an alkaline battery is called "alkaline".

**Potassium hydroxide is present in the form of a gel or paste (not liquid), and is involved in the anode reaction: $Zn(s) + OH^-(aq) \rightarrow 2\ e^- + Zn(OH)_2(s)$.**

**Alkali is a term describing strong bases like KOH, thus the expression "alkaline" battery.**

Problem 13.59 If alkaline batteries were not alkaline but rather acidic (as in the older dry cell batteries), what extra difficulties could you envision with corrosion, based on reactions that are part of the table of standard reduction potentials?

**Most metals will oxidize in the presence of hydrogen ion, as indicated by their negative reduction potential. This is true of zinc ($E^o_{red} = -0.763$ V) and iron/steel ($E^o_{red} = -0.763$ V); which means these materials may corrode quickly in the acidic environment. This can lead to fluid leaking from the battery.**

Problem 13.61 Battery manufacturers often assess batteries in terms of the specific energy (or energy capacity). The weight capacity of a battery is defined as $q \times V$/mass. Why would a battery manufacturer be interested in this quantity?

**Weight is one of the most important characteristics of a battery. The product of charge and voltage, $q \times V$, yields the total energy produced by the battery. Increasing the physical size of the battery can increase the energy produced, but also increases the weight. A better measure to access the performance of a battery would also take into account its mass, since smaller battery size is almost always desirable. The weight capacity, $q \times V$/mass, compares the energy produced per gram of battery mass, a better way to judge battery performance.**

Problem 13.63 What product forms from the lead components of a lead storage battery? Why does mechanical shock (bumps) ultimately degrade the performance of a lead storage battery?

**Both the oxidation of lead at the anode and the reduction of $PbO_2$ at the cathode produce the same product: $Pb^{2+}$. This cation combines with the sulfate ion present in the electrolyte solution from the sulfuric acid to form solid lead(II) sulfate. The $PbSO_4(s)$ "sticks" to the**

**surfaces of the electrodes, enabling the $Pb^{2+}$ to be converted back into Pb (anode) or $PbO_2$ (cathode) when the battery is charged.**

**Excessive mechanical shock can physically cause some of the $PbSO_4(s)$ to fall off the electrodes and settle to the bottom of the battery case. This represents a loss of the $Pb^{2+}$ ion, which can never be converted back to Pb and $PbO_2$. Over time, this will reduce the capacity of the battery to be recharged, leading to lower discharge voltages.**

Problem 13.65 You are an electrical engineer investigating different button batteries for possible use in a design. You obtain the discharge curves shown below for four different batteries under the same conditions and discharge load. The graph below shows discharge curves for four different button batteries under the same conditions and discharge load.
(a) Write a paragraph in which you interpret curve (a) and curve (d) to a customer seeking a battery for an electronic device.

**The curves represent the voltage produced by the battery as a function of time. The battery represented by curve (a) shows that it would produce approximately 1.5 volts for about 60 hours and then decline steadily to less then 1 volt over the next 20 hours. The battery represented by curve (d) will produce about 1.4 volts for a period of 200 hours and then rapidly decline to 0.5 volts.**

(b) Both (b) and (d) represent zinc/air cells. Explain why there is a difference in these two curves. (HINTS: Which electrode in the zinc/air cell has an unlimited life? Compare the approximate area of the other electrode for the two cells.)

**Since they are both zinc/air cells, they will produce approximately the same voltage. The difference in the discharge time of the batteries must be a physical one instead of chemical. The air is unlimited for both batteries, so that is not a factor in the difference. A larger surface area of the zinc electrode can provide a longer duration of discharge, thus the battery represented by (d) probably has a larger zinc anode.**

Problem 13.67 On the Internet, access the website of a major battery company, such as Rayovac, Duracell, or Eveready (Energizer). Search the site for a battery type not covered in this chapter. Find out all you can about the battery from engineering data or specification (spec) sheets. Print spec sheets that include graphs of discharge characteristics, operating temperature data, etc. Summarize and interpret your findings in a one-page report to which you attach the spec sheets.

**Answers will vary according to the websites found.**

Problem 13.69 Why is it easier to force an oxidation-reduction reaction to proceed in the nonspontaneous direction than it is to force an acid-base reaction to proceed in the nonspontaneous direction?

**Electrons are transferred in an oxidation-reduction reaction. It is easier to reverse the direction of flow of electrons to cause a non-spontaneous oxidation-reduction reaction to occur than it is to break the bonds of a stable product to reverse the direction of some other type of reaction.**

Problem 13.71 When aluminum is refined by electrolysis from its oxide ores, is the process used active or passive electrolysis? Explain your answer.

**Passive electrolysis occurs when the electrodes used in electrolysis are chemically inert materials that only provide a path for the electrons. Active electrolysis occurs when the electrodes used are actually part of the reaction, as in plating metal on the surface of an object.**
**In the electrolysis of aluminum oxide ores, graphite inert electrodes are used, making this process a passive electrolysis.**

Problem 13.73 In an electroplating operation, the cell potential is sometimes 0 V. Why is a zero potential possible in electrolysis but not in a galvanic cell?

**A galvanic cell is a spontaneous reaction, so a zero cell potential would not be possible to achieve a reaction. Electroplating is a nonspontaneous reaction driven by an external source of electrons, so cell potential is not crucial. Many electroplating operations are run at a very low voltage; zero potential can actually be advantageous.**

Problem 13.75 In barrel plating, why is the barrel containing the small parts rotated?

**The barrel is rotated during the plating process so that all the small parts it contains obtain electrical contact with the cathode, becoming part of the cathode themselves during the time of contact. As the parts become part of the cathode, reduction occurs and metal atoms are deposited on the surface of the part. The rotation and random mixing ensures that parts are plated for the same period of time, resulting in even coatings.**

Problem 13.77 Use the Internet to find electroplating companies that carry out silver plating. Popular impressions are that silver plating is done cosmetic reasons, making objects (like silverware) more attractive. Based on your Internet research, do you believe this popular impression accurately reflects the plating industry?

**Answers will vary according to the websites found. In addition to jewelry and flatware being plated with silver, silver-plated band instruments are common, as are electronics, communications, and electrical parts.**

Problem 13.79 If a current of 15 A is run through an electrolysis cell for 2.0 hours, how many moles of electrons have moved?

**Amperes and time are related to Coulombs of charge, which is related to electrons.**

$$15\text{ A} \times 2.0\text{ hrs} \times \frac{3600\text{ seconds}}{1\text{ hour}} \times \frac{1\text{ C}}{1\text{ A}\cdot\text{s}} \times \frac{1\text{ mol e}^-}{96485\text{ C}} = 1.1\text{ mol e}^-$$

Problem 13.81 If a barrel plating run uses 200.0 A for exactly 6 hours for an electroplating application at 0.30 V, how many kilowatt-hours have been used in the run? If the voltage is 0.90 V, what is the power usage (in kWh)?

**Kilowatts measure energy used per second, 1000 J/s. Kilowatt hours give the total amount of energy used. Charge is found by multiplying current and time, energy equals charge times voltage.**

$$200.0\text{ A} \times 6\text{ hours} \times \frac{3600\text{ seconds}}{1\text{ hour}} = 4.32 \times 10^6\text{ C}$$

$$4.32 \times 10^6\text{ C} \times 0.30\text{ V} \times \frac{1\text{ J}}{1\text{ C}\cdot\text{V}} = 1.30 \times 10^6\text{ J}$$

$$1.30 \times 10^6\text{ J} \times \frac{1\text{ kWh}}{3.60 \times 10^6\text{ J}} = 0.36\text{ kWh at 0.30 V}$$

**For 0.90 V,**

$$4.32 \times 10^6\text{ C} \times 0.90\text{ V} \times \frac{1\text{ J}}{1\text{ C}\cdot\text{V}} \times \frac{1\text{ kWh}}{3.60 \times 10^6\text{ J}} = 1.08\text{ kWh at 0.90 V}$$

Problem 13.83 In a copper plating experiment in which copper metal is deposited from a copper(II) ion solution, the system is run for 2.6 hours at a current of 12.0 A. What mass of copper is deposited?

**From the amperes and time, coulombs can be found. From C, moles of electrons are determined and that give moles of copper.**

$$Cu^{2+}(aq) + 2\,e^- \rightarrow Cu(s)$$

$$12.0\text{ A} \times 2.6\text{ hrs} \times \frac{3600\text{ seconds}}{1\text{ hour}} \times \frac{1\text{ C}}{1\text{ A}\cdot\text{s}} \times \frac{1\text{ mol e}^-}{96845\text{ C}} \times \frac{1\text{ mol Cu}}{2\text{ mol e}^-} \times \frac{63.55\text{ g Cu}}{1\text{ mol Cu}} = 37\text{ g Cu}$$

Problem 13.85 Tin-plated steel is used for "tin" cans. Suppose that in the production of sheets of tin-plate steel, a line at a factory operates at a current of 100.0 A for exactly 8 hours on a continuously fed sheet of unplated steel. If the electrolyte contains tin(II) ions, what is the total mass of tin that has plated out in this operation?

**The tin is deposited according to the half reaction:** $Sn^{2+} + 2e^- \rightarrow Sn(s)$

**The current and time gives us the coulombs of charge, which gives moles of electrons. The moles of electrons allows grams of tin to be calculated.**

$$100.0\text{ A} \times 8\text{ hours} \times \frac{3600\text{ s}}{1\text{ hr}} \times \frac{1\text{ C}}{1\text{ A}\cdot\text{s}} \times \frac{1\text{ mol e}^-}{96485\text{ C}} \times \frac{1\text{ mol Sn}}{2\text{ mol e}^-} \times \frac{118.71\text{ g Sn}}{1\text{ mol Sn}} = 1770\text{ g Sn}$$

Problem 13.87 If a plating line that deposits nickel (from $NiCl_2$ solutions) operates at a voltage of 0.40 V with a current of 400.0 A and a total mass of 49.0 kg of nickel is deposited, what is the minimum number of kWh consumed in this process?

**Nickel is deposited according to the half reaction:** $Ni^{2+} + 2e^- \rightarrow Ni(s)$
**This gives us the relationship between moles of nickel deposited and moles of electrons needed. Moles of electrons gives coulombs of charge and from charge and voltage, the joules used can be determined.**

$$49{,}000\text{ g Ni} \times \frac{1\text{ mol Ni}}{58.69\text{ g Ni}} \times \frac{2\text{ mol e}^-}{1\text{ mol Ni}} \times \frac{96{,}485\text{ C}}{1\text{ mol e}^-} \times 0.40\text{ V} \times \frac{1\text{ J}}{1\text{ C}\cdot\text{V}} \times \frac{1\text{ kWh}}{3.60\times10^6\text{ J}} = 18\text{ kWh}$$

Problem 13.89 When a lead storage battery is recharged by a current of 12.0 A for 15 minutes, what mass of $PbSO_4$ is consumed?

**Lead(II) sulfate is produced when a car battery discharges, and this material "sticks" to the electrodes. When the battery is recharged, the lead(II) ion is converted back into Pb, regenerating the electrode:** $PbSO_4(s) + H^+ + 2e^- \rightarrow Pb(s) + HSO_4^-(aq)$

**Using the time and current, the amount of charge in coulombs is found, then the number of moles of electrons, and finally the mass of** $PbSO_4(s)$**.**

$$12.0\text{ A} \times 15\text{ minutes} \times \frac{60\text{ s}}{1\text{ min}} \times \frac{1\text{ C}}{1\text{ A}\cdot\text{s}} \times \frac{1\text{ mol e}^-}{96485\text{ C}} \times \frac{1\text{ mol PbSO}_4}{2\text{ mol e}^-} \times \frac{303.3\text{ g PbSO}_4}{1\text{ mol PbSO}_4} =$$
$$= 17\text{ g PbSO}_4\text{ consumed}$$

Problem 13.91 ■ An engineer is designing a mirror for an optical system. A piece of metal that measures 1.3 cm by 0.83 cm will have a coating of rhodium plated on its surface to serve as the mirror. The rhodium thickness will be 0.00030 mm, and the electrolyte contains $Rh^{3+}$ ions. If the operating current of the electrolysis is 1.3 A, how long must it be operated to obtain the desired coating? What mass of rhodium is deposited? (The density of rhodium is 12.4 g $cm^{-3}$.)

**The first step in this problem is to calculate the volume of the rhodium coating.**
**Volume = surface area × coating thickness = 1.3 cm × 0.83 cm × 0.000030 cm =** $3.2 \times 10^{-5}$ $cm^3$**.**

**Using the density, the mass of rhodium deposited can be found.**

$$3.2 \times 10^{-5}\ \text{cm}^3 \times \frac{12.4\ \text{g Rh}}{\text{cm}^3} = 4.0 \times 10^{-4}\ \text{g Rh}$$

**Finally, using the mass of rhodium deposited according to this half-reaction, $Rh^{3+} + 3e^- \rightarrow Rh$, and the current, the time required can be determined.**

$$4.0 \times 10^{-4}\ \text{g Rh} \times \frac{1\ \text{mol Rh}}{102.91\ \text{g Rh}} \times \frac{3\ \text{mol e}^-}{1\ \text{mol Rh}} \times \frac{96{,}485\ \text{C}}{1\ \text{mol e}^-} \times \frac{1\ \text{A}\cdot\text{s}}{1\ \text{C}} \times \frac{1}{1.3\ \text{A}} = 0.87\ \text{seconds.}$$

Problem 13.93 Explain why lithium-ion batteries tend to be relatively lightweight.

**In a lithium-ion battery, a very lightweight metal, lithium, is oxidized instead of a much heavier metal such as cadmium or zinc. This gives the lithium-ion battery a very high energy density, which is one of its most attractive features.**

Problem 13.95 Looking at Figure 13.23, describe how the operation of a lithium-ion battery does not lead to a charge build up on one side of the battery or the other?

**The oxidation and reduction half reactions are separated, as they must be in any galvanic cell, but the separator is permeable to lithium ions. These ions are produced at the anode as Li is oxidized but pass through the separator to the cathode where they react with the lithium cobalt oxide. Therefore there is no net increase in charge as the cell operates.**

Problem 13.97 For a voltage-sensitive application, you are working on a battery that must have a working voltage of 0.85 V. The materials to be used have a standard cell potential of 0.97 V. What must be done to correct the voltage? What information would you need to look up?

**Assuming that the temperature for this application is somewhere close to 25°C, the variable that affects the voltage would be concentrations and/or pressures of the reactants and products in the reaction, see the Nernst equation, Equation 13.3:**

$$E^\circ = E^\circ_{\text{cell}} - \frac{2.303RT}{nF} \log [Q]$$

***R*, *n*, and *F* are all constants, the only variable is *Q*, the reaction quotient. This is where varying the concentrations and pressures would affect the voltage of the cell. Depending on the form of the overall equation, these concentrations and/or pressures could be adjusted to give a working voltage (nonstandard cell potential) of 0.85 V.**

Problem 13.99 An oxidation-reduction reaction using Sn(s) to remove $N_2O(g)$ from a reaction vessel has been proposed, and you need to find its equilibrium constant. You cannot find any thermodynamic information on one product of the reaction, $NH_3OH^+(aq)$. How could you estimate the equilibrium constant of the reaction?

**You could create a galvanic cell using the oxidation of Sn for the anode reaction and the reduction of $N_2O$ for the cathode reaction. From the measurement of the cell potential, the equilibrium constant could be calculated: $\ln K = \dfrac{nFE^{\circ}_{cell}}{RT}$.**

**Alternatively, one could look up the relevant potentials and calculate the cell potential.**

**cathode rxn:** $N_2O(g) + 6H^+(aq) + H_2O + 4e^- \rightarrow 2\ NH_3OH^+(aq)$ $E^{\circ}_{red} = -\ 0.05\ V$
**anode rxn:** $2\ \{Sn(s) \rightarrow Sn^{2+}(aq) + 2\ e^-\}$ $\underline{E^{\circ}_{ox} = -\ (-\ 0.14\ V)}$
**overall rxn:** $2Sn(s) + N_2O(g) + 6H^+(aq) + H_2O \rightarrow 2Sn^{2+}(aq) + 2NH_3OH^+(aq)$
$E^{\circ}_{cell} = 0.09\ V$

Problem 13.101 You need to make a gold-plated connector in a design to take advantage of the conductivity of and corrosion resistance of gold. To justify this choice to the project director, you must devise a cost projection for several levels of plating. What variables could be varied? What information would you have to look up?

**Assuming that the size of the connector is fixed, the surface area of the part must be determined. The total amount of gold used to plate the connector depends on the thickness of the layer. Several thicknesses of gold could be proposed and the total volume of gold deposited would be calculated using the surface area and thickness of the layer. The density of gold would have to be looked up, and then mass of gold deposited determined. By looking up the current price of gold per gram, cost projections for the various layers of gold could be presented.**

**Reduction potentials for the half-reactions involved would also have to be looked up.**

Problem 13.103 A current is passed through a solution of copper(II) sulfate long enough to deposit 14.5 g of copper. What volume of oxygen is also produced if the gas is measured at 24°C and 0.958 atm of pressure?

**The cathode reaction is $Cu^{2+}(aq) + 2e^- \rightarrow Cu(aq)$. At the anode, water is oxidized: $2H_2O(\ell) \rightarrow O_2(g) + 4H^+(aq) + 4e^-$. From the amount of copper produced, the moles of electrons transferred can be found. From the moles of $e^-$, the amount of oxygen formed can be determined.**

$$14.5\ \text{g Cu} \times \frac{1\ \text{mol Cu}}{63.55\ \text{g Cu}} \times \frac{2\ \text{mol e}^-}{1\ \text{mol Cu}} = 0.456\ \text{moles electrons transferred}$$

$$0.456\ \text{moles electrons} \times \frac{1\ \text{mol O}_2}{4\ \text{mol e}^-} = 0.114\ \text{moles O}_2\ \text{formed}$$

**Now we need to use the ideal gas law to find the volume.**

$$PV = nRT \text{ or } V = \frac{nRT}{P} = \frac{(0.114\text{ mol})(0.08206\text{ L}\cdot\text{atm/mol K})(297\text{ K})}{0.958\text{ atm}} = 2.90\text{ L } O_2$$

Problem 13.105 ■ A button case for a small battery must be silver coated. The button is a perfect cylinder with a radius of 3.0 mm and a height of 2.0 mm. For simplicity, assume that the silver solution used for plating is silver nitrate. (Industrial processes often use other solutions.) Assume that the silver plating is perfectly uniform and is carried out for 3.0 min at a current of 1.5 A. **(a)** What mass of silver is plated on the part? **(b)** How many atoms of silver have plated on the part? **(c)** Calculate an estimate of the thickness (in atoms) of the silver coating. (Silver has a density of 10.49 g/cm$^3$ and an atomic radius of 160 pm.)

**(a) From the current and time we can calculate the coulombs of charge, then moles of electrons used, finally moles and grams of silver.**

$$1.5\text{ A} \times 3.0\text{ min} \times \frac{60\text{ s}}{1\text{ min}} \times \frac{1\text{ C}}{1\text{ A}\cdot\text{s}} \times \frac{1\text{ mol e}^-}{96{,}485\text{ C}} \times \frac{1\text{ mol Ag}}{1\text{ mol e}^-} \times \frac{107.87\text{ g Ag}}{1\text{ mol Ag}} = 0.30\text{ g Ag}$$

**(b)** $$0.30\text{ g Ag} \times \frac{1\text{ mol Ag}}{107.87\text{ g Ag}} \times \frac{6.022\times10^{23}\text{ atoms}}{1\text{ mol Ag}} = 1.7\times10^{21}\text{ atoms Ag}$$

**(c) To find this we first calculate the surface area of the "button". From the mass and density of the silver, we calculate the volume. Dividing volume by surface area gives the thickness. Finally, dividing the thickness by the diameter per silver atom will yield the number of atoms thick that the coating is.**

$$\text{Surface area} = (\pi \times r^2 \times 2) + (2 \times \pi \times r \times h) = (\pi \times (3.0\text{ mm})^2 \times 2) + (2 \times \pi \times 3.0\text{ mm} \times 2.0\text{ mm}) = 94.2\text{ mm}^2$$

$$\text{Volume} = \frac{\text{mass}}{\text{density}} = \frac{0.30\text{ g Ag}}{\frac{10.49\text{ g}}{\text{cm}^3}} = 0.0286\text{ cm}^3$$

$$\text{Thickness} = \frac{0.0286\text{ cm}^3}{94.2\text{ mm}^2 \times \frac{1\text{ cm}^2}{100\text{ mm}^2}} = 0.0304\text{ cm}$$

$$0.0304\text{ cm} \times \frac{1\times10^{10}\text{ pm}}{1\text{ cm}} \times \frac{1\text{ Ag atom}}{2\times160\text{ pm}} = 9.5\times10^5\text{ Ag atoms} = \text{thickness of the layer in Ag atoms.}$$

## CHAPTER OBJECTIVE QUIZ

This quiz will test your understanding of the basic text chapter objectives and give you additional practice problems. You should work this quiz after completing the end-of-chapter questions. The solutions to these questions are found at the end of this chapter.

1. TRUE or FALSE: Corrosion is a spontaneous reaction.

2. Which of the following statements about corrosion is false?
    a. Metals are oxidized in corrosion.
    b. Rusting is a type of uniform corrosion.
    c. Two different metals in contact can create corrosion.
    d. Crevice corrosion is common in large machines.
    e. All metals corrode at essentially the same rate.

3. When one or more atoms gain one or more electrons, it is called _____________. Loss of electrons is called _____________.

4. Magnesium is added to a solution containing chromium (III) ions. Write and balance the half-reactions and the overall reaction.

5. Which of the following statements does **not** describe galvanic cells.
    a. Galvanic cells are spontaneous oxidation-reduction reactions.
    b. A galvanic cell creates a flow of electrons.
    c. Oxidation occurs at the cathode of a galvanic cell.
    d. Electrons are lost by some species at the anode.
    e. A salt bridge separates the two half-cells.

6. Identify which of the following statements describe electrolytic cells.
    I. Electrolytic cells require an outside source of electrons.
    II. Electrolytic cells are non-spontaneous.
    III. The cathode is positive.
    IV. Electroplating is a type of electrolysis.
    V. Batteries are examples of electrolytic cells.

    a. I, II, IV
    b. III, V
    c. I, II, III, IV
    d. I, III, IV
    e. All of these describe electrolytic cells.

7. Express the following reaction using galvanic cell notation.
    $$3\ Pb(s) + 2\ Au^{3+}(aq) \rightarrow 2\ Au(s) + 3\ Pb^{2+}(aq)$$

8. Calculate the standard cell potential for the following galvanic cell.
   $Cd(s) \mid Cd^{2+}(aq) \parallel Ag^{+}(aq) \mid Ag(s)$

9. In a galvanic cell, which of these half reactions would represent the reaction at the cathode? What would be the anode reaction? What would be the standard cell potential?
   $FeS(s) + 2e^{-} \rightarrow Fe(s) + S^{2-}(aq)$ $E^{o}_{red} = -1.01$ V
   $Cr^{2+}(aq) + 2e^{-} \rightarrow Cr(s)$ $E^{o}_{red} = -0.91$ V

10. Consider a galvanic cell created when an aluminum electrode is immersed in a solution of 0.5 M $Al_2(SO_4)_3$ and is connected to an iron electrode in 1.0 M $Fe(NO_3)_2$. A salt bridge connects the two half-cells. (a) Write the anode and cathode half-reactions. (b) Identify which electrode is positive and negative. (c) Describe which direction the electrons are flowing. (d) Calculate the cell potential.

11. Which of the following statements about a standard reduction potential table is **false**?
    a. Reduction potentials are tabulated relative to the standard hydrogen electrode (SHE, $E^{o}_{red} = 0.00$ V, $E^{o}_{ox} = 0.00$ V)
    b. A negative reduction potential indicates a greater tendency to be reduced relative to hydrogen.
    c. A positive reduction potential indicates that species tend to prefer to exist in the reduced state.
    d. Species with large negative reduction potentials are good reducing agents.
    e. Species with large positive reduction potentials are good oxidizing agents.

12. Calculate the cell potential of the following galvanic cell:
    $Zr(s) \mid Zr^{4+}(aq)(0.0030\ M) \parallel Pd^{2+}(aq)(0.15\ M) \mid Pd(s)$

13. Calculate the standard reduction potential for the following reaction and indicate if it is spontaneous.
    $Ni(s) + Co^{2+}(aq) \rightarrow Co(s) + Ni^{2+}(aq)$

14. Calculate the Gibb's free energy, $\Delta G^{o}_{rxn}$, of the cell described in the previous problem.

15. Calculate the standard cell potential if a galvanic cell has an equilibrium constant, $K = 3.4 \times 10^{4}$. Assume there is one mole of electrons transferred and the temperature is 298 K.

16. How many coulombs of charge are produced by 15.0 A of current flowing for 0.35 hours?

17. How much silver is deposited on a ring in electrolysis when 3.0 A flows for 27 minutes. The electrolyte is $AgNO_3$.

18. If 10.5 grams of tin is deposited on the surface of a "tin" can during electrolysis with 4.22 A, how many minutes did this take? The electrolyte is $Sn(NO_3)_2$.

19. What characteristic distinguishes primary and secondary batteries?

20. Write the anode and cathode reactions for:
    a. An alkaline dry cell.
    b. A nickel-cadmium cell.
    c. A lead-acid storage battery

21. Which of the following does not describe the lithium-ion battery?
    a. Lithium atoms are incorporated into both the anode and cathode.
    b. The energy density is very high due to the large reduction potential for $Li^+/Li$ and low mass of lithium.
    c. High temperatures cause lithium-ion batteries to degrade fairly quickly.
    d. Lithium-ion batteries catch fire very often.
    e. They are widely used in cell phones, laptops, and other portable devices.

## ANSWERS TO THE CHAPTER OBJECTIVE QUIZ

1. **TRUE**

2. **e**

3. **reduction. oxidation.**

4. **The magnesium will be oxidized and the chromium (III) ions will be reduced.**

   oxidation: $3\{ Mg(s) \rightarrow Mg^{2+}(aq) + 2e^- \}$
   reduction: $2\{ Cr^{3+}(aq) + 3e^- \rightarrow Cr(s) \}$
   overall: $3 Mg(s) + 2 Cr^{3+}(aq) \rightarrow 3 Mg^{2+}(aq) + 2 Cr(s)$

5. **c**

6. **a**

7. **The anode reactant is Pb(s), anode product is $Pb^{2+}(aq)$. The cathode reactant is $Au^{3+}(aq)$ and cathode product is Au(s). The salt bridge is symbolized by ‖ and separates the anode reaction on the left from the cathode reaction on the right.**

   $$Pb(s) \mid Pb^{2+}(aq) \parallel Au^{3+}(aq) \mid Au(s)$$

8. **From the galvanic cell notation, the half-reactions can be written and the standard reduction and oxidation potentials found in Appendix I. The standard cell potential is the sum of the reduction and oxidation potentials: $E^o_{cell} = E^o_{red} + E^o_{ox}$**

   cathode: $2\{Ag^+(aq) + e^- \rightarrow Ag(s)\}$ $E^o_{red} = 0.7994$ V
   anode: $Cd(s) \rightarrow Cd^{2+}(aq) + 2e^-$ $E^o_{ox} = -(-0.403$ V)
   overall: $2 Ag^+(aq) + Cd(s) \rightarrow 2 Ag(s) + Cd^{2+}(aq)$ $E^o_{cell} = 1.202$ V

9. **The more positive reduction potential ($E^o_{red} = -0.91$ V) indicates the reaction that has a greater tendency to be reduced and therefore be the cathode reaction:**
   $Cr^{2+}(aq) + 2e^- \rightarrow Cr(s)$
   **The other reaction must be an oxidation reaction, therefore it would be reversed:**
   $Fe(s) + S^{2-}(aq) \rightarrow FeS(s) + 2e^-$ $E^o_{ox} = +1.01$ V
   **The cell potential is the sum of the reduction and oxidation potentials:**
   $E^o_{cell} = -0.91\text{ V} + (+1.01\text{ V}) = 0.10\text{ V}$

10. **(a) Consulting a standard reduction potential table allows us to identify which electrode is the anode and cathode. Iron's reduction potential is more positive; it is the cathode. The half-reactions are:**

    cathode: $Fe^{2+}aq) + 2e^- \rightarrow Fe(s)$ $E^o_{red} = -0.44$ V
    anode: $Al(s) \rightarrow Al^{3+}(aq) + 3e^-$ $E^o_{ox} = -(-1.66$ V)

**(b) In a galvanic cell, the anode is negative due to the electrons being produced through oxidation occurring at that electrode. The cathode is positive (electrons are being consumed in reduction). Therefore, the iron electrode is positive and the aluminum electrode is negative.**

**(c) Electrons travel from negative to positive, so they move from the aluminum electrode to the iron electrode.**

**(d)** $E^o_{cell} = E^o_{red} + E^o_{ox} = -0.44\text{ V} + -(-1.66\text{ V}) = 1.22\text{ V}$

**11. b**

**12. The Nernst equation is used to calculate nonstandard cell potentials. At 298 K, the Nernst equation is:** $E_{cell} = E^o_{cell} - \frac{0.0257\text{ V}\cdot\text{mol e}^-}{n}\ln[Q]$**, where** $Q$ **is the reaction quotient, and n is the moles of** $e^-$ **transferred in the overall oxidation-reduction reaction.**

| | |
|---|---|
| cathode: $2\{Pd^{2+}(aq) + 2e^- \rightarrow Pd(s)\}$ | $E^o_{red} = 0.987\text{ V}$ |
| anode: $Zr(s) \rightarrow Zr^{4+}(aq) + 4e^-$ | $E^o_{ox} = -(-1.53\text{ V})$ |
| overall: $2Pd^{2+}(aq) + Zr(s) \rightarrow 2Pd(s) + Zr^{4+}(aq)$ | $E^o_{cell} = 2.517\text{ V}$ |

n = 4 mol $e^-$ $\quad Q = \frac{[Pd^{2+}]^2[1]}{[1][Zr^{4+}]} = \frac{[0.15]^2[1]}{[1][0.0030]} = 7.5$

**(Recall the "concentrations" of solids are set = 1 in the reaction quotient.)**

$$E_{cell} = E^o_{cell} - \frac{0.0257\text{ V}\cdot\text{mol e}^-}{n}\ln[Q] = 2.517\text{ V} - \frac{0.0257\text{ V}\cdot\text{mol e}^-}{4\text{ mol e}^-}\ln[7.5] = 2.504\text{ V}$$

**13. A spontaneous galvanic cell must have a positive reduction potential.**

| | |
|---|---|
| cathode: $Co^{2+}(aq) + 2e^- \rightarrow Co(s)$ | $E^o_{red} = -0.28\text{ V}$ |
| anode: $Ni(s) \rightarrow Ni^{2+}(aq) + 2e^-$ | $E^o_{ox} = -(-0.25\text{ V})$ |
| overall: $Co^{2+}(aq) + Ni(s) \rightarrow Co(s) + Ni^{2+}(aq)$ | $E^o_{cell} = -0.03\text{ V}$ |

**The cell potential is negative; the reaction is not spontaneous.**

**14. The Gibb's free energy change is related to the standard cell potential:**

$\Delta G^o = -nFE^o$ **where n = 2 mol** $e^-$ **transferred,** $F = 96,485\text{ J/V}\cdot\text{mol e}^-$

$$\Delta G^o = -nFE^o = -(2\text{ mol e}^-)(96,485\text{ J/V}\cdot\text{mol e}^-)(-0.03\text{ V}) = 5790\text{ J} \approx 6000\text{ J}$$

**15. The standard cell potential is related to the equilibrium constant:**

$$E^{o}_{cell} = \frac{RT}{nF} \ln K$$

**Where $n$ = mol e⁻ transferred, $R$ = gas constant, 8.314 J/mol·K, $F$ = Faraday's constant, 96,485 J/V·mol e⁻, $T$ = temperature in Kelvin, and $K$ = the equilibrium constant.**

$$E^{o}_{cell} = \frac{(8.314\ \text{J/mol}\cdot\text{K})(298\text{K})}{(1\ \text{mol e}^-)(96{,}485\ \text{J/V}\cdot\text{mol e}^-)} \ln(3.4 \times 10^4) = 0.27\ \text{V}$$

**16. Coulombs of charge is equal to amperes times seconds: C = A · s.**

$$15.0\ \text{A} \times 0.35\ \text{hrs} \times \frac{3600\ \text{seconds}}{1\ \text{hour}} \times \frac{1\ \text{C}}{1\ \text{A}\cdot\text{s}} = 18{,}900\ \text{C}$$

**17. After the coulombs are calculated from the amps and time, the moles of electrons are determined and then moles of silver.**
**The half-reaction is: $Ag^+ + e^- \rightarrow Ag$.**

$$3.0\ \text{A} \times 27\ \text{min} \times \frac{60\ \text{seconds}}{1\ \text{min}} \times \frac{1\ \text{C}}{1\ \text{A}\cdot\text{s}} \times \frac{1\ \text{mol e}^-}{96{,}485\ \text{C}} \times \frac{1\ \text{mol Ag}}{1\ \text{mol e}^-} \times \frac{107.87\ \text{g Ag}}{1\ \text{mol Ag}} = 5.4\ \text{g Ag}$$

**18. From the mass of tin deposited, the moles of tin and then moles of electrons are found. With moles of electrons, coulombs of charge is determined. Since coulombs equal amps × seconds, dividing by amps gives time.**

$$10.5\ \text{g Sn} \times \frac{1\ \text{mol Sn}}{118.71\ \text{g Sn}} \times \frac{2\ \text{mol e}^-}{1\ \text{mol Sn}} \times \frac{96{,}485\ \text{C}}{1\ \text{mol e}^-} \times \frac{1\ \text{A}\cdot\text{s}}{1\ \text{C}} \times \frac{1}{4.22\ \text{A}} \times \frac{1\ \text{min}}{60\ \text{s}} =$$

**= 67.4 minutes**

**19. Primary batteries are ones that cannot be recharged; secondary batteries can be recharged. To create a secondary battery, the design must allow for the products formed during discharge to be available during recharge to recreate the batteries' reactants.**

**20. (a) alkaline dry cell:**

**cathode: $2\ MnO_2(s) + H_2O(\ell) + 2\ e^- \rightarrow Mn_2O_3(s) + 2\ OH^-(aq)$**
**anode: $Zn(s) + 2\ OH^-(aq) \rightarrow Zn(OH)_2(aq) + 2\ e^-$**

**(b) nickel-cadmium cell:**

**cathode: $NiO(OH)(s) + H_2O(\ell) + e^- \rightarrow Ni(OH)_2(s) + OH^-(aq)$**
**anode: $Cd(s) + 2\ OH^-(aq) \rightarrow Cd(OH)_2(aq) + 2\ e^-$**

**(c) lead-acid storage cell:**

**cathode: $PbO_2(s) + 3\ H^+(aq) + HSO_4^-(aq) + 2\ e^- \rightarrow PbSO_4(s) + 2\ H_2O(\ell)$**

**anode: $Pb(s) + HSO_4^-(aq) \rightarrow PbSO_4(s) + H^+(aq) + 2\ e^-$**

**21. d**

# CHAPTER 14

## *Study Goals*

The study goals outline specific concepts to be mastered in this section of the text chapter. Related problems at the end of the text chapter will also be noted. Working the questions noted should aid in mastery of each study goal and also highlight any areas that you may need additional help in.

### Section 14.1 *INSIGHT INTO* Cosmic Rays and Carbon Dating

1. Describe cosmic rays and some of the ways that they influence Earth and its atmosphere, including formation of carbon-14. *Work Problems 1–6.*

### Section 14.2 Radioactivity and Nuclear Reactions

2. Describe various modes of nuclear decay, including alpha decay, beta decay, positron emission, and electron capture. *Work Problems 7–10.*
3. Write, balance, and interpret equations for simple nuclear reactions. *Work Problems 11–17, 93, 96, 97*

### Section 14.3 Kinetics of Radioactive Decay

4. Use first-order integrated rate equations to interpret the kinetics of nuclear decay. *Work Problems 18–30, 89, 91, 95, 99, 101, 102*

### Section 14.4 Nuclear Stability

5. Understand how radioactive decay processes enhance stability of the nucleus. *Work Problems 31–38, 92.*

### Section 14.5 Energetics of Nuclear Reactions

6. Use Einstein's equation ($E = mc^2$) to describe the interconversion of matter and energy to calculate the binding energies of nuclei and the energy changes of nuclear reactions. *Work Problems 39–50, 86, 98.*

### Section 14.6 Transmutation, Fission, and Fusion

7. Define the term transmutation as it applies to nuclear reactions. *Work Problems 51–53.*
8. Describe nuclear fission and fusion and explain how both processes can be highly exothermic. *Work Problems 54–61.*
9. Describe the basic operation of a nuclear reactor and recognize the challenges associated with nuclear waste. *Work Problems 61–66.*

### Section 14.7 The interaction of Radiation and Matter

10. Recognize how penetrating power and ionization power determine the effect of radiation on matter, including living tissue. *Work Problems 67–80, 87, 90, 100, 103.*

**Section 14.8** ***INSIGHT INTO*** **Modern Medical imaging Methods**

11. Explain how radioisotopes can be used in medical imaging techniques. *Work Problems 81–85, 88.*

## *Solutions to the Odd-Numbered Problems*

Problem 14.1 Cosmic rays are sometimes referred to as *corpuscular rays.* How does this term distinguish them from other rays of the sun that reach the planet?

**One definition of corpuscle is a minute or elementary particle, such as a subatomic particle. This distinguishes them from other rays from the sun that are only energy, not particles.**

Problem 14.3 Use the web to find information on the composition of cosmic rays beyond hydrogen and helium.

**Answers may vary. According to the NASA website, http://imagine.gsfc.nasa.gov/docs/science/know_l1/cosmic_rays.html, 90% of cosmic rays are hydrogen, 9% are helium, with the balance other elements. Even in this one percent, there may be some very rare elements and isotopes in addition to the more common carbon, oxygen, and nitrogen isotopes.**

Problem 14.5 **(a)** How does carbon-14 enter a living plant?

**Carbon-14 forms in the upper atmosphere where nitrogen atoms can absorb a neutron from cosmic rays, forming $^{14}C$. Therefore, a small percentage of carbon in $CO_2$ is carbon-14 and enters a living plant during photosynthesis.**

**(b)** Write the equation for this reaction.

$$^{14}_{7}N + ^{1}_{0}n \rightarrow ^{14}_{6}C + ^{1}_{1}p$$

Problem 14.7 Refer to Figure 14.1 and describe the experiment in which Ernest Rutherford demonstrated that different types of radiation emanate from uranium.

**Rutherford placed a sample of uranium inside a shielded container, forcing all emitted radiation in one direction. He directed the radiation through an electric field, noticing that some emissions were deflected towards the positive side of the field. Other emissions were attracted to the negative side of the field. Still others were unaffected and passed straight through.**

Problem 14.9 Match the following forms of radioactive decay with the appropriate result.

**(a)** Alpha **(3)** 1. No change in mass number or atomic number

**(b)** Positron **(2)** 2. Atomic number decreases by 1
**(c)** Gamma **(1)** 3. Atomic number decreases by 2, mass number decreases by 4
**(d)** K capture **(2)** 4. Atomic number increases by 1
**(e)** Beta **(4)**

Problem 14.11 Complete each equation and name the particle ejected from the nucleus.

(a) ${}^{20}_{8}O \rightarrow ? + {}^{0}_{-1}\beta + \nu$

**? = $^{20}_{9}F$, beta particle**

(b) ? $\rightarrow {}^{232}_{92}U + {}^{4}_{2}He$

**? = $^{236}_{94}Pu$, alpha particle**

(c) ${}^{201}_{82}Pb \rightarrow {}^{201}_{83}Bi + ?$

**? = $^{0}_{-1}\beta$, beta particle**

Problem 14.13 Write equations for the following nuclear reactions.

**(a)** Alpha decay by $^{188}Bi$ $\quad {}^{188}_{83}Bi \rightarrow {}^{4}_{2}He + {}^{184}_{81}Tl$

**(b)** Beta emission by $^{87}Rb$ $\quad {}^{87}_{37}Rb \rightarrow {}^{0}_{-1}\beta + {}^{87}_{38}Sr$

**(c)** Positron emission by $^{40}K$ $\quad {}^{40}_{19}K \rightarrow {}^{0}_{+1}\beta + {}^{40}_{18}Ar$

**(d)** Electron capture by $^{138}La$ $\quad {}^{0}_{-1}e + {}^{138}_{57}Ba \rightarrow {}^{138}_{56}Ba$

Problem 14.15 **(a)** Explain why tritium, $^{3}H$, cannot undergo alpha decay.
**(b)** What type of decay would you expect tritium to undergo?
**(c)** Write the nuclear equation for the expected decay.

**(a) Tritium has a mass number equal to three, so it cannot emit a particle with a larger mass number (4).**
**(b) Beta emission**
**(c)** $^{3}_{1}H \rightarrow {}^{0}_{-1}\beta + {}^{3}_{2}He$

Problem 14.17 One way to convert lead into gold is to irradiate $^{206}Pb$ with neutrons. There are several steps in the process, and three neutrons are required for each lead atom. In addition to gold, three beta particles and three alpha particles are produced. (This

process requires a particle accelerator, and the cost is much higher than the value of any gold that could be produced.)
**(a)** What isotope of gold will this process produce?
**(b)** Write the overall nuclear reaction for the process.

**(a) Gold-197**

**(b)** $3\,[{}^{1}_{0}\mathrm{n}] + {}^{206}_{82}\mathrm{Pb} \rightarrow 3\,[{}^{4}_{2}\mathrm{He}] + 3\,[{}^{0}_{-1}\beta] + {}^{197}_{79}\mathrm{Au}$

Problem 2.19 $^{137}$Cs has a half-life of 30.2 years. How many years will it take for a 100.0-g sample to decay to 0.01 g?

**All nuclear decay is first-order:** $\ln[\frac{No}{N}] = kt$ **and** $t_{½} = \frac{\ln 2}{k}$

**The half-life will allow us to find the decay constant, *k*:** $k = \frac{\ln 2}{t_{1/2}} = \frac{0.693}{30.2\ \mathrm{y}} = 0.0299\ \mathrm{y}^{-1}$

**Now we solve for t:** $t = \ln[\frac{No}{N}] \div k \quad = \ln[\frac{100.0\ \mathrm{g}}{0.01\ \mathrm{g}}] \div 0.0299\ \mathrm{y}^{-1} = 402\ \mathrm{y}$

Problem 14.21 The half-life of Sb-110 is 23.0 s.
(a) Determine its decay constant in $s^{-1}$.
(b) Compute the activity of a 1.000-g sample of $^{110}$Sb in Bq and Ci.

**(a) See Problem 19.** $k = \frac{\ln 2}{t_{1/2}} = \frac{0.693}{23.0\ \mathrm{s}} = 0.0301\ \mathrm{s}^{-1}$

**(b) Activity in Bq is the** $\frac{\Delta N}{\Delta t}$ **in disintegrations per second. In one half-life, the number *N* will be reduced by one-half:**

$$1.000\ \mathrm{g}\ {}^{110}\mathrm{Sb} \times \frac{1\ \mathrm{mol}}{110\ \mathrm{g}} \times \frac{6.022\times10^{23}\ \mathrm{nulcei}}{\mathrm{mol}} \div 2 \div 23.0\ \mathrm{s} = 1.2 \times 10^{20}\ \mathrm{Bq}$$

**One Ci = $3.7 \times 10^{10}$ Bq:**

$$1.2 \times 10^{20}\ \mathrm{Bq} \times \frac{\mathrm{Ci}}{3.7\times10^{10}\mathrm{Bq}} = 3.2 \times 10^{9}\ \mathrm{Ci}$$

Problem 14.23 The half-life of $^{19}$O is 29s. Suppose that a scientist wishes to incorporate $^{19}$O into a molecule and then use its radioactivity to trace the fate of that molecule in a reaction. To make an accurate measurement, there must be at least 1.50 mg of $^{19}$O left by the end of the study, which will take 2.5 min each time the experiment is run. What is the minimum mass of $^{19}$O that must be used each time the experiment is conducted?

**See Problem 19. The final amount, $N$, is 1.50 mg. The time, t, is 2.5 minutes or 150 seconds. The rate constant is found from the half-life.**

$$k = \frac{\ln 2}{t_{1/2}} = \frac{0.693}{29\ \mathrm{s}} = 0.024\ \mathrm{s}^{-1}$$

**Solving for $N_o$:**

$$N_o = \frac{N}{e^{-kt}} = \frac{1.50\ \mathrm{mg}}{e^{-(0.024\mathrm{s}^{-1})(150\ \mathrm{s})}} = 54\ \mathrm{mg}\ {}^{19}\mathrm{O}$$

Problem 14.25 A wooden artifact is burned and found to contain 21 g of carbon. The $^{14}C$ activity of the sample is 105 disintegrations/min. What is the age of the artifact?

**The half-life of $^{14}C$ is 5730 years and so the decay constant is:** $k = \frac{\ln 2}{t_{1/2}} = \frac{0.693}{5730\ \mathrm{y}} =$ $= 1.21 \times 10^{-4}\ \mathrm{y}^{-1}$.

**From the Check Your Understanding problem on page 425, the $^{14}C$ activity of all living things is 0.255 Bq/g C. This is equal to $0.255 \times 60 = 15.3$ disintegrations per minute per gram C. The activity of the artifact is 105 disintegrations per minute per 21 grams C or 5.0 disintegrations per minute per gram C. Using the first-order kinetic equation:**

$$t = \ln\left[\frac{N_o}{N}\right] \div k \qquad = \ln\left[\frac{15.3\ \mathrm{dist.}}{5.0\ \mathrm{dist.}}\right] \div 1.21 \times 10^{-4}\ \mathrm{y}^{-1} = 9200\ \mathrm{y}$$

Problem 14.27 Tritium, $^{3}H$, is a radioactive isotope of hydrogen produced by cosmic rays in the atmosphere, where it oxidizes and becomes part of the water cycle. Its half-life is 12.33 yr. Which of the following could be dated using tritium? Explain your answer.

**(a)** 4000-year-old ice from Antarctica

**No, 4000 years represents over 300 half-lives of tritium and there would not be any measurable amounts of tritium left.**

**(b)** Modern French wine

**Yes, a modern wine would experience only several half-lives of tritium at most.**

**(c)** Ancient Egyptian beer from a tomb in the Valley of the Kings

**No, same as (a).**

Problem 14.29 A piece of a spear handle is found in an archaeological dig in Central America. It contains 12.5% as much $^{14}C$ as a tree living today. Based only on the half-life of carbon-14, how old is the spear handle?

**The half-life of $^{14}C$ is 5730 years. There are four half-lives occurring to reach 12.5%:**
**100% → 50% → 25% → 12.5%.**

**The piece of wood is 4 × 5730 y = 22,900 years.**

Problem 14.31 What is the *N*/*Z* ratio for each of the following nuclides?
**(a)** $^{14}_{7}N$ **(b)** $^{114}_{50}Sn$ **(c)** $^{234}_{90}Th$

**(a) Z = 7; N = 14 – 7 = 7;** $N/Z = \frac{7}{7} = 1.0$

**(b) Z = 50; N = 114 – 50 = 64;** $N/Z = \frac{64}{50} = 1.28$

**(c) Z = 90; N = 234 – 90 = 144;** $N/Z = \frac{144}{90} = 1.6$

Problem 14.33 Based on their positions relative to the band of stability, predict the type of decay that each of the following will undergo. Write equations for the expected decays.
**(a)** $^{73}_{36}Kr$ **(b)** $^{13}_{8}O$ **(c)** $^{126}_{50}Sn$

**(a) The N/Z ratio is $\frac{37}{36} = 1.03$, indicating the isotope is probably below the band of stability. The decay is probably beta emission.**
$$^{73}_{36}Kr \rightarrow {}^{0}_{-1}\beta + {}^{73}_{37}Rb$$
**(b) The N/Z ratio is $\frac{5}{8} = 0.63$, indicating the isotope is probably above the band of stability. The decay is probably positron emission.**
$$^{13}_{8}O \rightarrow {}^{0}_{+1}\beta + {}^{13}_{7}N$$
**(c) The N/Z ratio is $\frac{76}{50} = 1.52$, indicating the isotope is probably below the band of stability. The decay is probably beta emission.**
$$^{126}_{50}Sn \rightarrow {}^{0}_{-1}\beta + {}^{126}_{51}Sb$$

Problem 14.35 ■ The thorium-232 radioactive decay series, beginning with $^{232}_{90}Th$ and ending with $^{208}Pb$, occurs in the following sequence: α, β⁻, β⁻, α, α, α, α, β⁻, β⁻, α. Write an equation for each step in this series.

α: $^{232}_{90}Th \rightarrow {}^{4}_{2}He + {}^{228}_{88}Ra$

α: $^{220}_{86}Rn \rightarrow {}^{4}_{2}He + {}^{216}_{84}Po$

$\beta^-$: $^{228}_{88}Ra \rightarrow {}^{0}_{-1}\beta + {}^{228}_{89}Ac$

α: $^{216}_{84}Po \rightarrow {}^{4}_{2}He + {}^{212}_{82}Pb$

$\beta^-$: $^{228}_{89}Ac \rightarrow {}^{0}_{-1}\beta + {}^{228}_{90}Th$

$\alpha$: $^{228}_{90}Th \rightarrow {}^{4}_{2}He + {}^{224}_{88}Ra$

$\alpha$: $^{224}_{88}Ra \rightarrow {}^{4}_{2}He + {}^{220}_{86}Rn$

$\beta^-$: $^{212}_{82}Pb \rightarrow {}^{0}_{-1}\beta + {}^{212}_{83}Bi$

$\beta^-$: $^{212}_{83}Bi \rightarrow {}^{0}_{-1}\beta + {}^{212}_{84}Po$

$\alpha$: $^{212}_{84}Po \rightarrow {}^{4}_{2}He + {}^{208}_{82}Pb$

Problem 14.37 The $^{238}U$ series can be used to date minerals containing uranium by measuring the ratio of $^{238}U$ to $^{206}Pb$, but all of the decay products have to remain in the mineral through several of their half-lives so that essentially all the $^{238}U$ that decays becomes $^{206}Pb$. One of the decay products in this series is the noble gas, $^{222}Rn$. How would the experimentally determined age of the mineral be affected if some of the radon diffused from the rock?

**The level of $^{206}Pb$ would be low, which would indicate a lower original amount of $^{238}U$. This would cause the calculated age to be biased low.**

Problem 14.39 Calculate the binding energy of 1 mole of $^{14}C$ nuclei. The experimentally determined mass of a carbon-14 atom (including its six electrons) is 14.003242 u.

**Binding energy is calculated using the mass defect, Δm: $E_b = \Delta mc^2$. The mass defect is the difference between the calculated mass of the atom and the actual.**

$$\Delta m = [6p \times 1.007825\text{ u} + 8n \times 1.008665\text{ u} + 6e \times 0.00054858\text{ u}] - 14.003242\text{ u} =$$
$$= 0.116319\text{ u or } 0.116319\text{ g/mol}$$

$$E_b = \Delta mc^2 = (0.116319 \times 10^{-3}\text{ kg/mol})(2.9979 \times 10^{8}\text{ m/s})^2 = 1.04541 \times 10^{13}\ \frac{\text{kg}\cdot\text{m}^2}{\text{s}^2\cdot\text{mol}} =$$
$$= 1.04541 \times 10^{13}\ \frac{\text{J}}{\text{mol}}$$

Problem 14.41 Compute the binding energy of the $^{7}Li$ nucleus, whose experimentally determined mass is 7.016004 u.

**See Problem 39.**

$$\Delta m = [3p \times 1.007825\text{ u} + 4n \times 1.008665\text{ u}] - 7.016004\text{ u} = 0.042131\text{ u or } 0.042131\text{ g/mol}$$

$$E_b = \Delta mc^2 = (0.042131 \times 10^{-3}\text{ kg/mol})(2.9979 \times 10^{8}\text{ m/s})^2 = 3.78648 \times 10^{12}\ \frac{\text{kg}\cdot\text{m}^2}{\text{s}^2\cdot\text{mol}} =$$
$$= 3.78648 \times 10^{12}\ \frac{\text{J}}{\text{mol}}$$

Problem 14.43 Compute the binding energy of the $^{14}N$ nucleus, which has an experimentally determined mass of 14.003074 u.

**See Problem 39.**

$$\Delta m = [7p \times 1.007825\text{ u} + 7n \times 1.008665\text{ u}] - 14.003074\text{ u} = 0.112356\text{ u or } 0.112356\text{ g/mol}$$

$$E_b = \Delta mc^2 = (0.112356 \times 10^{-3}\ \text{kg/mol})(2.9979 \times 10^8\ \text{m/s})^2 = 1.009789 \times 10^{13}\ \frac{\text{kg·m}^2}{\text{s}^2\text{·mol}} =$$

$$= 1.009789 \times 10^{13}\ \frac{\text{J}}{\text{mol}}$$

Problem 14.45 It takes 360 kJ to keep a 100-watt lightbulb burning for 1 hour. Assuming that the lightbulb won't burn out, for how many years could the energy of the mass defect of one mole of $^{14}C$ keep a 100-W lightbulb burning?

**From Problem 39, the binding energy of a mole of $^{14}C$ is $1.04541 \times 10^{13}$ J.**

$$1.04541 \times 10^{13}\ \text{J} \times \frac{1\ \text{hr}}{360 \times 10^3\ \text{J}} \times \frac{\text{day}}{24\ \text{hr}} \times \frac{\text{yr}}{365.25\ \text{day}} = 3310\ \text{years}$$

Problem 14.47 Lead has a magic number of protons. A doubly magic nuclide should be exceptionally stable. Why isn't lead-164 stable?

**The N/Z ratio is too low; it will undergo positron emission.**

Problem 14.49 Use the web to identify the heaviest element that has been observed to form a chemical compound. What element is it and what compound does it form? Are heavier elements incapable of forming compounds?

**The Inorganic Radiochemistry of Heavy Elements (Ivo Zvara, Joint Institute for Nuclear Research, Dubna, Russian Federation, Springer Publishing) reports the formation of an oxide of element 108, hassium: $HsO_4$.**

Problem 14.51 Ernest Rutherford carried out the first artificial transmutation in 1919 when he bombarded $^{14}_{7}N$ with alpha particles. The result was an oxygen nucleus:

$$^{14}_{7}\text{N} + {}^{4}_{2}\text{He} \rightarrow {}^{18}_{9}\text{F}^* \rightarrow {}^{17}_{8}\text{O} + {}^{1}_{1}\text{p}$$

Radiochemists use a shorthand notation to represent transmutations. Omitting the intermediate $^{18}_{9}F^*$ nucleus, the reaction above can be represented as $^{14}N\ (\alpha, p)\ ^{17}O$.

Write balanced equations for each of the indicated nuclear bombardments.

**(a)** $^{27}Al\ (p, \gamma)\ ^{28}Si$

**(b)** $^{40}Ar\ (n, \beta^-)\ ^{41}K$

$$\textbf{(a)}\ {}^{27}_{13}\text{Al} + {}^{1}_{1}\text{p} \rightarrow \gamma + {}^{28}_{14}\text{Si}$$

$$\textbf{(b)}\ {}^{40}_{18}\text{Ar} + {}^{1}_{0}\text{n} \rightarrow {}^{0}_{-1}\beta + {}^{41}_{19}\text{K}$$

Problem 14.53 Slow neutrons can be absorbed by some nuclei to produce new elements. Why does a neutron make a better projectile for a transmutation reaction than an alpha or beta particle?

**Neutrons are neutral, the lack of a charge means they do not have to overcome any electrostatic repulsions.**

Problem 14.55 How much energy is released in the fission of 1 kg of $^{235}U$ according to the equation below? The experimentally determined masses of the products are 136.92532 u and 96.91095 u, respectively. (The masses of $^{235}U$ and the neutron are given in Example Problem 14.6.)

$$^{235}_{92}U + 3\ ^{1}_{0}n \rightarrow\ ^{18}_{9}F^* \rightarrow\ ^{137}_{52}Te +\ ^{97}_{40}Zr + 2\ ^{1}_{0}n$$

**The mass defect for this reaction is the difference between the masses of the products versus the reactants.**

**$\Delta m = [235.0439231 + 1.0086649\ u] - [136.92532\ u + 96.91095\ u] = 2.216318\ u$ or $2.216318$ g/mol U.**

$$E_b = \Delta mc^2 = (2.216318 \times 10^{-3}\ kg/mol)(2.9979 \times 10^8\ m/s)^2 = 1.94414 \times 10^{14}\ \frac{kg \cdot m^2}{s^2 \cdot mol} =$$

$$= 1.94414 \times 10^{14}\ \frac{J}{mol\ ^{235}U}$$

$$1.94414 \times 10^{14}\ \frac{J}{mol\ ^{235}U} \times \frac{235.0439231\ g}{mol\ ^{235}U} \times \frac{1\ kg}{1000\ g} = 4.56958 \times 10^{13}\ \text{J for one kg of}\ ^{235}U$$

Problem 14.57 Because the fissionable nucleus of uranium is $^{235}U$, the large quantity of $^{238}U$ in a nuclear reactor isn't used. However, it is possible to produce fissionable $^{239}Pu$ from $^{238}U$ by the reaction $^{238}U\ (n, \beta^-)\ ^{239}Np$. $^{239}Np$ decays by $\beta^-$ emission to $^{239}Pu$. A reactor used for this purpose is called a *breeder reactor*.

**(a)** What is the compound nucleus formed by $^{238}U$? $^{239}_{92}U$

**(b)** Write balanced equations for production of $^{239}Pu$.

$$^{238}_{92}U + ^{1}_{0}n \rightarrow\ ^{0}_{-1}\beta +\ ^{239}_{93}Np$$

$$^{239}_{93}Np \rightarrow\ ^{0}_{-1}\beta +\ ^{239}_{94}Pu$$

Problem 14.59 According to the U.S. Nuclear Regulatory Commission, in 2011 there were 104 commercial nuclear reactors in the United States. In 2011, $4.1 \times 10^6$ gigawatt-hours of electricity were generated for use, and 19.43% of that was generated by nuclear power.

**(a)** A nuclear technician assumed that the products of the fission of $^{235}$U by one neutron were $^{97}$Zr and $^{137}$Te (see Problem 14.55). Ignoring decay products, the technician estimated the total mass of zirconium and tellurium in metric tons that would have to be disposed of in 2011 as nuclear waste (1 watt-hour = 3600 joules; 1 metric ton = 1000 kg). What was the technician's result?

**Using the energy released when one mole of $^{235}$U fissions (Problem 14.55), calculate amount of uranium needed to produce the energy (and amount of fission products):**

$$0.1943 \times (4.1 \times 10^{15} \text{ watt-hrs elec.}) \times \frac{3600 \text{ J}}{1 \text{ watt-hr}} \times \frac{1 \text{ mol } ^{235}\text{U}}{1.94414 \times 10^{14} \text{ J}} =$$

$$1.475 \times 10^4 \text{ mol } ^{235}\text{U} = 1.509 \times 10^4 \text{ mol } ^{97}\text{Zr} = 1.509 \times 10^4 \text{ mol } ^{137}\text{Te}$$

$$1.475 \times 10^4 \text{ mol } ^{97}\text{Zr} \times \frac{97 \times 10^{-3} \text{ kg}}{\text{mol } ^{97}\text{Zr}} \times \frac{1 \text{ metric ton}}{1000 \text{ kg}} = 1.4 \text{ metric tons}$$

$$1.475 \times 10^4 \text{ mol } ^{137}\text{Te} \times \frac{137 \times 10^{-3} \text{ kg}}{\text{mol } ^{137}\text{Te}} \times \frac{1 \text{ metric ton}}{1000 \text{ kg}} = 2.0 \text{ metric tons}$$

**Total waste = 1.4 + 3.0 = 3.4 metric tons**

**(b)** Look up the half-lives of $^{97}$Zr and $^{137}$Te. When the waste was disposed of, do you think there was a measurable quantity of either isotope? Explain.

**The half-life of $^{97}$Zr is 16.8 hours (beta emission) and is 2.5 seconds for $^{137}$Te (beta emission). There would not be significant quantities of either isotope present after several days.**

**(c)** Look up the decay mode of each isotope and write the decay series to a stable isotope for each one. Which decay series has a radioactive isotope that will remain in the waste for many years?

$^{97}\text{Zr} \rightarrow {}^{0}_{-1}\beta + {}^{97}\text{Nb}$ $t_{1/2}$=2.5s

$^{97}\text{Y} \rightarrow {}^{0}_{-1}\beta + {}^{97}\text{Mo}$ $t_{1/2}$=1.23h

**Stable**

$^{137}\text{Te} \rightarrow {}^{0}_{-1}\beta + {}^{137}\text{I}$ $t_{1/2}$ = 2.5s

$^{137}\text{I} \rightarrow {}^{0}_{-1}\beta + {}^{137}\text{Xe}$ $t_{1/2}$ = 24.5s

$^{137}\text{Xe} \rightarrow {}^{0}_{-1}\beta + {}^{137}\text{Cs}$ $t_{1/2}$ = 3.83 m

$^{137}\text{Cs} \rightarrow {}^{0}_{-1}\beta + {}^{137}\text{Ba}$ $t_{1/2}$ = 30.2 y

**Stable**

**Cesium-137 has a half-life of over 30 years; it will remain for many years.**

**(d)** Using the stable isotope products in your calculation, how far off was the technician's estimate of total waste? Explain.

**The two stable decay products both have the same mass numbers as the original fission products (because the decay series is all beta emission). This mean the mass of stable waste would be about the same mass as the original fission products.**

Problem 14.61 A fusion reaction that has been proposed involves bombarding boron-11 with a proton and producing a single, stable product.
**(a)** Write the equation for the reaction.
**(b)** Which of the obstacles in the previous problem is overcome by this reaction?

**(a)** $^{1}_{1}\text{P} + {}^{11}_{5}\text{B} \rightarrow {}^{12}_{6}\text{C}$

**(b) Fusion reactions require enormous amounts of energy to join the nuclei and many fusion reactions produce high-energy neutrons as products. This reaction, however, does not produce any radiation products.**

Problem 14.63 In 1998, the Nuclear Regulatory Commission suggested that residents who live near a nuclear power plant should have a supply of potassium iodide on hand in the event of a nuclear plant emergency. Explain why the suggestion makes sense.

**Radioactive iodine is one of the fission/decay products of nuclear fuel. The human body absorbs iodine to produce thyroid hormone. By taking supplemental KI when there is a release of radiation, it prevents absorption of the radioactive iodine.**

Problem 14.65 A nuclear waste storage facility like the Yucca Mountain site must last for roughly 10,000 years. What engineering uncertainties are introduced for the design of such a facility compared to a design for a tunnel for transportation services, for example?

**Significant geological changes may occur after a period of 10,000 years as opposed to a transportation tunnel which may have a reasonable life of 100 years, not to mention political and social uncertainties for a ten-thousand year period.**

Problem 14.67 Use information from Section 6.7 to estimate which form of electromagnetic radiation is the lowest energy ionizing radiation.

**The longest wavelength (low frequency) forms of electromagnetic radiation are the lowest energy forms of ionizing radiation. Examples are radio and TV waves.**

Problem 14.69 Rank the most common forms of radioactivity in order of increasing penetrating power.

$\alpha < \beta < \gamma$

Problem 14.71 Polonium-210 is an alpha-emitting nuclide found in cigarette smoke. How is the fact that it is found in such smoke related to its health hazard?

**Alpha radiation is not normally very hazardous due to its low penetrating ability; just skin can block alpha particles. When an alpha-emitter is inhaled, the sensitive lung cells are not protected by skin and can suffer significant damage from these particles.**

Problem 14.73 How is the damage from the possible impact of ionizing cosmic rays on an electronic device in a satellite related to the materials from which electronic devices are constructed?

**Electronic devices rely on specific distributions of electrons and holes in the semiconductor material. Introduction of ions, from ionizing radiation, could have catastrophic results in the performance of the electronics in this environment.**

Problem 14.75 A Geiger counter is set to click once for every 100 disintegrations that it detects. The counter registers 20 clicks per second for a sample of an unknown isotope. What is the activity of the isotope in Ci? In $\mu$Ci?

$$\frac{20\ \text{clicks}}{\text{s}} \times \frac{100\ \text{dist.}}{\text{click}} = \frac{2000\ \text{dist.}}{\text{s}} \times \frac{1\ \text{Ci}}{3.7\times10^{10}\ \text{dist.}} = 5.4\times10^{-8}\ \text{Ci}$$

$$(5.4\times10^{-8}\ \text{Ci}) \times \frac{1\times10^{6}\ \mu\text{Ci}}{1\ \text{Ci}} = 0.054\ \mu\text{Ci}$$

Problem 14.77 An average person is exposed to about 360 millirem of background radiation a year from a variety of natural and man-made sources. About two-thirds of that typically comes from inhalation of $^{222}$Rn produced in soil by the decay of $^{238}$U. Given that $^{222}$Rn decays by alpha emission, estimate the absorbed dose in **(a)** joules and **(b)** grays for a 60-kg person from one year's inhalation of radon.

$$\text{(a)}\ \tfrac{2}{3} \times (360\times10^{-3}\ \text{rem}) \times \frac{Q(20)\times\text{rad}}{1\ \text{rem}} \times \frac{1\ \text{Gy}}{100\ \text{rad}} = 0.048\ \text{grays}$$

$$\text{(b)}\ \tfrac{2}{3} \times (360\times10^{-3}\ \text{rem}) \times \frac{Q(20)\times\text{rad}}{1\ \text{rem}} \times \frac{1\ \frac{\text{erg}}{\text{g}}}{1\ \text{rad}} \times \frac{0.01\ \text{J}}{1\ \text{erg}} \times 60.0\times10^{3}\ \text{g body mass} =$$

**2880 J**

Problem 14.79 Concern over radiation exposure has led some local, state, and federal agencies to propose a "zero dose limit" above background radiation (see Problem 14.77). A zero dose limit means that industry could not allow anyone to be exposed to radiation above the average background exposure per year. Why does background radiation change with altitude? Suggest a reason why the airline industry might oppose a zero dose limit.

**Higher altitudes cause more exposure to cosmic rays. It would be impossible to prevent airline employees from experiencing less than above-average exposure to radiation because of the altitude of airplanes.**

Problem 14.81 Write the equation for the nuclear decay of each of the following isotopes used in positron emission tomography (PET): **(a)** $^{18}$F **(b)** $^{11}$C **(c)** $^{13}$N

**(a)** $^{18}_{9}\text{F} \rightarrow {}^{0}_{+1}\beta + {}^{18}_{8}\text{O}$

**(b)** $^{11}_{6}\text{C} \rightarrow {}^{0}_{+1}\beta + {}^{11}_{5}\text{B}$

**(c)** $^{13}_{7}\text{N} \rightarrow {}^{0}_{+1}\beta + {}^{13}_{6}\text{C}$

Problem 14.83 Write the nuclear equation for the antimatter annihilation of an electron by a positron to produce two gamma rays.

$$^{0}_{+1}\beta + {}^{0}_{-1}e \rightarrow 2\,[\,{}^{0}_{0}\gamma\,]$$

Problem 14.85 Osteoporosis (bone loss) is a growing medical problem as the average age of the U.S. population increases. It can be diagnosed using the isotope gadolinium-153. Using the Internet and other resources, research and write a paragraph about using $^{153}$Gd to determine bone density.

**Answers will vary.**

Problem 14.87 Recall that the definition of the curie is 1 Ci = $3.70 \times 10^{10}$ disintegrations per second and that this value was originally chosen so that the activity of 1.0 g of $^{226}$Ra would be equal to 1 Ci. Given that $^{226}$Ra has a molar mass of 226.025 g/mole and a half-life of 1600 years, calculate Avogadro's number.

**1.0 g $^{226}$Ra = 1 Ci = $3.70 \times 10^{10}$ atoms disintegrating /s, so ½ (1.0 g $^{226}$Ra) will decay in 1600 years at that rate.**

$$\frac{3.70\times10^{10}\text{ atoms}}{1\text{ s}} \times 1600\text{ y} \times \frac{365\text{ d}}{\text{y}} \times \frac{24\text{ h}}{1\text{ d}} \times \frac{3600\text{ s}}{1\text{h}} = 1.867 \times 10^{21}\text{ atoms decayed}$$

$$1.867 \times 10^{21}\text{ atoms decayed} = 0.5\text{ g }^{226}\text{Ra} = \frac{0.5\text{ g }^{226}\text{Ra}}{226.025\text{ g/mol}} = 0.0022089\text{ moles}$$

$$N_{av} = \frac{1.867\times10^{21}\text{ atoms}}{0.002089\text{ mol}} = 8.45 \times 10^{23}\text{ atoms/mole}$$

**This discrepancy from the accepted value of $6.022 \times 10^{23}$ is probably due to inaccuracies in measuring the half-life of radium-226.**

Problem 14.89 Tellurium-128 undergoes an extremely rare form of radioactive decay know as double beta decay.

$$^{128}_{52}\text{Te} \rightarrow {}^{128}_{54}\text{Xe} + 2\ {}^{0}_{-1}\beta + 2\ \bar{\nu}$$

This is an extremely low probability event; its half-life of $2.2 \times 10^{24}$ years is among the longest half-lives that have been determined. Estimate the activity in a 128-g (1 mol) sample of $^{128}$Te. How many $^{128}$Te atoms in such a 1-mol sample would decay in a day?

**Based on the half-life, in $2.2 \times 10^{24}$ years, ½ of a mole will decay:** $\dfrac{\mathbf{1/2\ mol\ (128\ g/mol)}}{\mathbf{2.2 \times 10^{24}\ yrs}}$

**$= 2.9 \times 10^{-23}$ g/yr.**

$$2.9 \times 10^{-23}\ \frac{\text{g}}{\text{yr}} \times \frac{1\ \text{yr}}{365\ \text{d}} \times \frac{1\ \text{mol}}{128\ \text{g}} \times \frac{6.022 \times 10^{23}\ \text{atoms}}{1\ \text{mol}} = 3.7 \times 10^{-4}\ \text{atoms/day}$$

Problem 14.91 Two radioactive isotopes of iron, $^{52}$Fe and $^{53}$Fe, are both positron emitters with half-lives of 8.28 hours and 8.51 minutes, respectively. A particular sample of iron initially emits positrons at a rate of 210 counts per minute. What would be the activity of the sample after 30 minutes if the sample contained the following?
(a) only $^{52}$Fe? (b) only $^{53}$Fe? (c) a mixture of 50% of each isotope

$$t_{½} = \frac{\ln 2}{k} \text{ so } k = \frac{\ln 2}{t_{1/2}} \quad k_{52\text{Fe}} = \frac{0.693}{8.28\ \text{hr}} = 0.0837\ \text{h}^{-1} \text{ and } k_{53\text{Fe}} = \frac{0.693}{8.51\ \text{min}} = 0.0814\ \text{min}^{-1}$$

**(a) Activity = [A] = $[A_o]e^{-kt}$ = (210 counts/min)$e^{-(0.0837\text{h-1})(0.5\text{h})}$ =**
**= (210 counts/min)$e^{-(0.0419)}$ = 201 counts/min**

**(b) Activity = [A] = $[A_o]e^{-kt}$ = (210 counts/min)$e^{-(0.0814\text{min-1})(30\text{min})}$ =**
**= (210 counts/min)$e^{-(2.44)}$ = 18 counts/min**

**(c) ½ (210) + ½ (18) = 110 counts /min**

Problem 14.93 Tailings from uranium mining contain radium, which decays to produce radon. The radon is also radioactive and undergoes further decay. Assume the radium present is all $^{226}$Ra, and write the pair of reactions for these two radioactive events. Use Figure 14.4 to help choose the most plausible pathway for the decay of radon.

$$^{226}_{88}\text{Ra} \rightarrow {}^{4}_{2}\text{He} + {}^{222}_{86}\text{Rn}$$

$$^{222}_{86}\text{Ra} \rightarrow {}^{4}_{2}\text{He} + {}^{218}_{84}\text{Po}$$

Problem 14.95 Both $^{192}$Ir and $^{137}$Cs are used in cancer treatments. The half-life of $^{192}$Ir is 74 days and that of $^{137}$Cs is 137 days. Suppose you work for a company that produces medical isotopes. If the radiation dosage needed from either of these isotopes is the same for a given treatment, which product would require a larger sample to be produced prior to shipment to a hospital that is 2 days away from your plant? Explain your reasoning.

**$^{192}$Ir. Its half-life is almost half that of $^{137}$Cs so it decays almost twice as fast.**

Problem 14.97 An interesting application of nuclear decay is the radioisotope thermoelectric generator, or RTG. In such a device, energy from radioactive decay is initially converted to heat in a manner similar to the decay heat discussed in Section 14.6. The heat is then converted to electrical energy using thermocouples in what is known as the Seebeck effect. The RTG is a long-lived source of low levels of electrical energy, and has been used to power various space probes, including *Cassini*, which is currently orbiting Saturn. The most commonly used isotope in RTGs is $^{238}$Pu. (a) Write a nuclear equation for the alpha decay of $^{238}$Pu. (b) The energy output from the decay of $^{238}$Pu is about 0.54 kW per kilogram of plutonium. Suppose the fuel for a particular RTG is 7.8 kg of $^{238}PuO_2$, and that the conversion of heat to electrical energy is about 6.5% efficient. Estimate the electrical power output in Watts for such an RTG.

**The mass of $^{238}$Pu in the $^{238}PuO_2$ is: $\frac{238 \text{ g/mol Pu}}{270 \text{ g/mol } PuO_2} \times 7.8 \text{ kg } ^{238}PuO_2 = 6.88 \text{ kg } ^{238}Pu$**

**$6.88 \text{ kg } ^{238}Pu \times \frac{0.54 \text{ kW}}{\text{kg}} \times 0.065 \text{ efficiency} = 0.24 \text{ kW} = 240 \text{ W}$**

Problem 14.99 The half-life of $^{237}$Np is $2.14 \times 10^6$ years, and Earth is $4.54 \times 10^9$ years old. If there was some amount of $^{237}$Np present when Earth was formed, and no $^{237}$Np has been formed since, what fraction of the original $^{237}$Np remains today?

**The fraction of $^{237}$Np remaining can be represented by $\frac{A}{A_o}$ where $A$ = the present $^{237}$Np and $A_o$ = the original $^{237}$Np. The decay can be represented by $\ln[\frac{A}{A_o}] = -kt$.**

**The decay constant can be found by $k = \frac{\ln 2}{t_{1/2}} = \frac{0.693}{2.14 \times 10^6 \text{ yrs}} = 3.23 \times 10^{-7} \text{ yrs}^{-1}$.**

**$\frac{A}{A_o} = e^{-kt} = e^{-(3.23 \times 10{-7} \text{ yrs}{-1})(4.54 \times 10{-9} \text{ yrs})} = e^{-1470} \approx 0$**

**This time period represents about 2100 half-lives, so the fraction would be $\frac{1}{2^{(2100)}}$.**

Problem 14.101 Thallium-201 has a long history of use in nuclear medicine. Suppose that a customer requires 140 mg of the isotope, but delivery will take 18 hours by an express courier service. You are assigned to determine how much isotope you need to produce. What information would you need to look up to answer the question?

**The half-life of the isotope must be known to determine how fast it will decay. The amount needed to ship out could then be calculated to ensure delivery of 140 mg.**

Problem 14.103 Suppose that you want to use radioactive iodine to run an experiment to treat a tumor. You can choose either $^{123}$I or $^{125}$I for this work. What do you need to know about these isotopes? What design parameters for the experiment factor into your decision?

**You would want to know the mode of decay. Gamma radiation is most effective in treating tumors. You would also want to know the half-lives – if one were very short it would not be practical to use that isotope.**

## CHAPTER OBJECTIVE QUIZ

This quiz will test your understanding of the basic text chapter objectives and give you additional practice problems. You should work this quiz after completing the end-of-chapter questions. The solutions to these questions are found at the end of this chapter.

1. Why does carbon-14 form in the upper atmosphere but not on our planet's surface?
2. Write and balance the following reactions:
   (a) Positron emission of $^{87}Zr$
   (b) Electron capture of $^{102}Rh$
   (c) Beta emission of $^{3}H$
3. What does *nuclear decay* mean? Why is nuclear decay always first-order?
4. What is the decay constant for tritium ($^{3}H$), whose half-life is 12.32 years?
5. Tritium is sometimes used to make glow-in-the-dark watch faces and gun sights. If a watch originally contained 0.55 mg of tritium, how much would remain after thirty years?
6. An ant trapped in amber (prehistoric tree sap) is analyzed and found to contain 18% of the carbon-14 activity in living ants. How old is the ant? Half-life = 5730 years.
7. Predict the type of nuclear decay given the following neutron-to-proton ratios (N/Z).
   (a) Z = 3, N/Z = 1.01
   (b) Z = 26, N/Z = 1.51
   (c) Z = 70, N/Z = 0.75
8. Find the mass defect and binding energy of $^{40}Ca$, actual mass = 39.96259 u.

9. Why is transmutation not possible in chemical reactions?

10. Using Figure 14.7 and the concept of binding energy, explain why smaller atoms more readily undergo fusion while heavier atoms would more likely fission?

11. About 21% of the world's electricity is generated using nuclear fission. Describe the function of control rods inside a nuclear reactor.

12. Alpha radiation has a relatively high amount of energy, yet is usually considered harmless. Why?

13. The average estimated annual dose of radiation for an American is 360 mrem. (Source: American Nuclear Society). If you live within 50 miles of a nuclear power plant, you are estimated to receive and extra 0.01 mrem per year, but if you live in Denver, you can expect an extra 63 mrem of radiation exposure each year. For a 180 pound person, how many extra joules does a person living in Denver absorb each year?

## ANSWERS TO THE CHAPTER OBJECTIVE QUIZ

1. **In the upper atmosphere, there exist many high-energy cosmic rays which can strike nuclei and initiate nuclear reactions. For example, a high-energy neutron can be absorbed by a nitrogen nucleus, resulting in a $^{13}C$ nucleus and a proton. This process could then occur a second time to produce a carbon-14 nucleus. For all practical purposes these cosmic rays don't reach the surface of the earth (sea-level) so this reaction is restricted to the upper atmosphere.**

2.

(a) $^{87}_{40}Zr \rightarrow {}^{0}_{+1}\beta + {}^{87}_{39}Y$

(b) $^{0}_{-1}e + {}^{102}_{45}Rh \rightarrow {}^{102}_{44}Ru$

(c) $^{3}_{1}H \rightarrow {}^{0}_{-1}\beta + {}^{3}_{2}He$

3. **Nuclear decay describe a spontaneous change in the nucleus of a radioactive atom. If there is an unfavorable combination of neutrons and protons in the nucleus, it will spontaneously undergo decay to attempt to improve the combination of nucleons. Nuclear decay is always first-order because these changes involve a single atom and are not affected by the presence or absence of any other atoms.**

4. **The decay constant is:** $k = \frac{\ln 2}{t_{1/2}} = \frac{0.693}{12.32\ y} = 5.625 \times 10^{-2}\ y^{-1}$

5. **See Problem 4 for the decay constant.**

   **All nuclear decay is first-order:** $\ln[\frac{N_0}{N}] = kt$ **or:**

   $N = N_o\ e^{-kt} = 0.55\ mg\ e^{-(5.625 \times 10-2\ y-1)(30\ y)} = 0.55\ mg\ e^{-1.6875} = 0.10$ **mg left**

6. **The decay constant is:** $k = \frac{\ln 2}{t_{1/2}} = \frac{0.693}{5730\ y} = 1.21 \times 10^{-4}\ y^{-1}$

   **All nuclear decay is first-order:** $\ln[\frac{N_0}{N}] = kt$;

   $N_o = 100\%$ **of the** $^{14}C$, $N = 18\%$

   $\ln[\frac{100\%}{18\%}] = 1.21 \times 10^{-4}\ y^{-1}(t)$ $\quad t = \ln[\frac{100\%}{18\%}] \div 1.21 \times 10^{-4}\ y^{-1} = 14{,}200\ y$

   **The ant is approximately 14,200 years old.**

7. **Consult Figure 14.4. The "band of stability is the N/Z ratios which yield stable isotopes. For small atoms, a 1:1 ratio is favorable, and this ratio gradually increases with greater numbers of protons in the nucleus.**

   (a) Z = 3, N/Z = 1.01

**For this small of an atom, a 1:1 ratio of neutrons and protons is ideal. The isotope would most likely be stable.**

(b) Z = 26, N/Z = 1.51
**This isotope has a high number of neutrons for its 26 protons. It would lie below the band of stability on Figure 14.4. Most likely decay would be beta emission.**

(c) Z = 70, N/Z = 1.21
**This atom is large where the N/Z ratio needs to be higher to promote stability. It would lie above the band of stability on Figure 14.4. Most likely decay would positron emission.**

8. **Binding energy is calculated using the mass defect, Δm: $E_b = \Delta mc^2$. The mass defect is the difference between the calculated mass of the atom and the actual.**

   **For $^{40}$Ca:**

   $$\Delta m = [20p \times 1.00783\ \text{u} + 20n \times 1.00867\ \text{u} + 20e \times 0.00054858\ \text{u}] - 39.96259\ \text{u} =$$
   $$= 0.37838\ \text{u or } 0.37838\ \text{g/mol}$$

   $$E_b = \Delta mc^2 = (0.37838 \times 10^{-3}\ \text{kg/mol})(2.9979 \times 10^8\ \text{m/s})^2 = 3.40007 \times 10^{13}\ \frac{\text{kg}\cdot\text{m}^2}{\text{s}^2\cdot\text{mol}} =$$
   $$= 3.40007 \times 10^{13}\ \frac{\text{J}}{\text{mol}}$$

9. **Transmutation is only possible when changes in the nucleus happen, i.e., the number of protons. A chemical reaction never involves the nucleus, only the electrons (and outer ones at that). A chemical reaction can never change one atom type into another.**

10. **The greater the binding energy, the more stable the isotope. Binding energies for atoms reach a maximum at iron-56. Therefore, when smaller atoms fuse together, the resulting atom is more stable; it has higher binding energy. When large atoms fission the same is true. The resulting fission products have higher binding energies than the original atom and are more stable.**

11. **A nuclear power plant uses the heat from a fission reaction to create steam and use the steam to run an electrical generator. The fuel (usually uranium) is in the form of cylindrical rods. Once the reaction is initiated with neutrons, the reaction is self-sustaining do to the release of neutrons during fission. Control rods are cylinders placed in between the fuel rods to moderate the reaction by absorbing neutrons. They are usually made of boron or cadmium. To slow down the reaction the control rods are placed further in between the fuel rods. To speed it up, they are withdrawn.**

**12. An alpha particle is a relatively massive form of radiation and so possesses a large amount of kinetic energy. However, they don't travel very fast and have low penetrating ability. That is why alpha radiation is usually harmless (see Problem 14.71).**

**13. Radiation units: 1 rem = 0.01 Sievert (Sv) 1 Sv = 1 J/ kg tissue**

$$180\ \text{lb} \times \frac{0.454\ \text{kg}}{1\ \text{lb}} = 81.7\ \text{kg}$$

$$63\ \text{mrem} \times \frac{1\ \text{rem}}{1000\ \text{mrem}} \times \frac{0.01\ \text{Sv}}{1\ \text{rem}} \times \frac{\frac{1\ \text{J}}{\text{kg}}}{1\ \text{Sv}} \times 81.7\ \text{kg} = 0.051\ \text{J}$$